EIGHTH EDITION

How to Solve General Chemistry Problems

R. S. Boikess

Rutgers, the State University of New Jersey

Upper Saddle River, New Jersey

Library of Congress Cataloging-in-Publication Data

Boikess, Robert S.
How to solve general chemistry problems / R.S. Boikess.—8th ed.
p. cm.
Includes bibliographical references and index.
ISBN-13: 978-0-13-154273-0 (alk. paper)
ISBN-10: 0-13-154273-7 (alk. paper)
1. Chemistry—Problems, exercises, etc. I. Title.
QD42.B64 2009
540.76—dc22

2008004254

Editor-in-Chief, Science: Nicole Folchetti
Senior Editor: Kent Porter Hamann
Assistant Editor: Jessica Neumann
Marketing Manager: Elizabeth Averbeck
Marketing Assistant: Che Flowers
Production Manager: Kathy Sleys
Creative Director: Jayne Conte
Cover Design: Yellow Dog Studios
Full-Service Project Management/Composition: Kavitha Kuttikan, Integra Software Services
Printer/Binder: Courier Companies, Inc.

Credits and acknowledgments borrowed from other sources and reproduced, with permission, in this textbook appear on appropriate page within text.

Pearson Education Ltd., London
Pearson Education Singapore, Pte. Ltd
Pearson Education Canada, Inc.
Pearson Education–Japan
Pearson Education Australia PTY, Limited
Pearson Education North Asia, Ltd., Hong Kong
Pearson Educación de Mexico, S.A. de C.V.
Pearson Education Malaysia, Pte. Ltd.
Pearson Education Upper Saddle River, New Jersey

10 9 8 7 6 5 4 3 2 1
ISBN-13: 978-0-13-154273-0
ISBN-10: 0-13-154273-7

To the students who will use their knowledge of science to make our world a better place to live.

CONTENTS

PREFACE

In this eighth edition, I have tried to preserve the self-teaching nature of previous editions, while responding to suggestions from users of the seventh edition. I have rewritten many sections for improved clarity, introduced many new problems and reworked many of the existing ones, and modernized certain aspects of chemical form and notation.

Features such as solutions for illustrative problems of almost every type, suggestions or clues for the solution of more difficult problems, and answers to all the problems have been retained.

The sequence of chapters has been changed to make it coincide more closely with the sequence in most popular general chemistry textbooks. Some long chapters have been divided and others restructured. The aim has been to make it easier for today's student to use this problem book as a supplement to any general chemistry textbook.

Among the changes that have been made are: the bringing together of the stoichiometry chapters; the placing of the material on thermochemistry close to that on stoichiometry to emphasize their similarities; the placing of gas laws after stoichiometry; and the division of the material on acids and bases into two chapters.

The beginning chemistry student of today will be required to master a broad range of problem solving skills. I believe that many students will need a self-teaching text from which they can obtain a mastery of these skills in an active way. My goal in preparing the eighth edition has been to make available the examples, problems, and explanations the student will need for success in the study of beginning chemistry. The student only need use the text with diligence.

I particularly wish to thank my colleagues at Rutgers for their valuable suggestions and criticisms.

Robert S. Boikess
boikess@rutchem.rutgers.edu

Chapter 1

How to Solve a Problem

Every problem you encounter, whether in chemistry or elsewhere, is solved in essentially the same fashion. *First*, you size up the situation or read the problem carefully, and decide *what* you are supposed to do and what you have to do it with. *Second*, having determined what you are supposed to do and what you have to do it with, you figure out *how* to do it. *Third*, you go ahead and *solve it* according to plan. Finally, you look at what you have done. The first two steps represent the *analysis* of the problem. The third step represents the *calculations*. Some problems are knottier than others, but they are all solved by these three fundamental steps. The last step is equally important. It will help you to avoid many errors.

To be more specific, when you go about solving any problem in this or any other book or in any test or examination:

1. *Read the problem carefully.* Note exactly what is given and what is sought. Recognize that most chemistry problems contain more information than is explicitly given. For example, if a problem specifies a given mass of water, it is also specifying a given amount of water in moles, molecules, and atoms. Note any and all special conditions. Be sure you understand the meaning of all terms and units and that you are familiar with all chemical principles relevant to the problem. Every problem in this book is designed to illustrate some principle, some relationship, some law, some definition, or some fact. If you understand the principle, relationship, law, definition, or fact, then you should have no difficulty solving the problem. The one big reason why students have difficulty with chemistry problems is failure to understand, exactly and well, the relevant chemical principles and the meaning and value of all terms and units that are used in the problem.

2. *Plan, in detail, just how the problem is to be solved.* Get into the habit of visualizing the entire solution before you execute a single step. Insist on knowing what you are going to do and why you are going to do it. Aim to learn to solve every problem in the most efficient manner, which generally means doing it the shortest way, with the fewest steps.

3. *Specify definitely what each number represents and the units in which it is expressed when you actually carry out the mathematical operation of solving the problem.* Don't just write

$$\frac{192}{32} = 6,$$

write

$$\frac{192 \text{ g sulfur}}{32 \text{ g sulfur per mole of sulfur}} = 6 \text{ mol sulfur}$$

or whatever other units are appropriate. Always divide and multiply the *units* as well as the *numbers*. This procedure is one way to give exactness to your thought process and is a very good way to avoid errors. You should jot down the unit or units in which your answer is to be expressed as the first step in the actual solution. For instance, if you are solving for the mass of oxygen in 200 g of silver oxide, you should write that the answer will be "= .... g of oxygen." Many problems are worked backward, since you will often first focus your attention on the units in which the answer is to be expressed and then plan the solution with those units in mind.

4. *Having solved the problem, examine the answer to see if it is reasonable and sensible.* A student who reports that 200 g of silver oxide contains 1380 g of oxygen should know that such an answer is not sensible. Get into the habit of checking the answer to see if it makes sense.

5. *If you do not understand how to solve a problem, have it explained to you at the very earliest possible time.* To be able to solve the later problems you must understand the earlier ones. After a problem has been explained to you, fix the explanation in your mind by working that problem and *other similar problems* at once, or at least within a few hours, while the explanation is still fresh in your mind. Test yourself to be sure that you can apply your understanding of the solution. In solving a problem you must not only understand how it is done, but be able to do it yourself.

Chapter 2

Units and Measurement

Fundamental to any quantitative examination of our surroundings is measurement. Science relies on quantitative data; measurement is generally how we obtain these data. It is the comparison of the quantity of interest with a standard quantity called a unit. In order to communicate and compare, we need everyone to agree on the same unit. The metric system is an example of an early agreement about units.

In this book we will assume that you are familiar with the common units of measure in the metric system and have a fair idea of the volume represented by 1 L, 100 cm^3, and 1 cm^3, the mass represented by 10 g, 100 g, or 1 kg, and the length represented by 760 mm, 10 cm, and 1 m. Also, we will assume that you are familiar with the Celsius (centigrade) temperature scale.

You should recall that the metric system employs multiplying prefixes that change the magnitude of the unit by powers of 10. For example, *micro-* means one-millionth (10^{-6}), *milli-* means one-thousandth (10^{-3}), *centi-* means one-hundredth (10^{-2}), and *deci-* means one-tenth (10^{-1}), while *kilo-* means one thousand times (10^3) and *mega-* means one million times (10^6).

These prefixes allow us to define relationships among units in the metric system. For example, 1 centimeter is one-hundredth of a meter, and 1 kilometer is one thousand meters.

Conversions of metric units (grams, liters, cubic centimeters, centimeters, etc.) to other units (pounds, quarts, inches, feet, etc.) are not often required. But when they are, we must know the conversion factors. The following list of conversion factors provides most of the ones we will need.

1 meter (m) = 39.37 inches (in) = 1.09 yards (yd)

1 kilogram (kg) = 2.2046 pounds (lb)

1 liter (L) = 0.264 US gallons (gal) = 1.06 US quarts (qt)

THE SI

Over the years many improvements have been introduced into the metric system. The result is a new and improved system of units that has been adopted by the international scientific community. It is called the International System of Units, or the SI from the initials of its name in French. Building on the metric system, the SI starts with seven *base units*, which correspond to a set of independent physical quantities. These base units and their symbols are listed in Table 2.1. The first five are commonly used in every branch of chemistry; the last two are more specialized.

The units for all other physical quantities are called *derived units* and are combinations of the appropriate base units without any numerical factors. For example, the unit for area is the square meter (m^2), derived from the relationship between length and area. The unit for volume is the cubic meter (m^3), from the relationship between length and volume. Such a system of units is called a *coherent system* and has an obvious advantage: we do not need to learn conversion factors. Some other important derived units are given in Table 2.2 and some common non-SI units for derived quantities are listed in Table 2.3.

TABLE 2.1

Physical Quantity	Unit	Symbol
Length	meter	m
Mass	kilogram	kg
Time	second	s
Amount	mole	mol
Temperature	kelvin	K
Electric current	ampere	A
Luminous intensity	candela	cd

TABLE 2.2

Physical Quantity	Unit	Symbol	Definition
Energy	joule	J	kg-m^2/s^2
Force	newton	N	kg-m/s^2 (or J/m)
Pressure	pascal	Pa	kg/m-s^2 (or N/m^2)
Electric charge	coulomb	C	A-s

TABLE 2.3

Physical Quantity	Unit	Symbol	Definition
Energy	calorie	cal	4.184 J
Pressure	atmosphere	atm	1.01×10^5 Pa
Pressure	millimeters of mercury	mmHg	1/760 atm
Pressure	torr	torr	1/760 atm
Force	dyne	dyn	10^{-5} N

TABLE 2.4

Fraction	Prefix	Symbol	Multiple	Prefix	Symbol
10^{-1}	deci	d	10	deka	da
10^{-2}	centi	c	10^2	hecto	h
10^{-3}	milli	m	10^3	kilo	k
10^{-6}	micro	μ	10^6	mega	M
10^{-9}	nano	n	10^9	giga	G

The SI also includes a set of multiplying prefixes that are used to form decimal fractions and decimal multiples of the base and derived SI units. Prefixes corresponding to all powers of 10 are not provided. Those smaller than 10^{-2} or larger than 10^2 must all have exponents that are divisible by 3. Some of these prefixes are listed in Table 2.4. Not all features of the SI have been adopted by chemists. In some cases the older systems are much more convenient. In this book we will try to use the SI to the extent consistent with chemical convenience.

CONVERSION OF UNITS

One very common type of scientific calculation is the conversion of the value of a physical quantity given in one unit into a value measured in another unit. The first step in such a calculation is to find a relationship between the two units. The relationship may merely be a power of 10 that is indicated by the prefix. For example:

$$1 \text{ m} = 100 \text{ cm}$$
$$1 \text{ m}^3 = (100 \text{ cm})^3 = 10^6 \text{ cm}^3$$

The relationship may instead require knowledge of a numerical factor, such as:

$$1 \text{ kg} = 2.205 \text{ lb}$$
$$1 \text{ atm} = 760 \text{ mmHg} = 760 \text{ torr}$$

Conversion factors that are used to perform unit conversions can be obtained by rewriting these types of relationships between units as fractions. For example:

$$\frac{1\ \text{m}}{100\ \text{cm}} = 1 \quad \text{or} \quad \frac{100\ \text{cm}}{1\ \text{m}} = 1$$

These two conversion factors are used to convert between meters and centimeters. The choice of conversion factor can be based on the logic of the conversion or on the canceling of units. Thus we convert a length of 3.2 m to centimeters using

$$3.2\ \cancel{\text{m}} \times \frac{100\ \text{cm}}{1\ \cancel{\text{m}}} = 320\ \text{cm}$$

While to convert a length of 53.7 cm to meters

$$53.7\ \cancel{\text{cm}} \times \frac{1\ \text{m}}{100\ \cancel{\text{cm}}} = 0.537\ \text{m}$$

The conversion factor used in each case is the one which has the unit of the original measurement in the denominator and the unit of the result in the numerator. As you can see, the original units cancel, leaving the units of the result. You should always cancel units as part of the use of conversion factors because doing so provides an additional check of your calculation.

You should always check such conversions to see if the answer makes sense. Since a centimeter is a smaller length than a meter, there are more centimeters in a given length than meters. Thus the conversion from meters to centimeters should result in a numerically larger answer. For example, 3.2 m is equivalent to 320 cm. Also, 53.7 cm is 0.537 m, a numerically smaller answer for the reverse conversion. If you had used the wrong conversion factor for the following conversion,

$$3.2\ \text{m} \times \frac{1\ \text{m}}{100\ \text{cm}} = 0.032\ \frac{\text{m}^2}{\text{cm}}$$

not only would the numerical value be illogical, indicating a smaller number of centimeters than meters in a given length, but the units would not cancel and the result would have units that make no sense in the context of the problem.

PROBLEMS

2.1 The mass of a paperback text is 410 g. Express this mass in kilograms and in pounds.

Solution: We start by expressing the relationship between the units of the conversion.

$$1 \text{ kg} = 1000 \text{ g}$$

We use this relationship to derive the necessary conversion factor. The factor is the one that has the unit of the answer in the numerator. In this case it is

$$\frac{1 \text{ kg}}{1000 \text{ g}}$$

The conversion is then carried out:

$$410 \cancel{\text{g}} \times \frac{1 \text{ kg}}{1000 \cancel{\text{g}}} = 0.410 \text{ kg}.$$

Note that when the units are canceled only the desired unit remains. The conversion to units of pounds is carried out in the same way.

$$1 \text{ lb} = 454 \text{ g}$$

$$410 \cancel{\text{g}} \times \frac{1 \text{ lb}}{454 \cancel{\text{g}}} = 0.903 \text{ lb}$$

2.2 Express the height of a 71-in person in units of feet, meters, and centimeters.

2.3 There are 32 fluid ounces in a US quart. Express the volume of a 750-cm^3 bottle of consumable liquid in units of ounces, quarts, and liters.

2.4 Express a volume of 1 cubic meter in units of liters.

2.5 There are 5280 feet in 1 mile. A highway speed limit is 55 mph. Express this speed in units of kilometers per hour and meters per second.

2.6 A rectangular building lot is 42 m × 35 m. Express its area in units of square feet.

2.7 Find the number of cubic nanometers in 1 km^3.

Solution: Both these units can be related to the meter, which is the base unit, and to the meter cubed, which is the derived unit for volume. To find relationships between derived SI units (in this case units of volume), we start with the relationship between the base unit and the unit with the prefix:

$$1 \text{ m} = 10^9 \text{ nm}$$

Then we carry out the operation that converts the base unit to the derived unit on both quantities

$$(1\ \text{m})^3 = (10^9\ \text{nm})^3$$

Similarly: $1\ \text{m} = 10^{-3}\ \text{km}$ and $(1\ \text{m})^3 = (10^{-3}\ \text{km})^3$

Setting equal things equal to each other gives

$$(10^9\ \text{nm})^3 = (10^{-3}\ \text{km})^3$$
$$10^{27}\ \text{nm}^3 = 10^{-9}\ \text{km}^3$$
$$10^{36}\ \text{nm}^3 = 1\ \text{km}^3$$

2.8 The density of water is 1 g/cm^3. Express the density of water in the following units: (a) kg/cm^3, (b) g/m^3, (c) kg/m^3, (d) g/L.

2.9 The mass of Phobos, the larger of Mars's two moons, is 2.7×10^{16} kg. Express this mass using three other SI mass units.

2.10 A diver at a depth of 50 m is subjected to a pressure of 6 atm. Express this pressure using three other units.

2.11 Dieters commonly encounter a unit called the "calorie." This calorie is related to the calorie that is used as the unit of energy by 1 "food calorie" = 1000 energy calories. A table of food composition lists the "calorie content" of 1 cup of roasted peanuts as 805 "calories." Convert this value to joules.

2.12 Express the data in today's weather report in SI units.

INTERCONVERSION OF CELSIUS (CENTIGRADE) AND FAHRENHEIT TEMPERATURE READINGS

The thermometers used in the laboratory are graduated in Celsius degrees, designated by the letter C. You should note that the correct term is *Celsius degrees* or *degrees Celsius* (rather than centigrade degrees or degrees centigrade) in honor of the Swedish scientist Anders Celsius, who devised the scale. The Celsius degree is related to the SI unit of temperature, the kelvin, by the equation

$$T = t + 273$$

where T is the temperature in kelvins (K) and t is the temperature in degrees Celsius (°C). In the SI, the unit of temperature is the kelvin; the term *degree* is not used. To convert between Celsius and SI we merely add or subtract 273.

Most household thermometers in the United States are graduated in Fahrenheit degrees, designated by the letter F. The fixed points on both the Celsius and Fahrenheit temperature scales are the boiling point and freezing point of water. On the Celsius scale the freezing point of water is 0 °C and the boiling point is 100 °C; the space between the fixed points is divided into 100 units and the space above 100 °C and below 0 °C is divided into the same size units. On the Fahrenheit scale the freezing point of water is 32 °F and the boiling point is 212 °F; the space between the fixed points is divided into 180 units and the space above 212 °F and below 32 °F is divided into the same size units. Since the space between the freezing point and boiling point of water is divided into 100° on the Celsius scale and 180° on the Fahrenheit scale, it follows that 100 Celsius degrees must represent the same temperature change as 180 Fahrenheit degrees. Thus 1 Celsius degree is equal to 1.8 Fahrenheit degrees; or expressing it in fractional form, 1 Celsius degree is equal to $\frac{9}{5}$ Fahrenheit degrees, and 1 Fahrenheit degree is equal to $\frac{5}{9}$ Celsius degree.

With these facts in mind we see that if we wish to find the Fahrenheit value, F, of a certain temperature in Celsius degrees, C, we first multiply the Celsius reading by $\frac{9}{5}$, which gives us $\frac{9}{5}$ C. Since the reference temperature (the freezing point of water) on the F scale is 32° above zero we must add 32 to $\frac{9}{5}$ C in order to get the actual reading on the Fahrenheit scale.

$$\text{Fahrenheit temperature} = \tfrac{9}{5}\ \text{Celsius temperature} + 32$$

or

$$F = \tfrac{9}{5}C + 32 \qquad \textbf{(2-1)}$$

Equation (2-1) can be transposed to the form

$$C = \tfrac{5}{9}(F - 32) \qquad \textbf{(2-2)}$$

Equation (2-2) tells us that to find the value, in degrees Celsius, of a Fahrenheit temperature, we first subtract 32 from the Fahrenheit temperature (because the Fahrenheit freezing point reference is 32° above zero) and then take $\frac{5}{9}$ of that answer.

The uses of Eqs. (2-1) and (2-2) are illustrated by the following problems.

a. Convert 144 °F to a Celsius reading.

In thinking our way through this problem we note that 144 °F is (144 − 32) or 112 °F above the freezing point of water. Since 1 Fahrenheit degree is equal to $\frac{5}{9}$ Celsius degree, 112 Fahrenheit degrees must be equal to $112 \times \frac{5}{9}$ or 62.2 Celsius degrees. That means that 144 °F is 62.2 Celsius degrees above the freezing point of water. Since the freezing point of water is 0 °C, 62.2 Celsius degrees above the freezing point of water is 62.2 °C.

b. Convert 80 °C to a Fahrenheit reading.

In thinking our way through this problem we note that 80 °C is 80 Celsius degrees above the freezing point of water. Since 1 °C equals $\frac{9}{5}$ °F, 80 °C is equal to $\frac{9}{5} \times 80$ or 144 Fahrenheit degrees above the freezing point of water.

But the freezing point of water on the Fahrenheit scale is 32°. Therefore, we must add 32 to our 144 to get the actual Fahrenheit temperature, 176 °F.

c. Convert 100 °C to SI.

To convert from Celsius to SI we add 273. Thus 100 °C is 373 K.

PROBLEMS

2.13 What temperature, in degrees Celsius, is represented by each of the following Fahrenheit temperatures?
a. 72.0 °F
b. −20.0 °F

2.14 What temperature in degrees Fahrenheit is represented by each of the following Celsius temperatures?
a. 12.0 °C
b. −50.0 °C

2.15 What temperature, in degrees Celsius and degrees Fahrenheit, is represented by each of the following kelvin temperatures?
a. 298 K
b. 0 K

2.16 At what temperature will the readings on the Fahrenheit and Celsius thermometers be the same?

2.17 Suppose you have designed a new thermometer called the X thermometer. On the X scale the boiling point of water is 130 °X and the freezing point of water is 10 °X. At what temperature will the readings on the Fahrenheit and X thermometers be the same?

2.18 On a new Jekyll temperature scale, water freezes at 17 °J and boils at 97 °J. On another new temperature scale, the Hyde scale, water freezes at 0 °H and boils at 120 °H. If methyl alcohol boils at 84 °H, what is its boiling point on the Jekyll scale?

MEASUREMENT

Expressions of the magnitude of a physical quantity have two parts, a number and a unit. So far we've talked about units. Let's look at the number.

Some numbers are exact; they result from counting each entity present. For example there are 14 students in the first row or 12 eggs in a dozen or 60 seconds in a minute. Some numbers are as exact as we want them to be, but not completely exact. For example we can write π or the fraction $\frac{2}{3}$ with as many digits after the decimal point as we want.

But many measurements of physical quantities are not exact. There is an error in the process of determining how many units equal the quantity we are measuring. We should write the result of the measurement, which is a number, in such a way as to convey the margin of error. There are two types of errors

1. *Accuracy*: How close is our value to the real or true value? We don't usually know the answer, because if we knew the true value, we would use it.
2. *Precision*: In a set of measurements of a quantity, how well do the individual measurements agree with each other? That is, how close is each value to the average? The closer they are, the more precise is the measurement. Precision is easily determined. Usually, but not always, high precision indicates high accuracy. Systematic errors can result in high precision, but low accuracy.

So when we express the result of a measurement, we write the number in a way that indicates something about the precision of the measurement.

SIGNIFICANT FIGURES

When a number is written for a measurement, we understand that there is uncertainty in the last digit. Therefore, the more digits to the right of the decimal point, the more precise is the measurement. As a rough rule, the number of digits in a number that reports a measurement is equal to the number of significant figures (SF). Examples: 1.85 m or 82.3 kg have 3 SF, while 1609 m has 4 SF. We need to know how many SF are in a number when we do calculations, as we shall see.

There is a complication that results from the use of zero to establish the magnitude of a number. Zeros are not always significant figures. We can state a few rules to cover all situations:

a. A zero between two nonzero digits is a significant figure. Examples: 100.1 K has 4 SF and 0.305 m has 3 SF.
b. A zero at the end of a number is a significant figure if it is to the right of a decimal point. Example: 91.10 g has 4 SF.
c. A zero that only positions the decimal point is not significant. Example: 0.0821 has 3 SF, but 1.082 has 4 SF.
d. A zero at the end of a whole number is ambiguous (in contrast to b). Thus 1000 m could have anywhere from 1 to 4 SF.

EXPONENTIAL NOTATION

To solve the problem of rule (d) above we can express numbers as the product of a small number (between 1 and 10) and a power of 10. Thus 400 is 4 times 100, and 100 is 10^2. So if the number has 1 SF we write 4×10^2. If it has 2 SF we write

4.0×10^2. And if it has 3 we write 4.00×10^2. This method works because zeros at the end of a number to the right of the decimal point are always significant (rule b). *Note:* The power of 10 has nothing to do with SF and the small number is always written with one digit to the left of the decimal point.

Another use for exponential notation is to make it easier to write very small or very large numbers. Suppose we have a measurement of 11 million (2 SF). Instead of 11,000,000, we can write 1.1×10^7. We know to use 7 because the decimal point has to move seven places from the end of the number to the left to make 1.1, or because 10^7 is 10 million and 1.1 times 10 million is 11 million.

Or more commonly, suppose we have a very small number, such as 0.00000000432. We write 4.32×10^{-9}. We know to use −9 because the decimal point has to move nine digits to the right to make 4.32. Similarly, if we see 5.43×10^{-5}, we know the number is 0.0000543, by moving the decimal point five digits to the left.

The use of exponents makes it quite easy to determine the correct number of digits in the answer to an operation involving multiplication and division of many numbers. Thus if the expression

$$\frac{417{,}000 \times 0.0036 \times 15{,}300{,}000}{0.000021 \times 293 \times 183{,}000}$$

is changed to the form

$$\frac{4.17 \times 10^5 \times 3.6 \times 10^{-3} \times 1.53 \times 10^7}{2.1 \times 10^{-5} \times 2.93 \times 10^2 \times 1.83 \times 10^5}$$

we can determine at a glance that the answer is approximately 2×10^7.

Likewise, when the expression

$$\frac{0.0045 \times 0.082 \times 600}{204 \times 23}$$

is changed to the form

$$\frac{4.5 \times 10^{-3} \times 8.2 \times 10^{-2} \times 6 \times 10^2}{2.04 \times 10^2 \times 2.3 \times 10^1}$$

we can see that the answer is approximately

$$48 \times 10^{-6} \quad \text{or} \quad 4.8 \times 10^{-5}$$

PROBLEMS

2.19 Express each of the following numbers in exponential form without using your calculator:

a. 21,000,000,000
b. 760
c. 0.0027
d. 0.0000018
e. 0.10

2.20 Estimate each of the following expressions without using your calculator by first writing each number in exponential form:

a. $\dfrac{136{,}000 \times 0.000322 \times 273}{0.082 \times 4200 \times 129.2}$

b. $\dfrac{120 \times 309 \times 800}{273 \times 600}$

2.21 Solve each of the following without using your calculator:

a. $\dfrac{1.76 \times 10^{-3}}{8.0 \times 10^{2}}$

Solution: In solving this problem it may be helpful to express the numerator as 17.6×10^{-4} rather than leaving it as 1.76×10^{-3}, in order to obtain the answer in the form 2.2×10^{-6} rather than 0.22×10^{-5}. Although these two numbers have the same value, the preferred expression is 2.2×10^{-6}.

b. $\dfrac{0.0234 \times 10^{-3}}{3.6 \times 10^{-4}}$

CALCULATIONS WITH SIGNIFICANT FIGURES

You cannot improve precision by calculation, since it comes from the way a measurement is done. So when you do calculations involving measurements, the result cannot be more precise than the inputted measurements. We take care of this issue by applying rules to arithmetical operations involving SF.

a. When doing multiplication or division, the result cannot have more SF than the number with the fewest SF. If rounding off is necessary, do it in the usual way. Example: $1.32 \times 8.1 = 11$ (2 SF)

b. When doing addition or subtraction, only consider digits (not SF) to the right of the decimal point. The result should have the same number of digits to the right of the decimal point as in the number with the fewest digits to the right of the decimal point. Example: $1.327 + 18.1 = 19.4$ (1 SF to the right of the decimal point)

PROBLEMS

2.22 Perform the following calculations and report the results with the correct number of significant figures.

a. $12.346 + 0.21 + 1.02$
b. $1.21 + 8.45 + 0.34$
c. $(123)(0.0037)$
d. $(1234)(4321)/5332$
e. $424 - 25$

2.23 There are exactly 60 seconds in a minute; there are exactly 60 minutes in an hour; there are exactly 24 hours in a mean solar day; and there are 365.24 solar days in a solar year. Find the number of seconds in a solar year. Be sure to give your answer with the correct number of significant figures.

Chapter 3

Atomic Weight, Amount, and the Mole

In order to study mass relationships in chemistry we need information about the masses of atoms. The approach we use is to designate the relative mass of an atom.

If we are going to designate the relative mass of atom, what is it relative to? By international agreement the standard for atomic mass to which we compare all other atomic masses is carbon-12, the most common isotope of carbon. An atom of carbon-12 is assigned a mass of *exactly* 12 and all other atomic masses are compared to it, by making very accurate measurements of the mass composition of pure substances, taking natural abundances of isotopes into account when necessary. For convenience, we assign a special unit for atomic masses, called the **atomic mass unit (amu)**, represented by the symbol **u**. By definition carbon-12 has a mass of 12 u. We can measure and find that 1 u $= 1.66054 \times 10^{-24}$ g.

When we use relative masses the values do not have units. For example, suppose we designate a student of height 5 feet as the standard. The relative height of someone 6 feet tall is then $\frac{6}{5}$, with no units. But you can do a unit conversion to get the actual height: $\frac{6}{5} \times 5$ ft $= 6$ ft.

The relative atomic mass of nitrogen-14 can be found by accurate measurement. It is 14.0030744. Since it is a relative atomic mass, it has no units. But to find the relative atomic mass of elements that are poly-isotopic (some elements have more than one stable isotope), we must know what isotopes exist on Earth and in what relative amounts—what we call their natural abundances.

Natural abundances can be expressed as the percent of each isotope or as the fraction (divide percent by 100) of each isotope in a naturally occurring sample of the element. We find these data in tables of natural abundances. We also need to know the relative atomic masses of the various isotopes to find the relative atomic mass of the element.

To find the relative atomic mass (sometimes inappropriately called the atomic weight) of a naturally occurring element on Earth, we take a weighted average of the relative masses of the existing isotopes, using their natural abundances. Taking a weighted average is the way you calculate your GPA. Multiply the grade in each course by the number of credits, add all the terms, and then divide by the total number of credits.

Here we get a weighted average (GPA) by multiplying the fractional natural abundance (credits) of each isotope (course) by its relative atomic mass (grade), adding the results and dividing by 1 (number of credits). Note the sum of all the fractions is 1. Because we get the relative atomic mass of an element by calculating a weighted average, the misnomer "atomic weight" is often used as the name for this quantity. It is wrong, but it can be considered slang for weighted average of relative atomic masses taking into consideration isotopes and their natural abundances. We are interested in mass, not weight, but the term *atomic weight* is still widely used.

EXAMPLE 1

Naturally occurring magnesium consists of three isotopes ^{24}Mg, ^{25}Mg, and ^{26}Mg, whose relative atomic masses are 23.985, 24.986, and 25.983 and whose natural abundances are 78.70%, 10.13%, and 11.17%, respectively. Calculate the relative atomic mass (or what some people call the atomic weight) of naturally occurring magnesium.

Solution: Find the weighted average. Either use the percents for natural abundances and divide by 100 at the end or first convert the percents to fractions.

$$(23.985)(0.7870) + (24.986)(0.1013) + (25.983)(0.1117) = 24.31$$

EXAMPLE 2

The relative atomic mass of boron is 10.81. It consists of two isotopes, ^{10}B and ^{11}B. The relative atomic mass of ^{10}B is 10.013 and its natural abundance is 19.78%. Find the relative atomic mass of ^{11}B.

Solution: Set up the calculation for the weighted average, letting x = relative atomic mass of ^{11}B.

$$(10.013)(0.1978) + (x)(1 - 0.1978) = 10.81.$$
$$x = 11.01, \text{ the relative atomic mass of } ^{11}B.$$

MASS—AMOUNT RELATIONSHIPS

There are two common ways to express the quantity of a material in chemistry. One way is as the mass, usually measured in grams (g), although the SI base unit of mass is the kilogram (kg). Another way is as the **amount**.

This word is a very familiar one that we use all the time in rather loose ways. In chemistry it is much more exactly defined and you should only use it when appropriate. *Amount is quantity expressed as a number of identifiable things or basic entities that we can count.*

Thus five students is an amount of students (the identifiable thing), but 450 kg of students is not an amount of students, but a mass of students. Six hamburgers is an amount of hamburgers (what we count), but 1300 g of hamburgers is a mass of hamburgers.

Note that you can only use amount to express quantity when there is an identifiable thing to count. Suppose you have a partly consumed keg of beer at a party. What amount of beer do you have? The question can't be answered because there is no identifiable thing to count. We could express the quantity as mass or more conveniently as volume, but not as amount in the chemical sense.

Also note that when you talk about amount, you must specify the entity you are counting—that is, what the amount is of. This rule is very important. You cannot say just five or six. You must say five students or six hamburgers. Otherwise no one will know what you are talking about.

Chemical formulas convey information about relative masses because of the availability of the Table of Atomic Weights, which gives the relative mass of every atom in the formula. Chemical formulas also convey information about amounts. The Atomic Theory gives us identifiable units to count: atoms, molecules, and ions. A chemical formula is first of all a statement about the amounts of the constituent elements in the compound. So in glucose, $C_6H_{12}O_6$, the formula tells us that there are 6 atoms of C, 12 atoms of H, and 6 atoms of O in the molecule. This interpretation of the formula is actually a statement about amounts. In calcium chloride the formula $CaCl_2$ tells us that there is one Ca^{2+} ion for every two Cl^- ions. This is a statement about relative amounts because there are no molecules of calcium chloride.

What is the relationship between mass and amount in these substances? The answer can be found using the data in a Table of Atomic Weights. One glucose molecule (amount) has a mass of 180.155 u. Ten glucose molecules have a mass of 1801.55 u.

But this relationship is not practical in the lab because we cannot work with a small number of molecules. We need to work with much larger amounts to get quantities big enough to have a measurable mass.

UNITS OF AMOUNT

Calculations with large amounts may be inconvenient, so we can devise units of amount to keep the numbers more manageable. You are familiar with this process. A unit of amount defines a number.

The most familiar unit of amount is the dozen. It is 12 things that we are counting. Another unit of amount is the gross. It is 144 things. Still another unit, used for paper, is the ream. It is 500 sheets of paper. It is easier to say 21 reams of paper than 10,500 sheets of paper.

None of these units of amount will work for atoms or molecules. The units are not big enough. We need a special unit for things on the atomic or molecular scale. It has to be very big, because atoms are so small. In fact it has to be so big that it we will be unable to count it. So what do we do?

We can define a unit of amount by reference to the number of things in a given quantity of a material. Suppose we define a ream as the number of sheets of paper in a package purchased at the bookstore. That works, even though we have not mentioned 500. Suppose we get another package of the same kind of paper somewhere else and suppose we measure the mass of both packages and they have the same mass. Then we know the second package is also a ream and has very close to or exactly the same number of sheets of paper. We still have not mentioned 500. But if we were asked how many sheets are in the packages, we could count them and say 500. But if we were not able to count past 20, then all we could say is "about" 500, if we had an indirect way of measuring the number of sheets (say by thickness or mass).

We do exactly the same thing in chemistry, but the numbers are bigger. To start with, we want to measure mass in grams. But atoms are incredibly small. The mass of a single atom of uranium-238 is 4.0×10^{-22} g.

So we shall define a unit of amount that corresponds to a convenient mass. Since we are making an arbitrary definition, we might as well make it as convenient for us as possible.

Just as we define relative atomic masses based on carbon-12, we also define the SI unit of amount based on carbon-12. The SI base unit of amount is the mole, abbreviated mol without a period. This unit of amount is so well accepted by chemists that they will often incorrectly use the phrase "number of moles" to mean amount. This phrase is the same as saying number of grams when you mean mass. It is incorrect but understandable.

The SI definition of a mole is: "the amount of substance that contains as many basic units as there are atoms in exactly 12 g of carbon-12." There is a very large number of atoms in 12 g of carbon-12. There are so many that in practice we cannot count them.

But the best measurements we have given a value of 6.0220978×10^{23} for the number. This number is called *Avogadro's number*, in honor of a 19th-century Italian chemist who did some important experiments on gases that turned out to be very useful in measuring atomic weights. Avogadro's number is usually represented by the symbol N. It is clearly not an exact number as is 12 in a dozen.

This definition may seem complicated, but it keeps quantitative relationships in chemistry very simple. The key is that both relative atomic mass (and therefore relative molecular mass and formula mass) and amount are defined in terms of carbon-12. By definition, the mass of one atom of carbon-12 is exactly 12 u; the mass of one mole of carbon-12 is exactly 12 g.

Since all the values in a Table of Atomic Weights are relative to carbon-12, the same is true of all elements and all pure substances. *The mass in grams of a mole of any pure substance is numerically equal to the mass in u of its basic unit*, whether a weighted average atom, a molecule, or a formula unit. We call the mass in grams of one mole of anything the *molar mass*; its units are g/mol. The substance under consideration must always be identified.

UNIT CONVERSIONS

We can convert between mass and amount and number of entities using the same method of unit conversions we have already seen.

For example 1.00 mol Cu = 63.5 g Cu (from the Table of Atomic Weights). Note this relationship only makes sense if we specify Cu; it is not true that 1.00 mol equals 63.5 g for any other substance. In the usual way we can get the two conversion factors to use in problem solving:

$$63.5 \text{ g Cu}/1.00 \text{ mol Cu and } 1.00 \text{ mol Cu}/63.5 \text{ g Cu}$$

Using amu, the Table of Atomic Weights also tells us that the mass of an imaginary "weighted average" Cu atom is 63.5 amu leading to the conversion factors

$$63.5 \text{ amu Cu}/1 \text{ atom Cu and } 1 \text{ atom Cu}/63.5 \text{ amu Cu}$$

The definition of the amu and Avogadro's number can lead to conversion factors, in the same way. These are the key conversion factors in many chemical calculations. Thus

$$1 \text{ g} = 6.022 \times 10^{23} \text{ amu and}$$
$$1 \text{ mol} = 6.022 \times 10^{23} \text{ units.}$$

Let's illustrate some mass–amount calculations, first with some ordinary objects and then with atoms.

EXAMPLE 3

The mass of an average orange is 65 g. Find the amount of oranges that has a mass of 5.4 kg.

Solution:

$$5.4 \text{ kg oranges} \times 1000 \text{ g}/1 \text{ kg} \times 1 \text{ orange}/65 \text{ g orange} = 83 \text{ oranges}$$

EXAMPLE 4

The mass of a molecule of human hemoglobin is 68,300 amu. Find the mass in grams of a single molecule and find the amount of hemoglobin that has a mass of 1.00 g.

Solution:

$$68{,}300 \text{ amu/hemoglobin molecule} \times 1 \text{ g}/6.02 \times 10^{23} \text{ amu} = 1.13 \times 10^{-19} \text{ g/hemoglobin molecule}$$

If the mass of one molecule of hemoglobin is 68,300 amu, then by definition the mass of one mole of hemoglobin is 68,300 g. From this relationship we get the necessary conversion factor for the second part of the example

$$1.00 \text{ g hemoglobin} \times 1 \text{ mol hemoglobin}/68{,}300 \text{ g hemoglobin} = 1.46 \times 10^{-5} \text{ mol hemoglobin}$$

EXAMPLE 5

The diameter of a red blood cell is 7.5 μm. Find the length of a row made up of 0.50 mol of red blood cells.

Solution:

$$0.50 \text{ mol cells} \times 6.02 \times 10^{23} \text{ cells/mol cells} \times 7.5 \text{ μm/cell} = 2.3 \times 10^{24} \text{ μm}$$

$$2.3 \times 10^{24} \text{ μm} \times 1 \text{ m}/10^{6} \text{ μm} = 2.3 \times 10^{18} \text{ m}$$

In miles this length is about 1,400,000,000,000,000 miles.

A mole has a large number of units.

The bottom line is that the mole is just like a dozen, but instead of 12 units it has Avogadro's number of units.

REVIEW

Since, as has been noted above, the atomic weight of an element in grams consists of 6.022×10^{23} atoms; since one mole of atoms means 6.022×10^{23} atoms; and since the number, 6.022×10^{23}, is the Avogadro number, N, it follows that, when applied to a particular element, the terms

One mole of atoms

One atomic weight in grams

6.022×10^{23} atoms

N atoms

all mean the same thing. As a specific example, 32.064 g of sulfur is

One mole of sulfur atoms

One atomic weight of sulfur in grams

6.022×10^{23} atoms of sulfur

N atoms of sulfur

Unless specifically stated otherwise it will always be understood that the unit **mole** *means that the mass of the quantity is measured in* **grams**. Therefore all amount–mass relationships are understood to be between grams and moles.

SYMBOL

The name of an element is commonly represented by an abbreviation, referred to as the *symbol* for that element. For example, the symbols for sodium and oxygen are, respectively, Na and O.

As one of its meanings, *the symbol for an element, when used in a chemical formula or equation, refers to one mole of atoms of that element.* **This is a very important fact to remember**; it is a basic fact in the solution of most chemical problems.

PROBLEMS

3.1 The ratio of the mass of a gold atom to the mass of a C-12 atom is 50:3. Find the mass of one mole of gold atoms.

3.2 The ratio of the mass of an oxygen atom to an atom of C-12 is 4:3. The ratio of sulfur and oxygen by mass in SO_2 is 1:1. Find the mass of one mole of S atoms.

3.3 Naturally occurring bromine consists of two isotopes, ^{79}Br, natural abundance 50.54%, and ^{81}Br, natural abundance 49.46%. Their relative atomic masses are 78.918 and 80.916, respectively. Calculate the atomic weight of bromine.

3.4 A sample of strontium of natural origin is found to consist of three major isotopes: Sr-86, natural abundance 9.86%; Sr-87, natural abundance 7.02%; and Sr-88, natural abundance 82.56%. Their relative atomic masses are 85.909, 86.909, and 87.906 respectively. Find the atomic weight of strontium.

3.5 Naturally occurring silver has an atomic weight of 107.87 and consists of only two isotopes: Ag-107, relative atomic mass 106.905, and Ag-109, relative atomic mass 108.905. Find the natural abundances of these two isotopes.

3.6* * A sample of chlorine consists of 80.00 mol % of ^{35}Cl, with atomic mass 35.00, and 20.00 mol % of ^{37}Cl, with atomic mass 37.00. Calculate the atomic weight of the chlorine in the sample.

Solution: The symbols ^{35}Cl and ^{37}Cl refer to the two isotopes of chlorine whose mass numbers are, respectively, 35 and 37 and whose atomic masses, according to the facts given in the problem, are 35.00 and 37.00, respectively. If the chlorine were pure ^{35}Cl, its atomic weight would be 35.00. If it were pure ^{37}Cl, its atomic weight would be 37.00. Since 80.00% of the atoms are ^{35}Cl and 20.00% are ^{37}Cl, the mass of the ^{35}Cl is 0.80×35.00 or 28.00 and the mass of the ^{37}Cl is 0.20×37.00 or 7.40. The atomic weight of the mixture is then $28.00 + 7.40$ or 35.40. This procedure is called *taking a weighted average.*

3.7* Deuterium (D) is a naturally occurring heavy isotope of hydrogen that is crucial in nuclear fusion. Its atomic mass is 2.0141. The atomic mass of pure hydrogen (^{1}H) is 1.0078 and the atomic weight of naturally occurring hydrogen is 1.0080. What is the mole percent (relative abundance) of the two isotopes in naturally occurring hydrogen, which is a mixture of ^{1}H and D only?

3.8* The atomic weight of boron is 10.81 and it consists of two isotopes, ^{10}B and ^{11}B. The atomic mass of ^{11}B is 11.01 and its natural abundance is 80.22%. Find the atomic mass of ^{10}B.

3.9 What amount of zinc atoms is 12.00 g of Zn?

Solution: The atomic weight of zinc is 65.37. Therefore, 65.37 g is 1 mol of Zn, and 12.00 g is 12.00/65.37 mol of Zn.

---

* The star following the number denotes the more sophisticated problems.

The calculation, in detail, is

$$12.00 \text{ g Zn} \div 65.37 \text{ g Zn/mol Zn} = \frac{12.00}{65.37} \text{ mol Zn}$$

Since any fraction represents a process of division, the calculation can take the form

$$\frac{12.00 \text{ g Zn}}{65.37 \text{ g Zn/1 mol Zn}} = \frac{12.00}{65.37} \text{ mol Zn}$$

Note that g Zn in the numerator cancels g Zn in the denominator. The answer is then in moles of zinc.

The solution can also be carried out as follows: We are given 12.00 g of zinc. We want to know what amount of zinc this mass represents. We reason that if we multiply

$$\text{g Zn} \times \frac{\text{mol Zn}}{\text{g Zn}}$$

then g Zn in the numerator and denominator will cancel and the answer will be in units of moles of Zn. Since 1 mol of Zn has a mass of 65.37 g, the actual calculation is

$$12.00 \text{ g Zn} \times \frac{1 \text{ mol Zn}}{65.37 \text{ g Zn}} = \frac{12.00}{65.37} \text{ mol Zn}$$

Note that in this and subsequent problems there are three quantities involved: a mass of substance, the amount represented by this mass, and the mass of one mole of this substance. These quantities are related:

$$\text{amount of substance} = \frac{\text{mass of substance}}{\text{mass of one mole}}$$

Given any two of these quantities we can find the third.

3.10 A mass of 35.59 g of potassium is what amount of potassium atoms?

3.11 A mass of 140.4 g of neon is what amount of Ne atoms?

3.12 Find the amounts of the following substances that have a mass of 101 g: (a) boron, (b) platinum, (c) uranium.

3.13 Elemental sulfur exists in a number of different forms at different temperatures. At room temperature it is primarily S_8, but as the temperature is raised it can be converted virtually completely to S_2 gas. At still higher temperature the S_2 can be converted completely to gaseous sulfur atoms. Find the amount of each of these three different forms of sulfur that has a mass of 64.1 g.

3.14 Calculate the mass of 2.32 mol of C atoms.

Solution: The atomic weight of carbon is 12.011. Therefore, 1 mol C has a mass of 12.011 g.

$$2.32 \text{ mol C} \times \frac{12.011 \text{ g C}}{\text{mol C}} = 27.9 \text{ g C}$$

Note that mol in the numerator and denominator cancel. The answer is in grams.

3.15 Calculate the mass of 5.52 mol of chlorine atoms.

3.16 Calculate the mass of 0.140 mol of lithium atoms.

3.17 Calculate the mass of 0.961 mol of cobalt.

3.18 Find the mass of 0.624 mol of the following substances: (a) xenon, (b) calcium, (c) lithium.

3.19 Calculate the total mass of 3.20 mol of silicon and of 3.20 mol of iron.

3.20 How many atoms are there in 20.00 g of boron?

Solution: We know that there are 6.022×10^{23} atoms in 1 mol of B. If we know how many *moles* of B there are in 20 g of B, we can then multiply moles of B by 6.022×10^{23} atoms per mole. The atomic weight of boron is 10.811. That means that there are 10.881 g in 1 mol of B. Therefore

$$20 \text{ g B} \times \frac{1 \text{ mol B}}{10.811 \text{ g B}} = \frac{20}{10.811} \text{ mol B}$$

$$\frac{20}{10.811} \text{ mol B} \times \frac{6.022 \times 10^{23} \text{ atoms}}{1 \text{ mol B}}$$

$$= \frac{20}{10.811} \times 6.022 \times 10^{23} \text{ atoms}$$

$$= 11 \times 10^{23} \text{ atoms}$$

$$= 1.1 \times 10^{24} \text{ atoms}$$

The entire calculation can be combined in one operation:

$$20 \text{ g B} \times \frac{1 \text{ mol B}}{10.811 \text{ g B}} \times \frac{6.022 \times 10^{23} \text{ atoms B}}{1 \text{ mol B}} = 1.1 \times 10^{24} \text{ atoms B}$$

Note that grams cancel grams and moles cancel moles.

3.21 What amount of Sn (in moles) is 4.63×10^{21} atoms of Sn?

Solution: 6.022×10^{23} atoms is 1 mol of Sn. Therefore, 4.63×10^{21} atoms is

$$\frac{4.63 \times 10^{21}}{6.022 \times 10^{23}} \text{ mol Sn}$$

The detailed calculation is

$$\frac{4.63 \times 10^{21}\text{ atoms}}{6.022 \times 10^{23}\text{ atoms/mol}} = \frac{4.63 \times 10^{21}}{6.022 \times 10^{23}}\text{ mol}$$
$$= 7.69 \times 10^{-3}\text{ mol Sn}$$

3.22 What mass of copper will contain 3.22×10^{24} atoms of copper?

Solution: The atomic weight of copper is 63.54. Therefore, there are 63.54 g of Cu in 1 mol of Cu. If we know how many *moles* of Cu are represented by 3.22×10^{24} atoms, we can multiply this number of moles by 63.54 g Cu/mol Cu.

We know that 1 mol contains 6.022×10^{23} atoms; therefore, 3.22×10^{24} atoms is

$$\frac{3.22 \times 10^{24}}{6.022 \times 10^{23}}\text{ mol Cu and}$$

$$\frac{3.22 \times 10^{24}}{6.022 \times 10^{23}}\text{ mol Cu} \times \frac{63.54\text{ g Cu}}{1\text{ mol Cu}} = 339\text{ g Cu}$$

The entire calculation can be carried out in one operation:

$$\frac{3.22 \times 10^{24}\text{ atoms}}{6.022 \times 10^{23}\text{ atoms/mol}} \times \frac{63.54\text{ g Cu}}{1\text{ mol Cu}} = 340\text{ g Cu}$$

3.23 How many atoms are there in 140 g of manganese?

3.24 Calculate the mass in grams of 8.00×10^{23} atoms of vanadium.

3.25 What mass of chromium will contain 6.12×10^{23} atoms of chromium?

3.26 How many atoms of carbon have a mass of 46.8 g?

3.27 What amount of uranium atoms (in moles) will contain 4.81×10^{20} atoms?

3.28 How many atoms are there in 125 lb of nickel?

3.29 How many atoms are there in 0.0260 ton of chromium?

3.30 Find the number of atoms in a 35.2-kg sample of uranium.

Chapter 4

Mass–Amount Relationships in Chemical Formulas

When atoms combine to form compounds, they always do so in definite proportions by mass. As a result, the composition of every pure compound is definite and constant. This statement is the *law of definite composition*. The definite composition of a particular compound is represented by its *chemical formula*.

When you write a chemical formula, whether corresponding to a molecule or to a collection of ions, you are conveying a great deal of information. You are indicating the elements that make up the compound and you are indicating the amount of each element. So when you write the formula for glucose as $C_6H_{12}O_6$ you are saying that glucose contains carbon, hydrogen, and oxygen and that a single unit of structure of glucose is a molecule containing 6 atoms of carbon, 12 atoms of hydrogen, and 6 atoms of oxygen. When you write the formula $CaCl_2$ for an ionic compound you are saying that the compound contains ions from calcium and chlorine in the ratio of 1:2. This type of quantitative information is fundamental to doing chemistry.

But formulas contain much more information as well. Some information is always considered available to the chemist. Among other things the Periodic Table is fundamental. So is a Table of Relative Atomic Masses (called in chemical slang a Table of Atomic Weights). So whenever you have to do a problem in real life chemistry you can presume that you have a Periodic Table and a Table of Atomic Weights available. Remember it is erroneously called a Table of Atomic Weights because the relative atomic masses of the elements are obtained by the calculation of a weighted average of their isotopes.

The combination of a formula, which gives the identities and amounts of the constituent elements, and the Table of Relative Atomic Masses, allows us to calculate the relative mass of the formula. We simply add the relative masses of the elements (that we look up in the Table), taking into account the amount of each element. If the formula is of a molecule then the result is the relative molecular mass, called in slang the molecular weight. If the formula is of a collection of ions then it is the relative formula mass or in slang the formula weight.

One molecular weight (in grams) of any substance contains 6.022×10^{23} *molecules*. The reasoning that leads to this conclusion can be illustrated in the case of a compound such as CS_2. This formula tells us that 1 atom of C combines with 2 atoms of S to form 1 molecule of CS_2. Therefore 6.022×10^{23} atoms (1 mol) of C must combine with $2 \times 6.022 \times 10^{23}$ atoms (2 mol) of S to yield 6.022×10^{23} molecules of CS_2. Therefore, 1 molecular weight of CS_2 must contain 6.022×10^{23} individual molecules. In a similar manner it can be reasoned that 1 *molecular weight (in grams) of any molecular substance contains* 6.022×10^{23} *molecules.*

But 6.022×10^{23} units is, by definition, 1 mol. *Since the formula represents one molecular weight, since one molecular weight consists of* 6.022×10^{23} *molecules, and since* 6.022×10^{23} *molecules is 1 mol, the chemical formula of a compound, when used in an equation, represents 1 mol of that substance. This is an important fact to remember.*

We define 1 mol of an ionic compound such as NaCl as that quantity of NaCl which contains 1 mol of Na^+ ions and 1 mol of Cl^- ions; since the symbol for an element represents 1 mol of atoms of that element, the formula NaCl, in fact, represents 1 mol of NaCl. Likewise, 1 mol of Na_2SO_4 is defined as that quantity of Na_2SO_4 which contains 2 mol of Na^+ ions and 1 mol of SO_4^{2-} ions; the formula Na_2SO_4 does, in fact, represent 1 mol of Na_2SO_4 The same line of reasoning can be applied to all compounds that, like NaCl and Na_2SO_4, exist as ions rather than as discrete neutral molecules. Accordingly, the statement that the formula of a compound used in a chemical equation represents 1 mol of that compound applies to all compounds even though they may not exist as discrete neutral molecules.

USE OF THE UNIT *MOLE*

In the present-day use of the unit *mole, the symbol or formula for any chemical substance*, whether it is an element, a compound, or an ion, when used in a formula or equation, *represents 1 mole of that substance*. So that there may be no question about the identity of the mole of substance, its symbol or formula should always be given. The following statements, each of which represents correct usage of the term *mole*, illustrate this point.

The formula $KClO_3$ tells us that 1 mol of $KClO_3$ contains 1 mol of K, 1 mol of Cl, and 3 mol of O.

One mole of $KClO_3$, when heated, will yield 1 mol of KCl and 1.5 mol of O_2.

One mole of Na_2SO_4 contains 2 mol of Na^+ ions and 1 mol of SO_4^{2-} ions.

In the reaction represented by the equation $C + O_2 \rightarrow CO_2$, 1 mol of C atoms combines with 1 mol of O_2 molecules to form 1 mol of CO_2 molecules. One mole of CO_2 molecules can be considered to consist of 1 mol of C atoms and 2 mol of O atoms.

In the reaction $Ag^+ + Cl^- \longrightarrow AgCl$, the Ag^+ ions and Cl^- ions combine in the ratio of 1 mol of Ag^+ ions to 1 mol of Cl^- ions to form 1 mol of AgCl.

MASS RELATIONSHIPS IN CHEMICAL FORMULAS

As we have seen, a chemical formula is a statement about amount. For example cryolite (used in glass making and metallurgy) has the formula Na_3AlF_6. The formula tells us that 1 mol of cryolite contains 3 mol of Na, 1 mol of Al, and 6 mol of F.

We can present this same result as the fraction by amount of each element. We find the fraction by dividing the quantity of interest by the total quantity. One mole of cryolite has a total of $3 + 1 + 6 = 10$ mol of atoms. Thus the fractions by amounts are Na, $3/10 = 0.3$, Al, $1/10 = 0.1$, and F, $6/10 = 0.6$. *Notice that the sum of all the fractions is 1.* We could also express this composition as amount in percents by multiplying all the fractions by 100%. Thus Na is 30% by amount and so on. *The sum of all the percents is 100%.*

Because a Table of Atomic Weights, which allows amount to mass conversions, is available, we can also find the composition of cryolite or any other substance by mass. That means converting the amount of each element to a mass of each element and then finding the fraction or percentage of each element in the substance by mass.

DETERMINING THE FORMULA OF A COMPOUND

We can also do this type of calculation in reverse. That is, given the composition by mass of an unknown substance, we can find amount relationships and get a formula. But the formula we get is limited. It is a formula that gives the relative amounts of the constituent elements. In the case of molecular compounds and some ionic compounds it may not give the actual formula. This type of formula is called the *empirical formula*. The subscripts are in lowest terms (no factors in common). Thus hydrogen peroxide, with the molecular formula H_2O_2, has the empirical formula HO; and ethylene, with the molecular formula C_2H_4, has the empirical formula CH_2.

Suppose we want to determine experimentally the empirical formula for water. First we synthesize pure water (prepare it from pure oxygen and pure hydrogen). We find, in an idealized experiment, that 15.999 g of oxygen combines with exactly 2.016 g of hydrogen to form exactly 18.015 g of water. We check our results by analyzing (breaking down) these 18.015 g of water, and we find that this quantity of water yields exactly 15.999 g of oxygen and 2.016 g of hydrogen. The atomic weight of oxygen is 15.999 and the atomic weight of hydrogen is 1.008. That means that 1 mol of O has combined with 2 mol of H to form 1 mol of water. The empirical formula for water is therefore H_2O. The subscript to the right of and below the H means that there are two atoms of hydrogen combined with one atom of oxygen. We obtained the 2, representing the number of atoms of H, by dividing the 2.016 g of hydrogen by 1.008 g (the atomic weight of hydrogen). That is,

$$2.016 \text{ g H} \times \frac{1 \text{ mol H}}{1.008 \text{ g H}} = 2 \text{ mol H}$$

$$15.999 \text{ g O} \times \frac{1 \text{ mol O}}{15.999 \text{ g O}} = 1 \text{ mol O}$$

Summarizing what we did in getting the formula for water, we proceed as follows in determining the empirical formula for any chemical compound.

1. Determine the exact composition of the compound, that is, the mass of each element that combines, or the percentage of each element in the compound, either by analysis or by synthesis.
2. Divide the mass of each element, or the percentage of each element, by its atomic weight. The simplest whole-number ratio between the quotients gives the empirical formula.

In Probs. 4.50–4.60, which are designed to show how formulas are determined, the results of the experimental analysis or synthesis are given. Only the subsequent calculations are required.

CALCULATIONS FROM THE FORMULA OF A COMPOUND

PROBLEMS

4.1 Calculate the molecular weight of glucose.

Solution:

Step 1 Write the formula. $C_6H_{12}O_6$.

Step 2 Look up the atomic weights of the elements in the Table. C = 12.011, H = 1.008, and O = 15.999.

Step 3 Multiply each atomic weight by the subscript of the element in the formula. $6 \times 12.011 = 72.066$; $12 \times 1.0079 = 12.095$; $6 \times 15.999 = 95.994$. Watch significant figures.

Step 4 Add the results, watching significant figures. $72.066 + 12.095 + 95.994 = 180.155$, the molecular weight of glucose.

This result can be regarded as the mass of a single "weighted average" glucose molecule in units of amu (u). It gives us the mass–amount relationships for glucose and therefore the conversion factors we need for calculations about glucose.

4.2 What is the mass of 1.00 mol of H_2SO_4?

Solution: A mole is the mass in grams of the atomic weights of the elements represented by the formula of a substance. Therefore, to find the mass of 1.00 mol of H_2SO_4 we simply find the sum of the atomic weights in grams of its constituent atoms.

$$\begin{aligned} 2\text{H} &= 2.0\text{ g} \\ 1\text{S} &= 32.1\text{ g} \\ 4\text{O} &= \underline{64.0\text{ g}} \\ 1.00\text{ mol of H}_2\text{SO}_4 &= 98.1\text{ g} \end{aligned}$$

4.3 Calculate the formula weight of calcium chloride.

Solution: For $CaCl_2$ we repeat the process of 4.1 and 4.2 with the appropriate numbers:

$$40.078 + (2)35.453 = 110.984.$$

This time the result corresponds to the relative mass of the weighted average formula and is commonly called the formula weight. It also gives us the mass–amount relationships for calcium chloride.

4.4 Calculate the formula weight of $Al_2(SO_4)_3$.

Solution: The chemical formula $Al_2(SO_4)_3$ indicates that aluminum sulfate is made up of aluminum, sulfur, and oxygen combined in the ratio of 2 atoms of aluminum to 3 atoms of sulfur to 12 atoms of oxygen. The SO_4^{2-} ion is enclosed in parentheses with a subscript 3 outside the parentheses. This notation means that the ion is taken 3 times. The subscript 2 applies only to the aluminum atom; the subscript 3 applies to the entire SO_4^{2-} ion.

$$\begin{aligned} 2\text{Al} &= 54.0 \\ 3\text{S} &= 96.3 \\ 12\text{O} &= \underline{192.0} \\ \text{Formula weight} &= 342.3 \end{aligned}$$

4.5 Find the mass of 2.32 mol of $NaHCO_3$, sodium bicarbonate (baking soda).

Solution: Start by using the Table of Atomic Weights to find the formula weight of the sodium bicarbonate. Remember the mass of 1 mol is the formula weight in grams. The calculation of formula weight is

$$1 \times 22.99 + 1 \times 1.008 + 1 \times 12.01 + 3 \times 16.00 = 84.01$$

So now we can write 1 mol $NaHCO_3$ = 84.01 g $NaHCO_3$ to get conversion factors.

We want the factor that cancels mol $NaHCO_3$ and leaves g $NaHCO_3$, which is what the problem is asking for.

$$\frac{2.32\ \cancel{\text{mol NaHCO}_3} \times 84.01\ \text{g NaHCO}_3}{\cancel{1\ \text{mol NaHCO}_3}} = 195\ \text{g NaHCO}_3$$

4.6 What amount is 7 g of CH_4?

Solution: One mole of CH_4 is 16 g (from the Table of Atomic Weights)

$$7\ \text{g CH}_4 \times \frac{1\ \text{mol CH}_4}{16\ \text{g CH}_4} = 0.4\ \text{mol CH}_4$$

Note: Grams of CH_4 cancel, giving the answer as moles of CH_4.

4.7 What amount is 11 g of NH_3?

4.8 A mass of 81 g of CO_2 (carbon dioxide) is what amount of CO_2?

Solution:

$$81\ \text{g CO}_2 \times \frac{1\ \text{mol CO}_2}{44.0\ \text{g CO}_2} = 1.8\ \text{mol CO}_2$$

4.9 What amount of $C_7H_5N_3O_6$ (TNT) is 214 g?

4.10 The mass of 0.234 mol of a substance is 53.7 g. Find the molecular weight of the substance.

4.11 The mass of 0.0863 mol of an acid HXO_4 is 12.5 g. Find the atomic weight of the element X. Identify X.

4.12 What amount of P is there in 2.4 mol of P_4O_{10}?

Solution: The formula P_4O_{10} shows that 1 mol of P_4O_{10} contains 4 mol of P.

$$2.4 \text{ mol } P_4O_{10} \times \frac{4 \text{ mol P}}{1 \text{ mol } P_4O_{10}} = 9.6 \text{ mol P}$$

Note: Moles of P_4O_{10} cancel, giving the answer as moles of P.

4.13 What amount of C is there in 0.190 mol of $C_{16}H_{18}N_2O_4S$ (penicillin)?

4.14 What amount of H is there in 5.6 mol of N_2H_4 (hydrazine)?

4.15 Find the amount of O contained in 0.85 mol of the following substances: (a) NO_2, (b) N_2O_4, (c) $Al_2(SO_4)_3$, (d) $Fe_2(CO)_9$.

4.16 What amount of CCl_4 contains 2.4 mol of Cl?

Solution:

$$2.4 \text{ mol Cl} \times \frac{1 \text{ mol } CCl_4}{4 \text{ mol Cl}} = 0.60 \text{ mol } CCl_4$$

Note: Moles of Cl cancel, giving the answer as moles of CCl_4.

4.17 Find the amount of penicillin, $C_{16}H_{18}N_2O_4S$, that contains 0.10 mol of C.

4.18 What amount of C_3H_8 (propane) contains 31 mol of H?

4.19 Find the amount of SiO_2 that contains 1.0×10^{20} atoms of O.

4.20 How many molecules are there in 2.70 mol of H_2S?

Solution: One mole of H_2S contains 6.022×10^{23} molecules.

$$2.70 \text{ mol } H_2S \times \frac{6.022 \times 10^{23} \text{ molecules } H_2S}{1 \text{ mol } H_2S} = 1.63 \times 10^{24} \text{ molecules}$$

Note: Moles of H_2S cancel, giving the answer as molecules.

4.21 How many molecules are there in 0.0569 mol of $C_9H_8O_4$ (aspirin)?

4.22 What amount of CH_4 contains 4.31×10^{25} molecules of CH_4?

Solution:

$$4.31 \times 10^{25} \text{ molecules } CH_4 \times \frac{1 \text{ mol } CH_4}{6.022 \times 10^{23} \text{ molecules } CH_4} = 71.6 \text{ mol } CH_4$$

4.23 What amount of NO is 7.89×10^{20} molecules of NO?

4.24 How many molecules of SO_2 are there in 200 g of SO_2?

Solution: The molecular weight of SO_2 is 64.1. Therefore, 200 g of SO_2 is 200/64.1 mol of SO_2.

$$\frac{200}{64.1} \text{ mol } SO_2 \times \frac{6.022 \times 10^{23} \text{ molecules } SO_2}{1 \text{ mol } SO_2} = 1.88 \times 10^{24} \text{ molecules } SO_2$$

4.25 Find the number of molecules of H_2 in 1.0 fg (10^{-15} g) of H_2.

4.26 How many molecules of CH_4 (methane) are there in 1.91 g of CH_4?

4.27 What mass of CO_2 contains 5.10×10^{24} molecules of CO_2?

Solution: One mole of CO_2 has a mass of 44.0 g and contains 6.022×10^{23} molecules.

$$5.10 \times 10^{24} \text{ molecules } CO_2 \times \frac{44.0 \text{ g } CO_2}{6.022 \times 10^{23} \text{ molecules } CO_2} = 373 \text{ g } CO_2$$

Note: Molecules of CO_2 cancel molecules of CO_2.

4.28 Find the mass of 2.41×10^{30} molecules of water.

4.29 Calculate the mass in grams of 5.03×10^{22} molecules of SO_3.

4.30 The mass of an average molecule of a substance is 2.59×10^{-22} g. Find the molecular weight of the substance.

4.31 Find the total number of atoms in 12.9 g of $(NH_4)_2Cr_2O_7$.

4.32 The mass of 2.60 mol of a compound is 312 g. Calculate the molecular weight of the compound.

Solution: The molecular weight is equal to the mass in grams of 1 mol.

$$\frac{312 \text{ g}}{2.60 \text{ mol}} = 120 \text{ g/mol}$$

Therefore, molecular weight = 120.

Note: Any fraction, when solved, expresses the value per one unit. Thus in the above calculation,

$$\frac{312\ \text{g}}{2.60\ \text{mol}} = 120\ \text{g per } \textit{one} \text{ mole.}$$

4.33 A 7.25-mol sample of a compound has a mass of 689 g. Calculate the molecular weight of the compound.

4.34 What mass of sulfur is in 2.20 mol of H_2S?

Solution: One mole of H_2S contains 1 mol of S. One mole of S has a mass of 32.1 g.

$$2.20\ \text{mol H}_2\text{S} \times \frac{1\ \text{mol S}}{1\ \text{mol H}_2\text{S}} \times \frac{32.1\ \text{g S}}{1\ \text{mol S}} = 70.6\ \text{g S}$$

4.35 What mass of sulfur is in 1.67 mol of P_4S_3?

4.36 What amount of SiO_2 (silica) contains 58.5 g of oxygen?

4.37 What amount of O is there in 182 g of $KClO_3$?

Solution:

$$182\ \text{g KClO}_3 \times \frac{1\ \text{mol KClO}_3}{122.6\ \text{g KClO}_3} \times \frac{3\ \text{mol O}}{1\ \text{mol KClO}_3} = 4.45\ \text{mol O}$$

4.38 What amount of S is there in 0.0254 g of CS_2?

4.39 What mass of Sb_2S_3 contains 3.14 mol of S?

4.40 What mass of phosphorus is in 160 g of P_4O_{10}?

Solution: One mole of P_4O_{10} contains 4 mol of P. Moles of P = 4 × moles of P_4O_{10}.

$$160\ \text{g P}_4\text{O}_{10} \times \frac{1\ \text{mol P}_4\text{O}_{10}}{284.0\ \text{g P}_4\text{O}_{10}} = \frac{160}{284}\ \text{mol P}_4\text{O}_{10}$$

$$\frac{160}{284}\ \text{mol P}_4\text{O}_{10} \times \frac{4\ \text{mol P}}{1\ \text{mol P}_4\text{O}_{10}} = 4 \times \frac{160}{284}\ \text{mol P}$$

$$\frac{4 \times 160}{284}\ \text{mol P} \times \frac{31.0\ \text{g P}}{1\ \text{mol P}} = 69.9\ \text{g P}$$

The solution, in one operation is

$$160 \text{ g } P_4O_{10} \times \frac{1 \text{ mol } P_4O_{10}}{284.0 \text{ g } P_4O_{10}} \times \frac{4 \text{ mol P}}{1 \text{ mol } P_4O_{10}} \times \frac{31.0 \text{ g P}}{1 \text{ mol P}} = 69.9 \text{ g P}$$

4.41 What mass of oxygen is contained in 2.51 g of $K_2Cr_2O_7$?

4.42 What mass of SO_3 contains 3.58 g of oxygen?

4.43 What mass in tons of Fe_2O_3 contains 12.0 tons of Fe?

4.44 Calculate the mass percent of carbon in CO_2.

Solution: By definition, mass percent means parts per 100 parts by mass. So all we need do in this problem is find how many grams of C there are in 100 g of CO_2; the answer will then be the percent of C in CO_2. The term *mass fraction* is more general. It is the ratio of the number of parts by mass of the particular thing you want to find to the parts by mass of the whole thing. That is,

$$\text{mass fraction of C in } CO_2 = \frac{\text{mass of the C in } CO_2}{\text{mass of the } CO_2}$$

A mole of CO_2 has a mass of 44.0 g and contains 12.0 g of C. Therefore,

$$\text{mass fraction of C in } CO_2 = \frac{12.0 \text{ g C}}{44.0 \text{ g } CO_2} = 0.273$$

To convert fraction to percent we simply multiply by 100%. Doing the whole calculation in one operation, we have

$$\text{mass percent of C in } CO_2 = \frac{12.0 \text{ g C}}{44.0 \text{ g } CO_2} \times 100\% = 27.3\%$$

The general form of this relationship is

percent of an element in a compound

$$= \frac{\text{mass of the element in the formula of the compound}}{\text{formula weight of the compound}} \times 100\%$$

4.45 Calculate the percent of oxygen in

a. $KMnO_4$

b. Na_2CO_3

4.46 For the compound $C_6H_5NO_2$ (nitrobenzene) calculate

a. amount of $C_6H_5NO_2$ in 131 g of $C_6H_5NO_2$
b. mass of C in 3.27 mol of $C_6H_5NO_2$
c. mass of C in 131 g of $C_6H_5NO_2$
d. mass of C per 10.0 g of N
e. amount of O in 230 g of $C_6H_5NO_2$
f. amount of $C_6H_5NO_2$ containing 5.0 g of N
g. mass of $C_6H_5NO_2$ containing 0.500 mol of C
h. molecules of $C_6H_5NO_2$ in 4.59 g of $C_6H_5NO_2$
i. atoms of C in 3.00 g of $C_6H_5NO_2$
j. the percent by mass of carbon
k. atoms of N per atom of C
l. mass of H per gram of C
m. mass of C per mole of H

4.47 Some important ores of zinc are zincblende (ZnS), franklinite ($ZnCO_3$), and willemite (Zn_2SiO_4). Find the percentage by mass of zinc in the ore that is richest in zinc.

4.48* Dealer A sells NaClO bleach at 80 cents per kilogram of NaClO content. Dealer B sells the identical bleach at $1.00 per kilogram of ClO content. Which dealer offers the better bargain?

4.49* The element M forms the chloride MCl_4. This chloride contains 65.0% chlorine by mass. Calculate the atomic weight of M, knowing that the atomic weight of chlorine is 35.5.

Hint: What must be the numerical value of the term below?

$$\frac{\text{\% of M/atomic weight of M}}{\text{\% of Cl/atomic weight of Cl}}$$

CALCULATION OF THE FORMULA OF A COMPOUND

PROBLEMS

4.50 It was found that 56 g of iron combined with 32 g of sulfur. Calculate the empirical formula of the compound that was formed.

Solution: The formula of a compound gives the amount in moles of atoms of each element in 1 mol of the compound. The atomic weight of iron is 56 and of sulfur is 32. Therefore, 56 g is 1 mol of Fe and 32 g is 1 mol of S. Therefore, Fe and S are combined in the ratio of 1 mol of Fe to 1 mol of S. Since 1 mol of iron atoms is represented by the symbol Fe and 1 mol of sulfur atoms by the symbol S, the empirical formula for the compound iron sulfide is FeS.

4.51 When heated, 433.22 g of a pure compound yielded 401.22 g of mercury and 32.00 g of oxygen. Calculate the empirical formula of the compound.

Solution: First we find the amount of atoms of each element present in the usual way.

$$401.22 \text{ g Hg} \times \frac{1 \text{ mol Hg}}{200.61 \text{ g Hg}} = 2 \text{ mol Hg}$$

$$32.00 \text{ g O} \times \frac{1 \text{ mol O}}{16.00 \text{ g O}} = 2 \text{ mol O}$$

Since we are interested in getting the simplest formula (the empirical formula), we find the simplest ratio. Since 2 is to 2 as 1 is to 1, the empirical formula is HgO.

4.52 A 10.0-g sample of a pure compound contains 2.82 g of K, 2.56 g of Cl, and 4.62 g of O. Calculate the empirical formula of the compound.

4.53 It is found that 50.7 g of S combines with 93.8 g of Sn to form one compound. Find the empirical formula of the compound.

4.54 A sample of mass 60.5 g of a compound containing only nickel, carbon, and oxygen is found to contain 20.8 g of Ni and 17.0 g of C. Find the empirical formula of the compound.

4.55 In the laboratory 2.38 g of copper combined with 1.19 g of sulfur. In a duplicate experiment 4.26 g of copper combined with 1.80 g of sulfur. Are these results in agreement with the law of definite composition?

4.56 When burned, 4.04 g of magnesium combined with 2.66 g of oxygen to form 6.70 g of magnesium oxide. Calculate the empirical formula of the oxide.

Solution: To find the amount of Mg in 6.70 g of magnesium oxide we use the mass in grams of the magnesium and the atomic weight of Mg, and to find the amount of O we use the mass in grams of the oxygen and the atomic weight of O.

$$4.04 \text{ g Mg} \times \frac{1 \text{ mol Mg}}{24.3 \text{ g Mg}} = 0.166 \text{ mol Mg}$$

$$2.66 \text{ g O} \times \frac{1 \text{ mol O}}{16.0 \text{ g O}} = 0.166 \text{ mol O}$$

Therefore, the magnesium and oxygen are combined in the ratio of 0.166 mol of Mg to 0.166 mol of O. But 0.166 is to 0.166 as 1 is to 1. Therefore, the simplest formula is MgO. The formula represents the simplest *whole-number* ratio of the moles. In this case the simplest ratio is 1:1.

4.57 A sample of 4.24 g of copper was heated in oxygen until no further change took place. The resulting oxide had a mass of 4.77 g. Calculate the empirical formula of the oxide.

4.58 A pure compound was found on analysis to contain 31.9% potassium, 28.9% chlorine, and 39.2% oxygen. Calculate its empirical formula.

Solution: To say that a compound contains 31.9% potassium, 28.9% chlorine, and 39.2% oxygen is equivalent to saying that 100 g of the compound contains 31.9 g of potassium, 28.9 g of chlorine, and 39.2 g of oxygen. Therefore, to find the amount of each element in a mole of the compound, we divide the percent of each element by its atomic weight.

$$31.9 \text{ g K} \times \frac{1 \text{ mol K}}{39.1 \text{ g K}} = 0.816 \text{ mol K}$$

$$28.9 \text{ g Cl} \times \frac{1 \text{ mol Cl}}{35.5 \text{ g Cl}} = 0.814 \text{ mol Cl}$$

$$39.2 \text{ g O} \times \frac{1 \text{ mol O}}{16.0 \text{ g O}} = 2.45 \text{ mol O}$$

The mole ratio of K to Cl to O is 0.816 to 0.814 to 2.45. To simplify this ratio, we divide all three numbers by the smallest.

$$\text{K} = \frac{0.816}{0.814} = 1 \qquad \text{Cl} = \frac{0.814}{0.814} = 1 \qquad \text{O} = \frac{2.45}{0.814} = 3$$

The simplest formula of the compound is, therefore, $KClO_3$.

4.59 A compound contains 36.5% sodium, 25.5% sulfur, and 38.1% oxygen. What is its empirical formula?

4.60 A compound was found on analysis to contain 27.7% magnesium, 23.6% phosphorus, and 48.7% oxygen. Calculate the empirical formula of the compound.

4.61 An unknown substance is found to contain 55.8% carbon, 11.6% hydrogen, and 32.6% nitrogen by mass. Find the empirical formula of the compound.

4.62* A binary compound of the unknown elements X and Y is $\frac{1}{3}$ X by mass. The atomic weight of element X is $\frac{3}{4}$ the atomic weight of element Y. Find the empirical formula of the compound.

4.63 A compound was found to have the empirical formula CH_2 and a molecular weight of 71. Calculate the molecular formula.

Solution: Calculate the formula weight of CH_2: $1 \times 12 + 2 \times 1 = 14$. Calculate how many times the empirical formula goes into the molecular formula by dividing the empirical formula weight into the molecular weight 71/14 = 5.07 and round the answer off to the nearest integer, which is 5. Multiply each subscript in the empirical formula by this integer to get the molecular formula: $CH_2 \times 5 = C_5H_{10}$.

4.64 A compound was found to contain 65.4% carbon, 9.15% hydrogen, and 25.4% nitrogen. Its molecular weight was found to be 220. Calculate the molecular formula of the compound.

4.65 Mannose is a sugar that contains only carbon, hydrogen, and oxygen and has a molecular weight of 180. A 2.36-g sample of mannose was found on analysis to contain 0.944 g of carbon and 0.158 g of hydrogen. Calculate the molecular formula.

Hint: Calculate the mass of oxygen in the sample by difference and then proceed as in Probs. 4.58 and 4.63.

4.66 A 1.20-g sample of a compound that contains only carbon and hydrogen is burned completely in excess O_2 to yield 3.60 g of CO_2 and 1.96 g of H_2O. Calculate the empirical formula of the compound.

Solution: The mass of carbon in the sample is the same as the mass of carbon in CO_2, and the mass of hydrogen in the sample is the same as the mass of hydrogen in H_2O.

$$\text{Mass of C} = \frac{12 \text{ g C}}{44 \text{ g } CO_2} \times 3.60 \text{ g } CO_2 = 0.982 \text{ g C}$$

$$\text{Mass of H} = \frac{2 \text{ g H}}{18 \text{ g } H_2O} \times 1.96 \text{ g } H_2O = 0.218 \text{ g H}$$

Now proceed as in Probs. 4.51 and 4.58.

4.67 A 2.36-g sample of a compound that contains only carbon, hydrogen, and oxygen is burned completely in excess O_2 to yield 5.76 g of CO_2 and 2.34 g of H_2O. Calculate the empirical formula of the compound.

Solution: As in Prob. 4.66, the mass of carbon and hydrogen in the original sample can be calculated.

$$\text{Mass of C} = \frac{12 \text{ g C}}{44 \text{ g } CO_2} \times 5.76 \text{ g } CO_2 = 1.57 \text{ g C}$$

$$\text{Mass of H} = \frac{2 \text{ g H}}{18 \text{ g } H_2O} \times 2.34 \text{ g } H_2O = 0.260 \text{ g H}$$

Since the total mass of the sample is the sum of the masses of C, H, and O, the mass of oxygen in the sample can be obtained by difference.

$$\text{Mass of O} = \text{mass of sample} - \text{mass of C} - \text{mass of H}$$
$$= 2.36\text{ g} - 1.57\text{ g} - 0.26\text{ g} = 0.53\text{ g O}$$

Now proceed as in Probs. 4.51 and 4.58.

4.68 A sample of a compound containing only carbon and hydrogen was burned in excess oxygen to yield 2.02 g of CO_2 and 0.551 g of H_2O. Calculate the empirical formula of the compound.

4.69 A 1.78-g sample of a compound containing only carbon and hydrogen was burned in excess oxygen to yield 5.79 g of CO_2. Calculate the empirical formula of the compound.

4.70 A 2.44-g sample of a compound containing only carbon, hydrogen, and nitrogen was burned in excess oxygen to yield 6.78 g of CO_2 and 1.35 g of H_2O. Calculate the empirical formula of the compound.

4.71 Complete combustion of a sample of a compound containing only carbon and hydrogen produced 1.404 g of CO_2 and 0.764 g of H_2O. Find the empirical formula of the compound.

4.72 A 1.46-g sample of a compound containing only carbon, hydrogen, and oxygen is burned in excess O_2 to form 3.57 g of CO_2 and 1.45 g of H_2O. Find the empirical formula of the compound.

4.73 A 1.05-g sample of a compound containing only carbon, hydrogen, and nitrogen is burned in excess O_2 to form 2.92 g of CO_2 and 0.581 g of H_2O. The molecular weight of the compound is found to be 238. Find the molecular formula of the compound.

4.74 A 1.48-g sample of a compound containing only carbon, hydrogen, nitrogen, and chlorine was burned in excess O_2 to yield 2.21 g of CO_2 and 0.452 g of H_2O. Another sample of this compound, of mass 2.62 g, was found to contain 1.05 g of Cl. Calculate the empirical formula of the compound.

Hint: The mass of Cl in the 1.48-g sample can be calculated from the mass of Cl in the 2.62-g sample.

$$\text{Mass of Cl} = \frac{1.05\text{ g Cl}}{2.62\text{ g sample}} \times 1.48\text{ g sample} = 0.593\text{ g Cl}$$

The mass of C and the mass of H in the 1.48-g sample are calculated as in Prob. 4.67. The mass of N in the sample = 1.48 g of sample – mass of C – mass of H – 0.593 g of Cl as in Prob. 4.67. Now proceed as in Probs. 4.51 and 4.58.

4.75 A 3.42-g sample of a compound containing only carbon, hydrogen, nitrogen, and oxygen was burned in excess O_2 to yield 2.47 g of CO_2 and 1.51 g of H_2O. Another sample of this compound of mass 5.26 g was found to contain 1.20 g of N. Calculate the empirical formula of the compound.

4.76* The following data were obtained from experiments to find the molecular formula of benzocaine, a local anesthetic, which contains only carbon, hydrogen, nitrogen, and oxygen. Complete combustion of a 3.54-g sample of benzocaine with excess O_2 formed 8.49 g of CO_2 and 2.14 g of H_2O. Another sample of mass 2.35 g was found to contain 0.199 g of N. The molecular weight of benzocaine was found to be 165. Find the molecular formula of benzocaine.

4.77* A 2.52-g sample of a compound containing only carbon, hydrogen, nitrogen, oxygen, and sulfur was burned in excess O_2 to yield 4.23 g of CO_2 and 1.01 g of H_2O. Another sample of the same compound, of mass 4.14 g, yielded 2.11 g of SO_3. A third sample, of mass 5.66 g, yielded 2.27 g of HNO_3. Calculate the empirical formula of the compound.

4.78* A compound of molecular weight 177 contains only carbon, hydrogen, bromine, and oxygen. Analysis shows that a sample of the compound contains 8 times as much carbon as hydrogen, by mass. Calculate the molecular formula of the compound.

4.79* The elements X and Y form a compound that is 40% X and 60% Y by mass. The atomic weight of X is twice that of Y. What is the empirical formula of the compound?

Hint: Assign some value, 1 for instance, for the atomic weight of Y. The atomic weight of X will then be 2. Then proceed as in Prob. 4.58.

4.80* Three pure compounds are formed when 1.00-g portions of element X combine with, respectively, 0.472 g, 0.630 g, and 0.789 g of element Z. The first compound has the formula X_2Z_3. What are the empirical formulas of the other two compounds? How does the atomic weight of X compare with that of Z?

Hint: Arbitrarily assign some convenient value, 0.50 for instance, for the atomic weight of X. Then proceed as in Prob. 4.49.

Chapter 5

Mass–Amount Relationships in Chemical Reactions

I. Stoichiometry

We have looked at some of the quantitative relationships in chemical formulas. Now let's do the same for chemical change.

CHEMICAL EQUATIONS

We describe chemical change, how substances are converted to other substances, with a chemical equation. A chemical equation is a precise description of the conversion of one or more substances called starting materials or *reactants* into one or more new substances called *products*.

Chemical equations are basically similar to algebraic equations, with chemical formulas replacing x and y. But there is some additional information built into a chemical equation. The most obvious is that there is an arrow instead of an equal sign. The arrow indicates the direction of chemical change. By tradition the reactants appear to the left of the arrow and the products to the right.

Two important laws govern ordinary chemical processes. First is the law of conservation of mass, meaning that the total mass of the products must equal the total mass of reactants. Second is that there is no transmutation of elements, that is the conversion of one element into another does not take place. This law means that the number and type of atoms in the products must equal the number and type of atoms in the reactants.

So when we write a chemical equation it must meet these two requirements. When we write a chemical equation we do two major things. First we list all the reactants and products, which tells us what the change is. Then we make sure these two requirements are met using a process called balancing. A balanced (redundant) chemical equation is obtained by using *coefficients* (not subscripts or superscripts), which are small whole numbers (fractions can also be used, but we generally prefer whole numbers) written preceding the formulas in the equation. A coefficient multiplies everything in the formula that follows it. When the coefficient is 1 we don't write it.

When we speak of an equation we will always mean a *balanced equation.* So, the phrase "balanced chemical equation" is actually redundant.

It is important to remember that the *chemical formula* of a substance, when it appears in a chemical reaction, *refers to* **one mole** *of that substance.*

The process by which the balanced chemical equation is used as a basis for making calculations is called *stoichiometry*. In a common use of the term, the *stoichiometry of a reaction* refers to the *mole relationships* represented by the equation for that reaction.

PROBLEMS

5.1 What amount of O_2 will be obtained by heating 3.50 mol of $KClO_3$?

Solution: The equation for the liberation of O_2 from $KClO_3$ is

$$\underset{2\ \text{mol}}{2KClO_3} \rightarrow \underset{2\ \text{mol}}{2KCl} + \underset{3\ \text{mol}}{3O_2}$$

In order to be able to solve this problem at all we must *know* that heating liberates all the oxygen from $KClO_3$. That is, we must *know the reaction* that occurs as written.

The equation tells us that 3 mol of O_2 is liberated from 2 mol of $KClO_3$ or that 1 mol of $KClO_3$ liberates 1.5 mol of oxygen.

Therefore, 3.50 mol of $KClO_3$ will liberate 3.50×1.5 or 5.25 mol of O_2. We can also write the equation in the form

$$KClO_3 \rightarrow KCl + \frac{3}{2}O_2$$

which tells us at a glance that the amount of O_2 liberated is 1.5 times the amount of $KClO_3$ that reacts.

5.2 When antimony is burned in oxygen, the following reaction occurs:

$$4Sb + 3O_2 \rightarrow 2Sb_2O_3$$

What amount of oxygen will be needed to burn 18 mol of antimony? What mass of Sb_2O_3 will be formed?

Solution: A glance at the equation tells us that 3 mol of O_2 is needed for every 4 mol of Sb burned. That means that it takes $\frac{3}{4}$ mol of O_2 to burn 1 mol of Sb. Therefore, $18 \times \frac{3}{4}$ mol 13.5 mol of O_2 will be needed to burn 18 mol of Sb.

To find the mass of Sb_2O_3 we first find the amount of Sb_2O_3. The equation tells us that the amount of Sb_2O_3 formed is half the amount of Sb burned. Therefore, 9 mol of Sb_2O_3 will be formed.

$$9 \text{ mol } Sb_2O_3 \times \frac{291.6 \text{ g } Sb_2O_3}{1 \text{ mol } Sb_2O_3} = 2600 \text{ g } Sb_2O_3$$

5.3 When C_3H_8 is burned in O_2, the products are CO_2 and H_2O. If 2.40 mol of C_3H_8 is burned in a plentiful supply of oxygen, what mass of H_2O and of CO_2 will be formed?

Solution: We can see the molar relationships here without writing out the full equation if we wish. Since one molecule of C_3H_8 contains 3 atoms of C and 8 atoms of H, while one molecule of CO_2 contains 1 atom of C and one molecule of H_2O contains 2 atoms of H, *1 mol of* C_3H_8 *will yield 3 mol of* CO_2 *and 4 mol of* H_2O. Therefore, 2.40 mol of C_3H_8 will yield 2.40×3 or 7.20 mol of CO_2 and 2.40×4 or 9.60 mol of H_2O. Since the molecular weight of CO_2 is 44, 1 mol of CO_2 has a mass of 44 g and 7.20 mol has a mass of 317 g.

Since the molecular weight of H_2O is 18, 1 mol of H_2O has a mass of 18 g and 9.60 mol has a mass of 173 g.

Each calculation can be carried out in one operation.

$$2.40 \text{ mol } C_3H_8 \times \frac{3 \text{ mol } CO_2}{1 \text{ mol } C_3H_8} \times \frac{44.0 \text{ g } CO_2}{1 \text{ mol } CO_2} = 317 \text{ g } CO_2$$

$$2.40 \text{ mol } C_3H_8 \times \frac{4 \text{ mol } H_2O}{1 \text{ mol } C_3H_8} \times \frac{18.0 \text{ g } H_2O}{1 \text{ mol } H_2O} = 173 \text{ g } H_2O$$

Note: In analyzing and solving this problem, we did not *write* the equation for the reaction. However we *thought* about the significant part of the reaction when we observed that since one *molecule* of C_3H_8 contains 3 atoms of C and 8 atoms of H, 1 mol of C_3H_8 will yield 3 mol of CO_2 and 4 mol of H_2O. This relationship was so obvious that it was not necessary to write it down. Had the problem asked us to calculate the amount or the mass of O_2 required to burn the 2.40 mol of C_3H_8, we probably would have needed to *write* the equation for the reaction.

5.4 When C_4H_{10} is burned in excess oxygen, the following reaction occurs:

$$2C_4H_{10} + 13O_2 \rightarrow 8CO_2 + 10H_2O$$

What mass of O_2 will be needed to burn 36.0 g of C_4H_{10}? What mass of CO_2 and what mass of H_2O will be formed?

Solution: The equation tells us that 6.5 mol of O_2 is consumed for every mole of C_4H_{10} burned.

The equation, or the formula, C_4H_{10}, tells us that 1 mol of C_4H_{10} yields 4 mol of CO_2 and 5 mol of H_2O. The molecular weights of C_4H_{10}, O_2, CO_2, and H_2O are 58.0, 32.0, 44.0, and 18.0, respectively. First we convert the mass of C_4H_{10} to amount, using the molecular weight:

$$36.0\text{ g } C_4H_{10} \times \frac{1\text{ mol } C_4H_{10}}{58.0\text{ g } C_4H_{10}} = 0.621\text{ mol } C_4H_{10}$$

$$0.621\text{ mol } C_4H_{10} \times \frac{6.5\text{ mol } O_2}{1\text{ mol } C_4H_{10}} \times \frac{32.0\text{ g } O_2}{1\text{ mol } O_2} = 129\text{ g } O_2$$

$$0.621\text{ mol } C_4H_{10} \times \frac{4\text{ mol } CO_2}{1\text{ mol } C_4H_{10}} \times \frac{44.0\text{ g } CO_2}{1\text{ mol } CO_2} = 109\text{ g } CO_2$$

$$0.621\text{ mol } C_4H_{10} \times \frac{5\text{ mol } H_2O}{1\text{ mol } C_4H_{10}} \times \frac{18.0\text{ g } H_2O}{1\text{ mol } H_2O} = 55.9\text{ g } H_2O$$

Note that in solving this problem we first determined the amount of C_4H_{10} burned. Then we determined the amounts of O_2, CO_2, and H_2O involved. Then we converted moles of O_2, CO_2, and H_2O to grams.

In general, this is the procedure that should be followed in all stoichiometric calculations. *The first questions should be: What amount of reactant do we have? What amount of product will be formed per* **mole** *of reactant?* Having determined the amount in *moles* we then convert to mass in grams.

5.5 What mass of copper will be formed when the hydrogen gas liberated from treatment of 41.6 g of aluminum with excess HCl is passed over excess CuO?

Solution: The reactions that occur are

$$2Al + 6HCl \rightarrow 3H_2 + 2AlCl_3$$
$$H_2 + CuO \rightarrow Cu + H_2O$$

We note that 2 mol of Al liberates 3 mol of H_2, which means that 1 mol of Al liberates 1.5 mol of H_2.

When 1 mol of H_2 reacts with CuO, it will produce 1 mol of Cu.

That means that the 1.5 mol of H_2 liberated by 1 mol of Al will produce 1.5 mol of Cu.

In short, 1 mol of Al will liberate enough hydrogen to produce 1.5 mol of Cu. The Al that reacts and the Cu that is produced are in the *ratio* of 1 mol of Al to 1.5 mol of Cu. The 41.6 g of Al is 41.6/27.0 mol.

Therefore, 1.5 × 41.6/27.0 mol of Cu will be produced.

$$\frac{1.5 \times 41.6}{27.0} \text{ mol Cu} \times \frac{63.5 \text{ g Cu}}{1 \text{ mol Cu}} = 147 \text{ g Cu}$$

More formally we could write

$$41.6 \text{ g Al} \times \frac{1 \text{ mol Al}}{27.0 \text{ g Al}} \times \frac{3 \text{ mol } H_2}{2 \text{ mol Al}} \times \frac{1 \text{ mol Cu}}{1 \text{ mol } H_2} \times \frac{63.5 \text{ g Cu}}{1 \text{ mol Cu}} = 147 \text{ g Cu}$$

5.6 At elevated temperatures, the compound NF_3 decomposes to N_2 and F_2. Find the amounts of N_2 and F_2 formed by the decomposition of (a) 3 mol of NF_3 and (b) 0.268 mol of NF_3.

5.7 When C_2H_6 is burned in excess oxygen, the following reaction occurs:

$$2C_2H_6 + 7O_2 \rightarrow 4CO_2 + 6H_2O$$

What amount of oxygen will be consumed when 1.5 mol of C_2H_6 is burned? What amount of CO_2 and what amount of H_2O will be produced?

5.8 A 0.262-mol amount of the compound As_2S_5 was subjected to a series of treatments by which all the sulfur in the As_2S_5 was converted to $BaSO_4$ and all the arsenic was converted to Ag_3AsO_4. What amount of $BaSO_4$ and Ag_3AsO_4, respectively, was formed?

Hint: Inspection of the formulas of the reactant, As_2S_5, and the products, $BaSO_4$ and Ag_3AsO_4, tells us that 1 mol of As_2S_5 will yield 5 mol of $BaSO_4$ and 2 mol of Ag_3AsO_4. We may, if we wish, write down the skeleton equation

$$1As_2S_5 \rightarrow 2Ag_3AsO_4 + 5BaSO_4$$

5.9 An amount of 3.16 mol of $KClO_3$ was heated until all the oxygen was liberated. This oxygen was then all used to oxidize arsenic to As_2O_5. What amount of As_2O_5 was formed?

Hint: An inspection of the formulas $KClO_3$ and As_2O_5 tells us that 1 mol of $KClO_3$ will liberate enough oxygen to produce 3/5 mol of As_2O_5.

5.10 A certain quantity of $FeCl_3$ was completely oxidized, all the chlorine being liberated as Cl_2 gas. This Cl_2 gas was all used to convert Si to $SiCl_4$. A total of 4.86 mol of $SiCl_4$ was produced. What amount of $FeCl_3$ was oxidized?

5.11 Decomposition of HBr produces H_2 and Br_2. The Br_2 is used to convert As to $AsBr_3$. Find the amount of HBr required to form 0.448 mol of $AsBr_3$.

5.12 The Cl_2 formed by the decomposition of 1.3 mol of PCl_3 is used to convert carbon to CCl_4. Find the amount of CCl_4 formed.

5.13* Hydrolysis of the compound B_5H_9 forms boric acid $B(OH)_3$. Fusion of boric acid with Na_2O forms a borate salt, $Na_2B_4O_7$. Find the amount of B_5H_9 that forms 0.75 mol of the borate salt by this reaction sequence.

Hint: You do not need to write complete equations to solve this problem.

5.14 What mass of oxygen will be required to prepare 200 g of P_4O_{10} from elemental phosphorus?

Solution: The molecular weight of P_4O_{10} is 284; thus 200 g of P_4O_{10} is 200/284 mol of P_4O_{10}.

Inspection of the formulas P_4O_{10} and O_2 tells us that 5 mol of O_2 will be needed to produce 1 mol of P_4O_{10}; therefore, $5 \times 200/284$ mol of O_2 will be needed. The molecular weight of O_2 is 32; therefore, $5 \times 200/284 \times 32 = 113$ g of O_2. The entire calculation, in one operation, is

$$200 \text{ g } P_4O_{10} \times \frac{1 \text{ mol } P_4O_{10}}{284 \text{ g } P_4O_{10}} \times \frac{5 \text{ mol } O_2}{1 \text{ mol } P_4O_{10}} \times \frac{32.0 \text{ g } O_2}{1 \text{ mol } O_2} = 113 \text{ g } O_2$$

5.15 What mass of pure zinc must be treated with an excess of dilute sulfuric acid to liberate 7.65 g of hydrogen?

5.16 What mass of potassium chlorate must be heated to give 60.0 g of oxygen?

Solution:

$$2KClO_3 \rightarrow 2KCl + 3O_2$$

or

$$KClO_3 \rightarrow KCl + 1.5O_2$$

1 mol of $KClO_3$ will yield 1.5 mol of O_2. That means that 1/1.5 or 0.667 mol of $KClO_3$ will yield 1 mol of O_2. In other words, moles of $KClO_3$ needed = 0.667 × moles of O_2 produced.

The formula weight of O_2 is 32.0; of $KClO_3$, 122.5. The entire calculation in one operation is

$$60.0 \text{ g } O_2 \times \frac{1 \text{ mol } O_2}{32.0 \text{ g } O_2} \times \frac{1 \text{ mol } KClO_3}{1.5 \text{ mol } O_2} \times \frac{122.6 \text{ g } KClO_3}{1 \text{ mol } KClO_3} = 153 \text{ g } KClO_3$$

5.17 Direct reaction of P with Cl_2 can form PCl_5 under suitable conditions. Find the mass of Cl_2 required to form 2.70 mol of PCl_5.

5.18 Find the mass of $AsCl_3$ formed by the reaction of 0.133 mol of Cl_2 with arsenic.

5.19 What mass in tons of sulfur must be burned to produce 12 tons of SO_2 gas?

Solution:

$$S + O_2 \rightarrow SO_2$$

Moles of SO_2 = moles of S.

It is not necessary to convert tons to grams in order to do this calculation. We can work with the molecular weight in tons and think of a "ton mole" as the mass in tons that has the numerical value of the molecular weight. Thus

$$12 \text{ tons } SO_2 \times \frac{1 \text{ ton mol } SO_2}{64 \text{ tons } SO_2} \times \frac{1 \text{ ton mol S}}{1 \text{ ton mol } SO_2} \times \frac{32 \text{ tons S}}{1 \text{ ton mol S}} = 6.0 \text{ tons S}$$

As in previous problems, we can if we wish include all the detailed steps in one operation. Because some of the steps are generally obvious from an inspection of the equation for the reaction or the formulas of the substances and can be done "in our heads," we will, it is hoped, fall into the practice of writing only the essential steps, thereby making the solution as brief and efficient as possible. In this problem, for example, we should be able to decide at a glance that 12/64 mol, hence, 12/64 × 32 tons, of S will be needed.

Note: In Probs. 5.19 and 5.20 the numerical solution is the same regardless of the units in which mass is expressed.

5.20 What mass in pounds of ZnO will be formed by the complete oxidation of 153 lb of pure Zn?

5.21 What mass of copper oxide (CuO) can be formed by the oxygen liberated when 122 g of silver oxide is decomposed?

Hint: Moles of CuO formed = moles of Ag_2O decomposed.

5.22 An important reaction that takes place in a blast furnace during the production of iron is the formation of iron metal and carbon dioxide from Fe_2O_3 and carbon monoxide. Find the mass of Fe_2O_3 required to form 910 kg of iron. Find the amount of carbon dioxide that forms in this process.

5.23 A sample of pure MgO was first dissolved in hydrochloric acid to give a solution of $MgCl_2$ that was then converted to a precipitate of pure dry $Mg_2P_2O_7$ having a mass of 7.02 g. Calculate the mass of the sample of MgO.

Hint: Moles of MgO used $= 2 \times$ moles of $Mg_2P_2O_7$ formed.

5.24 The formula weight of P_4S_3 is 220. The formula weight of Ag_3PO_4 is 419. A 13.2-g sample of P_4S_3 was first boiled with excess HNO_3 and eventually treated with excess $AgNO_3$. In the process all the phosphorus in the P_4S_3 was converted to insoluble Ag_3PO_4. What mass of Ag_3PO_4 was formed?

5.25 A sample of impure copper having a mass of 1.25 g was dissolved in nitric acid to yield $Cu(NO_3)_2$. It was subsequently converted, first to $Cu(OH)_2$, then to CuO, then to $CuCl_2$, and finally to $Cu_3(PO_4)_2$. There was no loss of copper in any step. The pure dry $Cu_3(PO_4)_2$ that was recovered had a mass of 2.00 g. Calculate the percent of pure copper in the impure sample.

Hint: The equations for the reactions that occur can be written in the following skeleton form:

$$3Cu \rightarrow 3Cu(NO_3)_2 \rightarrow 3Cu(OH)_2 \rightarrow 3CuO \rightarrow 3CuCl_2 \rightarrow 1Cu_3(PO_4)_2$$

This expression tells us at a glance that moles of pure Cu $= 3 \times$ moles of $Cu_3(PO_4)_2$.

$$\text{Percent of pure Cu} = \frac{\text{mass of pure Cu}}{\text{mass of sample}} \times 100\%$$

5.26 A 5.00-g sample of a crude sulfide ore in which all the sulfur was present as As_2S_5 was analyzed as follows: The sample was digested with concentrated HNO_3 until all the sulfur was converted to sulfuric acid. The sulfate was then completely precipitated as $BaSO_4$. The recovered $BaSO_4$ had a mass of 0.752 g. Calculate the percent of As_2S_5 in the crude ore.

5.27 What mass (in tons) of lead will be obtained from 1310 tons of ore containing 21.0% PbS, the yield of lead being 94.0% of the theoretical amount?

5.28 Carbon disulfide, CS_2, is a very flammable substance that reacts with O_2 to form CO_2 and SO_2. Find the mass of O_2 that is required to react with 9.34 g of CS_2. Find the mass of carbon dioxide and of sulfur dioxide formed in this reaction.

5.29 A certain oxide of lead is converted by H_2 to lead metal and water. One mole of the oxide reacts with 8.1 g of H_2 and forms 622 g of lead. Find the formula of the oxide.

5.30 A pure compound containing 63.3% manganese and 36.7% oxygen was heated, evolving oxygen, until no more reaction took place. The solid product was a pure compound containing 72% manganese and 28% oxygen. Write a chemical equation to represent the reaction that took place.

Hint: Calculate the formulas of the initial and final compounds as in Prob. 4.58.

5.31 After complete reduction of 0.800 g of a pure oxide of lead with excess hydrogen gas, there remained 0.725 g of lead. Write the chemical equation for the reaction that took place.

5.32 A compound contained 27.1% sodium, 16.5% nitrogen, and 56.4% oxygen. A 5.00-g sample of this compound was heated until no more reaction took place. A mass of 0.942 g of oxygen was given off. A pure chemical compound remained as a solid product. Write a chemical equation to represent the reaction that took place.

5.33 A mixture of 12.2 g of potassium and 22.2 g of bromine was heated until the reaction was completed. What mass of KBr was formed?

Solution: From the formula KBr and the equation

$$2K + Br_2 \rightarrow 2KBr$$

we see that potassium and Br_2 combine in the ratio of 2 mol of K to 1 mol of Br_2. The masses we start with are not in this molar ratio, since 12.2 g of K is 12.2/39.1 = 0.312 mol of K and 22.2 g of Br_2 is 22.2/159.8 = 0.139 mol of Br_2. Since 0.139 mol of Br_2 requires only 0.278 mol of K for complete reaction, the K is present in excess and Br_2 is said to be the *limiting reagent* because it limits the quantity of KBr that will form. We now ignore the K and note that the 0.139 mol of Br_2 will form 0.278 mol of KBr.

$$0.278 \text{ mol KBr} \times \frac{119 \text{ g KBr}}{1 \text{ mol KBr}} = 33.1 \text{ g of KBr}$$

5.34 What mass of AgCl will be formed when 35.4 g of NaCl and 99.8 g of $AgNO_3$ are mixed in water solution?

5.35* The fertilizer triple superphosphate of lime is prepared from equal parts $Ca_3(PO_4)_2$, which costs 95.1¢ per kilogram, and H_3PO_4, which costs 79.8¢ per kilogram. In order to achieve the maximum production of the fertilizer an excess of one of the reagents is used. Which reagent should be present in excess in order to produce the fertilizer as economically as possible?

5.36 What mass of Bi_2S_3 will form when 12.3 g of H_2S is mixed with 126 g of $Bi(NO_3)_3$ in water solution?

5.37 Quantities of 11.1 g of H_2 and 33.3 g Cl_2 are mixed. Find the mass of hydrogen chloride that forms.

5.38 Reaction of tungsten with Cl_2 forms WCl_6. Find the mass of the unreacted starting material when 12.6 g of tungsten is treated with 13.6 g Cl_2. Find the mass of WCl_6 that forms.

5.39 Treatment of gold with BrF_3 and KF leads to the formation of Br_2 and the salt $KAuF_4$. Find the limiting reagent when equal masses of the three reagents are mixed. Find the mass of the gold salt formed from 24 g of such a mixture.

5.40* Metallic aluminum reacts with MnO_2 at elevated temperatures to form manganese metal and aluminum oxide. A mixture of the two reagents is 39% Al by mass. Find the limiting reagent in this mixture. Find the mass of manganese that forms from the reaction of 250 g of this mixture.

5.41 The nitrogen in $NaNO_3$ and $(NH_4)_2SO_4$ is all available to plants as fertilizer. Which is the more economical source of nitrogen, a fertilizer containing 30% $NaNO_3$ and costing $9.00 per 100 lb or one containing 20% $(NH_4)_2SO_4$ and costing $8.10 per 100 lb?

5.42 When HgO is heated it decomposes to mercury and O_2. A 75.8-g sample of the oxide decomposes to form 62.7 g of mercury. Find the percent yield of the reaction.

5.43 The reaction of carbon with calcium oxide produces carbon monoxide and CaC_2. A total of 2.45 g of CaC_2 is isolated from the reaction of 5.00 g of CaO with 2.50 g of carbon. Find the percent yield of CaC_2.

5.44 Large quantities of formic acid HCOOH can be prepared in 76% yield from the reaction of carbon monoxide with sodium hydroxide. Find the mass of CO needed to prepare 125 kg of formic acid.

5.45* Exactly 3.00 mol of chromium is reacted with an excess of element Q; all the Cr is converted to Cr_2Q_3. The Cr_2Q_3 is then treated with excess strontium metal; all the Q in the Cr_2Q_3 is converted to SrQ. The SrQ is then reacted with excess sodium metal; all the SrQ is converted to Na_2Q; 782 g of Na_2Q is formed. What is the atomic weight of element Q? Atomic weight of Na = 23.0.

Hint:

$$2Cr \rightarrow 1Cr_2Q_3 \rightarrow 3SrQ \rightarrow 3Na_2Q$$

Therefore, moles of $Na_2Q = 1.5 \times$ moles of Cr.

$$\text{Moles of } Na_2Q = 1.5 \times 3.00 \text{ mol} = 4.5 \text{ mol } Na_2Q.$$

Knowing that the atomic weight of Na is 23.0 and that 782 g is 4.5 mol of Na_2Q, the atomic weight of Q can be calculated.

5.46* When 2.451 g of pure, dry MXO_3 is heated, 0.9600 g of oxygen gas is liberated. The other product is 1.491 g of solid MX. When this MX is treated with excess $AgNO_3$, all of it reacts with the $AgNO_3$ to form solid AgX; 2.869 g of AgX is formed. Knowing that the atomic weights of O and Ag are, respectively, 16.00 and 108.0, calculate the atomic weights of M and X.

Solution:

$$MXO_3 \rightarrow MX + 1.5O_2$$

$$MX + AgNO_3 \rightarrow AgX + MNO_3$$

$$0.9600 \text{ g } O_2 \times \frac{1 \text{ mol } O_2}{32.00 \text{ g } O_2} = 0.03000 \text{ mol } O_2$$

$$\text{Moles of AgX} = \text{moles of MX} = \frac{1}{1.5} \times \text{mol } O_2$$

$$= \frac{1}{1.5} \times 0.03000 \text{ mol } O_2 = 0.02000 \text{ mol AgX}$$

$$\text{Molecular weight of AgX} = \frac{2.869 \text{ g}}{0.02000 \text{ mol}} = 143.5 \text{ g/mol.}$$

$$\text{Atomic weight of X} = 143.5 - 108.0 = 35.5.$$

$$\text{Molecular weight of MX} = \frac{1.491 \text{ g}}{0.02000 \text{ mol}} = 74.55 \text{ g/mol.}$$

$$\text{Atomic weight of M} = 74.55 - 35.5 = 39.1.$$

5.47* A 5.68-g sample of pure P_4O_{10} was completely converted to H_3PO_4 by dissolving it in water. This H_3PO_4 was completely converted to Ag_3PO_4 by treatment with excess $AgNO_3$. The Ag_3PO_4 was then completely converted to AgCl by treatment with excess HCl. The AgCl had a mass of 34.44 g. The molecular weight of P_4O_{10} is known to be 284.0, and the atomic weight of chlorine is known to be 35.5. Calculate the atomic weight of Ag.

5.48* A 2.000-g sample of NH_3 was neutralized by HCl, NH_4Cl being formed as a product. In a separate experiment 20.00 g of $AgNO_3$ was formed by the action of excess HNO_3 on 12.70 g of silver. All the NH_4Cl formed in the first experiment was exactly sufficient to react with all the $AgNO_3$ formed in the second experiment. Knowing that the atomic weight of H is 1.008 and of O, 15.999, calculate the atomic weights of N and Ag.

Hint:

$$NH_3 + HCl \rightleftarrows NH_4Cl$$

$$Ag \rightarrow AgNO_3 \qquad NH_4Cl + AgNO_3 \rightleftarrows AgCl + NH_4NO_3$$

Moles of NH_3 = moles of NH_4Cl = moles of $AgNO_3$ = moles of NO_3^- = moles of Ag.

Let N represent the atomic weight of nitrogen.

$$\text{Moles of } NH_3 = \frac{\text{g of } NH_3}{\text{mol. wt. of } NH_3} = \frac{2.000 \text{ g}}{(N + 3.024)\text{g/mol}}$$

$$\text{Moles of } NO_3^- = \frac{\text{g of } NO_3^-}{\text{mol. wt. of } NO_3^-} = \frac{7.30 \text{ g}}{(N + 48.00)\text{g/mol}}$$

Since moles of NH_3 equals moles of NO_3^-, we can set them equal to each other and solve for N.

Having calculated N we can then solve for the atomic weight of Ag, knowing that moles of Ag equals moles of NH_3.

5.49 In one version of the Thermit reaction, metallic aluminum and manganese dioxide react at elevated temperatures to form manganese metal and Al_2O_3. Find the mass of manganese metal that forms when 175 g of a mixture that contains 39.1% by mass of aluminum undergoes this reaction.

5.50* A certain compound contains C, H, and O. One molecule of the compound is known to contain 2 atoms of O, and the number of atoms of H per molecule is 2 times the number of atoms of C per molecule. When a quantity of the compound is burned completely in O_2 to form CO_2 and H_2O as the only products, 0.375 mol of O_2 is consumed and 0.300 mol of H_2O is formed. Calculate the chemical formula of the compound.

5.51 The oxide of an unknown element is believed to have the formula XO_4. Heating a 7.56-g sample of the oxide leads to its complete decomposition to O_2 and metallic X. The mass of O_2 is 1.92 g. Find the atomic weight of the metal.

5.52* The chloride of an unknown metal is believed to have the formula MCl_3. A 2.395-g sample of the chloride is dissolved in water and treated with excess silver nitrate solution. The mass of the AgCl precipitate formed is found to be 5.168 g. Find the atomic weight of M, the unknown metal.

5.53 A mixture of CS_2 and excess O_2 gas contains a total of 1.0 mol. When the mixture is ignited by a spark, all the CS_2 is oxidized to CO_2 and SO_2. There is now a total of 0.80 mol of gas present. Calculate the mass of CS_2 in the original mixture.

Hint: If we examine the equation

$$CS_2 + 3O_2 \rightleftarrows CO_2 + 2SO_2$$

we note that 4 mol of reactants yields 3 mol of products. There is a decrease of 1 mol. Most significant is the fact that *the decrease in the total amount is equal to the amount of* CS_2 *that reacts*. To solve this problem we simply calculate the decrease in the total amount of gases, which gives us the amount of CS_2 that was burned.

5.54* When solid $CrCl_3$ is heated with H_2 gas, reduction occurs; HCl is the only gaseous product, the other possible products ($CrCl_2$, CrCl, and Cr) being nonvolatile solids. A reaction bomb contained 0.2000 g of anhydrous $CrCl_3$ and 0.1218 mol of H_2. When the temperature was raised to 327 °C, a reduction reaction took place. The total amount of gas (H_2 and HCl) when the reaction was completed was 0.1237 mol. Write the equation for the reaction that took place.

Solution: See Prob. 5.53. Three reactions are possible:

a. $CrCl_3(s) + 0.5H_2(g) \rightleftarrows CrCl_2(s) + HCl(g)$

b. $CrCl_3(s) + H_2(g) \rightleftarrows CrCl(s) + 2HCl(g)$

c. $CrCl_3(s) + 1.5H_2(g) \rightleftarrows Cr(s) + 3HCl(g)$

For each of these reactions how does the increase in the amount *of gas* compare with the amount *of solid* $CrCl_3$ that reacts? In the experiment that is reported how does the increase in the amount of gas compare with the amount of $CrCl_3$ consumed?

Chapter 6

Mass–Amount Relationships in Chemical Reactions

II. Mixtures

The combination of the formula of a compound, the Table of Atomic Weights, and an understanding of the amount relationships that we discussed in Chapter 5 provide powerful tools for the solutions of many complex problems in stoichiometry. A seemingly formidable type of problem is one that deals with the composition of mixtures of two or more substances. With what appears to be limited information, we will find that in fact we can solve many problems of this type. The approach used in solving these problems is essentially the same as that we used in solving the problems in Chapter 5, but you may find that many of these problems are more challenging. Strict attention to and recognition of the *amount* relationships is the key to the solution of each problem.

PROBLEMS

6.1 A mixture of C and S, when burned, yielded a mixture of CO_2 and SO_2 in which the amount of the two gases was equal. Calculate the mole percent of C in the original mixture.

Solution:

$$C + O_2 \rightarrow CO_2$$
$$S + O_2 \rightarrow SO_2$$

When two (or more) substances in a mixture react with another substance, as when the C and S in a mixture both react with O_2, *two separate equations should always be written*. The overall reaction should *not* be written as

$$C + S + 2O_2 \rightarrow CO_2 + SO_2$$

This equation would be correct only if the C and S were present in a 1-to-1 mol ratio. The relationships given by the two equations are

$$\text{amount of } CO_2 \text{ formed} = \text{amount of C burned}$$
$$\text{amount of } SO_2 \text{ formed} = \text{amount of S burned}$$

Since amount of C equals amount of CO_2 and amount of S equals amount of SO_2, and since amount of CO_2 equals amount of SO_2, it follows that amount of C equals amount of S.

That is, the mixture consists of 50 mol % C and 50 mol % S.

6.2 A mixture of C and S was burned to CO_2 and SO_2. Twice as many moles of CO_2 as SO_2 were formed. What was the mass percent of C in the mixture?

6.3 A mixture of C and S was burned to CO_2 and SO_2. The mass of the initial mixture was 10.0 g. The amount of CO_2 and of SO_2 formed was equal. What mass of carbon was burned?

6.4 A 10.0-g mixture of H_2S and CS_2 was burned in oxygen to form a mixture of H_2O, SO_2, and CO_2. The dried mixture on being separated into its pure components yielded 0.275 mol of SO_2 and 0.0774 mol of CO_2. What mass of H_2S was in the original mixture?

Solution: From the equations

$$H_2S + 1.5O_2 \rightarrow H_2O + SO_2$$
$$CS_2 + 3O_2 \rightarrow CO_2 + 2SO_2$$

we note that CO_2 and SO_2 are formed from CS_2 in the ratio of 1 mol of CO_2 to 2 mol of SO_2. Since 0.0774 mol of CO_2 is present, and this CO_2 was derived from the CS_2, $2(0.0774) = 0.155$ mol of SO_2 must also have come from the CS_2. Therefore, $(0.275 - 0.155) = 0.120$ mol of SO_2 must have come from the H_2S. The mass of H_2S must have been 0.120 mol $\times$ 34 g H_2S/mol H_2S = 4.1 g H_2S.

6.5 A 13-g mixture of carbon and sulfur, when burned in air, yielded a mixture of CO_2 and SO_2 in which the amount of CO_2 was one-half the amount of the SO_2. What mass of carbon was contained in the mixture?

Solution: Let X = mass of C and Y = mass of S. Then $X + Y = 13$.

$$C + O_2 \rightarrow CO_2$$
$$S + O_2 \rightarrow SO_2$$

We note that

1. Moles of CO_2 = moles of C burned
2. Moles of SO_2 = moles of S burned
3. $X/12$ = moles of C = moles of CO_2 (atomic weight of C = 12)
4. $Y/32$ = moles of S = moles of SO_2 (atomic weight of S = 32)

Since amount of $CO_2 = \frac{1}{2}$ amount of SO_2,

$$\frac{X}{12} = \frac{1}{2}\left(\frac{Y}{32}\right)$$

We now have two equations in two unknowns. The first is derived from the mass relationship: The total mass is the sum of the masses of the components. The second is derived from the amount relationship and the fact that moles = mass/atomic weight (or molecular weight). Many mixture problems can be solved by setting up and solving two such equations. In this case, solving gives $X = 2.1$ g of carbon.

6.6 A 2.0-g mixture of sulfur and carbon was burned to give a mixture of SO_2 and CO_2 of mass 6.0 g. What mass of carbon was contained in the original mixture?

Solution: In order to illustrate many possible ways to such a mixture problem, we note that each of the following equations holds true:

(a) X = mass of C (in g) Y = mass of S (in g)

(b) $X + Y = 2.0$ g

(c) $C + O_2 \rightarrow CO_2$

(d) $S + O_2 \rightarrow SO_2$

(e) Amount of CO_2 = Amount of C

(f) Amount of SO_2 = Amount of S

(g) Amount of C + amount of S = amount of O_2

(h) Moles of C $= \dfrac{X}{12}$

(i) Moles of S $= \frac{Y}{32}$

(j) Moles of $CO_2 = \frac{X}{12}$

(k) Moles of $SO_2 = \frac{Y}{32}$

(l) Mass of CO_2 + mass of $SO_2 = 6.0$ g

(m) Mass of ($CO_2 + SO_2$) − mass of (C + S) = mass of $O_2 = 4.0$ g

(n) Moles of $O_2 = \frac{4.0}{32}$

(o) Mass of CO_2 = moles of $CO_2 \times \frac{\text{mass of } CO_2}{1 \text{ mol of } CO_2} = \frac{X}{12} \times 44$

(p) Mass of SO_2 = moles of $SO_2 \times \frac{\text{mass of } SO_2}{1 \text{ mol of } SO_2} = \frac{Y}{32} \times 64$

(q) Mass of O_2 in CO_2 + mass of O_2 in $SO_2 = 4.0$ g

(r) Mass of O_2 in $CO_2 = \frac{32}{12} \times$ mass of C $= \frac{32}{12}X$

(s) Since 1 mol of CO_2 contains 1 mol of C and 1 mol of O_2, it follows that in CO_2

$$\frac{\text{mass of the } O_2}{\text{mass of the C}} = \frac{\text{mol. wt. of } O_2}{\text{mol. wt. of C}} = \frac{32}{12}$$

Therefore,

$$\text{mass of the } O_2 = \frac{32}{12} \times \text{mass of the C}$$

$$\text{mass of the } O_2 \text{ in } SO_2 = \frac{32}{12} \times \text{mass of S} = \frac{32}{32} \times Y$$

$$\text{mass of the } CO_2 = \frac{44}{12} \times \text{mass of C} = \frac{44}{12}X$$

(t) Since 1 mol of CO_2 contains 1 mol of C it follows that

$$\frac{\text{mass of the } CO_2}{\text{mass of the C}} = \frac{\text{mol. wt. of } CO_2}{\text{mol. wt. of C}} = \frac{44}{12}$$

(u) Therefore,

$$\text{mass of the } CO_2 = \frac{44}{12} \times \text{mass of the C}$$

$$\text{mass of } SO_2 = \frac{64}{32} \times \text{mass of S} = \frac{64}{32}Y$$

To solve for the value of X we need two equations in two unknowns. One is given by Eq. (b); this is the mass equation. A second equation can be constructed by various combinations of the 21 relationships. In one way or another any second equation will incorporate an amount relationship.

Solution 1: From Eqs. (*g*), (*h*), (*i*), and (*n*),

$$\text{moles of C} + \text{moles of S} = \text{moles of } O_2 \qquad \textbf{(g)}$$

$$\frac{X}{12} + \frac{Y}{32} = \frac{4.0}{32} \quad (X = 1.2)$$

Solution 2: From Eqs. (l), (o), and (p),

$$\text{mass of } CO_2 + \text{mass of } SO_2 = 6.0 \text{ g} \qquad \textbf{(1)}$$

$$\frac{X}{12} \times 44 + \frac{Y}{32} \times 64 = 6.0 \text{ g} \quad (X = 1.2)$$

Solution 3: From Eqs. (q), (r), and (s),

$$\text{mass of } O_2 \text{ in } CO_2 + \text{mass of } O_2 \text{ in } SO_2 = 4.0 \text{ g} \qquad \textbf{(q)}$$

$$\frac{32}{12}X + \frac{32}{32}Y = 4.0 \text{ g} \quad (X = 1.2)$$

Solution 4: From Eqs. (l), (t), and (u),

$$\text{mass of } CO_2 + \text{mass of } SO_2 = 6.0\text{g} \qquad \textbf{(1)}$$

$$\frac{44}{12}Y + \frac{64}{32}Y = 6.0 \text{ g} \quad (X = 1.2)$$

Note: Although solutions 2 and 4 appear the same, they represent different approaches to the problem.

In most mixture problems more than one method of solution is possible. Of course, you need not probe all the methods as we did in this problem. You should select the simplest, shortest, and most obvious method.

6.7 A mixture of Mg and Zn having a mass of 1.000 g was burned in oxygen. It gave a mixture of MgO and ZnO that had a mass of 1.409 g. What mass of Zn was in the original mixture?

Solution: $X =$ mass of Zn. $Y =$ mass of Mg. $X + Y = 1.000$. By examining the formulas MgO and ZnO we see that 1 mol of Mg will yield 1 mol of MgO and 1 mol of Zn will yield 1 mol of ZnO. Since X g of Zn $= X/65.4$ mol of Zn and Y g of Mg $= Y/24.3$ mol of Mg, $X/65.4$ mol of ZnO and $Y/24.3$ mol of MgO have formed. The total mass of ZnO and MgO is known, so we convert these amounts to masses by multiplying by the appropriate molecular weight: mass of ZnO $= 81.4(X/65.4)$ and mass of MgO $= 40.3(Y/24.3)$. The sum of these masses is given as 1.409 g. This result gives the second equation,

$$\frac{81.4}{65.4}X + \frac{40.3}{24.3}Y = 1.409 \text{ g} \quad (X = 0.603 \text{ g Zn})$$

6.8 A mixture of NaBr and NaI has a mass of 1.620 g. When treated with excess $AgNO_3$, it yields a mixture of AgBr and AgI which has a mass of 2.822 g. What mass of NaI was present in the original mixture?

Solution: $X =$ mass of NaI; $Y =$ mass of NaBr.

$$X + Y = 1.620 \quad \text{(the mass equation)}$$

The formula weights are: NaBr, 102.9; NaI, 149.9; AgBr, 187.8; and AgI, 234.8. Since 1 mol of NaBr and 1 mol of AgBr each contain 1 mol of Br, and since 1 mol of NaI and 1 mol of AgI each contain 1 mol of I, then 1 mol of NaBr will yield 1 mol of AgBr and 1 mol of NaI will yield 1 mol of AgI. It follows, therefore, that

$$\frac{\text{mass of AgBr}}{\text{mass of NaBr}} = \frac{\text{mol. wt. of AgBr}}{\text{mol. wt. of NaBr}} = \frac{187.8}{102.9}$$

$$\text{mass of AgBr} = \frac{187.8}{102.9} \times \text{mass of NaBr}$$

Likewise,

$$\text{mass of AgI} = \frac{234.8}{149.9} \times \text{mass of NaI} = \frac{234.8}{149.9}X$$

$$\text{mass of AgBr} = \frac{187.8}{102.9} \times \text{mass of NaBr} = \frac{187.8}{102.9}Y$$

$$\text{mass of AgI} + \text{mass of AgBr} = 2.822 \text{ g}$$

$$\frac{234.8}{149.9}X + \frac{187.8}{102.9}Y = 2.822\text{ g} \quad \text{(the amount relationship equation)}$$

$$X = 0.520\text{ g of NaI}$$

6.9 A mixture of CO_2 and SO_2 has a mass of 2.952 g and contains a total of 5.300×10^{-2} mol. What amount of CO_2 is in the mixture?

Molecular weights: $CO_2 = 44.01$, $SO_2 = 64.06$.

Solution: X = mass of CO_2; Y = mass of SO_2.

$$X + Y = 2.952 \quad \text{(the mass equation)}$$

$$\frac{X}{44.01} + \frac{Y}{64.06} = 5.300 \times 10^{-2} \quad \text{(the mole equation)}$$

$$X = 0.9728\text{ g}$$

$$\text{mol } CO_2 = 0.9728\text{ g } CO_2 \times \frac{1\text{ mol } CO_2}{44.01\text{ g } CO_2}$$

$$= 2.210 \times 10^{-2}\text{ mol } CO_2$$

6.10 A mixture of pure AgCl and pure AgBr contains 66.35% silver. What is the mass percent of bromine in the mixture?

Solution: We will assume that we have 100 g of mixture. The answer, in grams, will then equal, numerically, the mass percent. Let

X = mass of Br, Y = mass of Cl, mass of Ag = 66.35
$X + Y = 100 - 66.35$ (the mass equation)

$$\text{moles of AgBr} = \text{moles of Br} = \frac{X}{79.91}$$

$$\text{moles of AgCl} = \text{moles of Cl} = \frac{Y}{35.45}$$

$$\text{moles of (AgCl + AgBr)} = \text{moles of Ag} = \frac{66.35}{107.9}$$

Now we can write the mole relationship equation:

$$\frac{X}{79.91} + \frac{Y}{35.45} = \frac{66.35}{107.9} \quad (X = 21.3\text{ g} = 21.3\%)$$

Other variations on the second equation are possible.

6.11 A mixture of $BaCl_2$ and $CaCl_2$ contains 43.1% chlorine. Calculate the mass percent of barium in the mixture.

6.12 A mixture of pure Na_2SO_4 and Na_2CO_3 has a mass of 1.200 g and yields a mixture of $BaSO_4$ and $BaCO_3$ having a mass of 2.077 g. Calculate the mass percent of Na_2SO_4 in the original mixture.

6.13 A mixture of CO_2 and CS_2 contains 20.0 mass % carbon. What mass of SO_2 will be formed by the complete oxidation of 10.0 g of the mixture to CO_2 and SO_2?

6.14 A mixture of NaCl and NaBr contains twice the mass of NaCl as of NaBr. When treated with excess $AgNO_3$, this mixture yields 100 g of a mixture of AgCl and AgBr. What mass of NaCl was present in the original mixture?

6.15 A 2.00-g sample of a mixture of NaCl and NaBr is found to contain 0.75 g of Na. Calculate the mass fraction of NaCl in the mixture.

6.16 A 3.20-g sample of a mixture of NaCl and $CaCl_2$ is dissolved in water and treated with excess silver nitrate solution. All the Cl^- in the mixture is converted into a precipitate of silver chloride, AgCl. The precipitate is found to have a mass of 7.94 g. Calculate the mass of sodium chloride in the mixture.

6.17 A mixture of $CaCO_3$ and $(NH_4)_2CO_3$ is 61.9% by mass CO_3. Find the mass percent of $CaCO_3$ in the mixture.

6.18* A mixture of 50.0 g of S and 100.0 g of Cl_2 reacts completely to form S_2Cl_2 and SCl_2 and no other products. Find the mass of S_2Cl_2 formed.

6.19* A mixture of C_3H_8 and C_2H_2 has a mass of 2.0 g. It is burned in excess O_2 to form a mixture of water and carbon dioxide that contains 1.5 times as many moles of CO_2 as of water. Find the mass of C_2H_2 in the original mixture.

6.20* A mixture of As_2S_3 and CuS having a mass of 8.00 g was roasted in air until completely oxidized to SO_2, As_2O_3, and CuO. The SO_2 gas was oxidized to sulfate, which was then completely precipitated as $BaSO_4$; 21.5 g of $BaSO_4$ was formed. Calculate the mass of Cu in the initial mixture.

6.21* When 50.0 g of mercury and 50.0 g of iodine are heated together, they are completely converted into a mixture of Hg_2I_2 and HgI_2. What mass of Hg_2I_2 is present in the mixture?

6.22* A mixture of Al and Mg contained 3 times the mass of Al as of Mg. When the mixture was treated with excess HCl the hydrogen that was liberated reduced 119.25 g of CuO to Cu. What mass of Al was present in the mixture?

6.23* When a mixture of H_2S and CS_2 was burned in oxygen to give H_2O, CO_2, and SO_2, the mass in grams of the SO_2 that was formed was 4 times the mass of the CO_2. Calculate the mass percent of CS_2 in the mixture of H_2S and CS_2.

Hint:

$$H_2S + 1.5O_2 \rightarrow H_2O + SO_2$$
$$CS_2 + 3O_2 \rightarrow CO_2 + 2SO_2$$

Since no specific quantities are given and since the required answer (percent of CS_2) represents a *relative* value, let us assume that 1 mol (44 g) of CO_2 was formed. The total mass of SO_2 formed would then be 176 g. Since the burning of CS_2 forms 2 mol of SO_2 for each mole of CO_2 formed, 2×64 g or 128 g of SO_2 would have been formed from the CS_2. The remaining 48 g of SO_2 ($176 - 128 = 48$) must have been formed from the H_2S. With this information, the relative amount of CS_2 and H_2S and the mass percent of CS_2 in the original mixture can be calculated.

6.24* To a mixture of C_2H_6 and C_2H_6S gases contained in a constant-volume reaction vessel was added the exact amount of O_2 gas required to burn it completely to CO_2, SO_2, and H_2O. When a spark was passed, complete combustion of the gases to CO_2, SO_2, and steam (H_2O) occurred. The mole fraction of the SO_2 in the gaseous mixture of CO_2, SO_2, and H_2O was 0.1237. (There was no liquid present.) Calculate the mass percent of the C_2H_6 in the original mixture of C_2H_6 and C_2H_6S.

Hint: Let X = the mole fraction of the C_2H_6 in the original mixture of C_2H_6 and C_2H_6S. The mole fraction of C_2H_6S in the mixture will then be $1 - X$.

Chapter 7

The Gas Laws

BOYLE'S LAW

If pressure is applied to a gas, the volume of the gas decreases. As the pressure goes *up* the volume goes *down*. The relationship between pressure and volume is a simple one if we do not measure too accurately. For example, if we double the pressure, keeping the temperature constant, the volume will be reduced by half. We can state that at constant temperature *the volume of a given quantity of gas is inversely proportional to the pressure*. This statement is *Boyle's law*. If we call P_1 the original or first pressure and V_1 the original or first volume, then if the pressure is increased to a second value P_2, the volume will be decreased to a second value V_2. We can represent the change as follows:

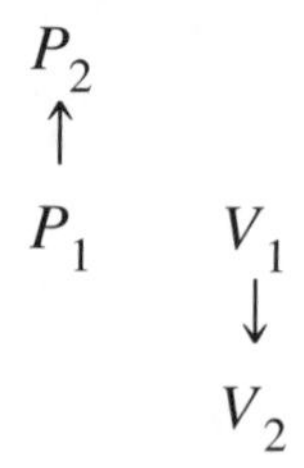

The pressure goes up, from P_1 to P_2; the volume goes down, from V_1 to V_2. The relationship is written

$$P_1V_1 = P_2V_2$$

or

$$\frac{P_2}{P_1} = \frac{V_1}{V_2}$$

This formula is called Boyle's law. It tells us that at constant temperature *the volume of a given quantity of gas is inversely proportional to the pressure.*

STANDARD PRESSURE

The pressure of a gas is usually expressed in units of millimeters of mercury* or atmospheres. A pressure of 740 mmHg means the pressure that would be exerted by a column of liquid mercury 740 mm high. At sea level the average pressure of the atmosphere is 1 atm or 760 mmHg. A pressure of 760 mmHg or 1 atm is, therefore, called *standard pressure*.

The following problems illustrate the application of Boyle's law.

PROBLEMS

7.1 A quantity of gas has a volume of 500 cm^3 at a pressure of 700 mmHg. What volume will it occupy if the pressure is increased to 800 mmHg, at constant temperature?

Solution: One way of solving is to substitute in the Boyle's law equation,

$$\frac{V_1}{V_2} = \frac{P_2}{P_1}$$

The first volume V_1 is 500 cm^3.
The second volume V_2 is what we want to find.
The first pressure P_1 is 700 mmHg.
The second pressure P_2 is 800 mmHg.
Substituting these values in the formula, we have

$$\frac{500 \text{ cm}^3}{V_2} = \frac{800 \text{ mmHg}}{700 \text{ mmHg}}$$

---

* Another common unit of pressure is the torr. It is almost exactly equal to the millimeter of mercury: 1 torr = 1 mmHg.

Solving, we obtain

$$V_2 = \frac{500\ \text{cm}^3 \times 700\ \text{mmHg}}{800\ \text{mmHg}} = 438\ \text{cm}^3$$

7.2 The volume of a gas is 520 L at 750 mmHg and 20 °C. What volume in liters will it occupy at 710 mmHg and 20 °C?

7.3 A tank of compressed helium has a volume of 1.4 L. The helium is at a pressure of 42 atm. All the helium is then used to fill balloons at a pressure of 1.0 atm. Find the total volume of the balloons that are filled with the helium, assuming no change in temperature.

7.4 A mass of hydrogen gas has a volume of 1200 L at a pressure of 1.1 atm. To what value in atmospheres must the pressure be changed if the volume is to be reduced to 2.0 L at constant temperature?

7.5 A cylinder is filled with a mixture of gasoline vapor and air at a pressure of 0.82 atm. The piston of the cylinder is pushed down so that the volume of the cylinder is decreased to 75% of the original volume. Calculate the pressure of the gases at this volume.

7.6 A cylinder of compressed gas has a volume of 0.35 L and the gas it contains is at a pressure of 48 atm. All the gas in the cylinder is then used to fill a balloon. The final pressure of gas in the balloon is 0.98 atm. Find the volume of the balloon.

7.7 A cylinder of an automobile engine has a volume of 0.83 L at the point in its operating cycle when it fills with the mixture of gasoline vapor and air. The pressure of the mixture of gases is 0.62 atm. The piston then moves into the cylinder compressing the mixture of gases to a volume of 0.12 L before it is ignited by the spark plug. Assume ideal gas behavior and find the pressure of the gases at this volume.

THE EFFECT OF CHANGE OF TEMPERATURE

When a gas is heated, it expands. That is, when the temperature goes up, the volume increases. We can say that *the volume of a gas varies directly as the temperature*: When the temperature increases, the volume increases. When the temperature decreases, the volume decreases.

ABSOLUTE ZERO

Suppose we were to take 1.00 L of some gas, helium for example, at exactly 0 °C and at some constant pressure, say 750 mmHg. If we were to cool this gas, keeping the pressure constant, and note how the volume changes, we would observe that the volume decreases by 1/273 of the volume at 0 °C for each °C drop in temperature. If the volume were to keep on shrinking at the same rate, then at −273 °C, the volume would be zero. Since cooling below −273 °C would give a volume less than zero, and since it is reasonable to assume that there is no such thing as negative volume, it can be concluded that −273 °C is the lowest temperature theoretically possible. Therefore, −273 °C is designated as *absolute zero*. (The exact value is −273.15 °C. The value −273 °C is sufficiently accurate for all calculations in this book.) Absolute zero also defines the zero point on the absolute or Kelvin temperature scale. Thus −273.15 °C = 0 K, where K is the abbreviation for the kelvin, which is the unit of temperature on the absolute scale. The kelvin is equal to the degree Celsius; only the zero point is different between the Kelvin and Celsius temperature scales. To convert from Celsius to Kelvin we simply add 273. The relationship is

$$T(\text{K}) = t(^\circ\text{C}) + 273$$

CHARLES'S LAW

If we had taken the 1.00 L of helium gas at 0 °C and heated it, we would have found that at 273 °C, the volume would have increased to 2.00 L. That is, at 0 °C the volume is 1.00 L, while at 273 °C it is 2.00 L, or just twice as great. But 0 °C is the same as 273 K (absolute) and 273 °C is the same as 546 K. When we double the absolute temperature (from 273 K to 546 K), we double the volume (from 1.00 L to 2.00 L). From these data you can see that *the volume of a quantity of gas at constant pressure is directly proportional to the absolute temperature. This statement is Charles's law.* The formula for Charles's law is

$$\frac{V_1}{V_2} = \frac{T_1}{T_2}$$

STANDARD CONDITIONS

Zero degrees Celsius is referred to as *standard temperature*. Standard temperature (0 °C) and standard pressure (760 mmHg) are commonly referred to by the notation STP.

The following problems illustrate the application of Charles's law.

PROBLEMS

7.8 A gas occupies a volume of 200 cm^3 at 0 °C and 760 mmHg. What volume will it occupy at 100 °C and 760 mmHg?

Solution: Since the pressure is constant, this is a problem in which only the temperature is changed. We can solve by substituting in the Charles's law formula, provided we convert the Celsius temperatures to absolute temperatures. Only the absolute temperatures can be used in calculations involving gases.

$$V_1 = 221 \text{ cm}^3; T_1 = 273 \text{ K } (0 + 273); T_2 = 373 \text{ K } (100 + 273).$$

$$\frac{V_1}{V_2} = \frac{T_1}{T_2}$$

$$\frac{221 \text{ cm}^3}{V_2} = \frac{273 \text{ K}}{373 \text{ K}}$$

$$V_2 = \frac{221 \text{ cm}^3 \times 373 \text{ K}}{273 \text{ K}} = 302 \text{ cm}^3$$

7.9 The volume of a gas is 740 cm^3 at 12 °C. What volume will it occupy at 0 °C, the pressure remaining constant?

7.10 The volume of a gas is 13.1 ft^3 at −20 °C and 750 mmHg. What volume will it occupy at 20 °C and 750 mmHg?

7.11 A mass of helium gas occupies a volume of 100 L at 20 °C. If the volume occupied by the gas is tripled, to what must the temperature be changed in order to keep the pressure constant?

7.12 The gas inside a balloon of volume 12.3 m^3 is heated from 25 °C to 125 °C. Find the volume of the balloon.

7.13 A natural gas storage tank is a cylinder with a movable top whose volume can change only as its height changes. Its radius remains fixed. The height of the cylinder is 22.6 m on a day when the temperature is 22 °C. The next day the height of the cylinder increases to 23.8 m as the gas expands because of a heat wave. Find the temperature, assuming that the pressure and the amount of gas in the storage tank have not changed.

CHANGE OF PRESSURE AND TEMPERATURE

By combining the formulas for Boyle's law and Charles's law we get the following general formula expressing the change of volume with change of pressure and change of temperature:

$$\frac{V_1}{V_2} = \frac{T_1 \times P_2}{T_2 \times P_1} \quad \textbf{(7-1)}$$

Equation (7-1) can be transposed to give Eq. (7-2):

$$\frac{V_1P_1}{T_1} = \frac{V_2P_2}{T_2} \tag{7-2}$$

This equation is very useful because it gives us the relationship between any two sets of conditions of a given quantity of gas. Each set of conditions is specified by three variables (P, V, and T), so this equation involves six variables. Given any five of these variables the sixth can be calculated. For example, if we know the volume of a quantity of gas at one temperature and pressure, we can calculate its volume at any other temperature and pressure. Or if we know the temperature of a quantity of gas that occupies a certain volume at a certain pressure, we can calculate its temperature when it occupies any volume at any pressure, and so on.

You should note that no gas obeys Boyle's and Charles's laws perfectly over all ranges of temperature and pressure; that is, no gas is "perfect." The higher the temperature and the lower the pressure, the more nearly perfect every gas is in its response to changes of temperature and pressure. It will be assumed in the problems in this book that all gases are perfect (ideal) and that Boyle's and Charles's laws are obeyed at the temperatures and pressures encountered.

PROBLEMS

7.14 The volume of a gas is 200 L at 12 °C and 750 mmHg. What volume will it occupy at 40 °C and 720 mmHg?

Solution: We first convert the temperature to the Kelvin scale:

$$12\ °\text{C} = 285\ \text{K}$$
$$40\ °\text{C} = 313\ \text{K}$$

We then solve by substituting in the general formula

$$\frac{V_1P_1}{T_1} = \frac{V_2P_2}{T_2}$$

$$\frac{200\ \text{L} \times 750\ \text{mmHg}}{285\ \text{K}} = \frac{V_2 \times 720\ \text{mmHg}}{313\ \text{K}}$$

$$V_2 = 200\ \text{L} \times \frac{313\ \text{K} \times 750\ \text{mmHg}}{285\ \text{K} \times 720\ \text{mmHg}} = 229\ \text{L}$$

7.15 The volume of a dry gas is 50.0 L at 20 °C and 742 mmHg. What volume will it occupy at STP?

7.16 At what temperature will a mass of gas whose volume is 150 L at 12 °C and 750 mmHg occupy a volume of 200 L at a pressure of 730 mmHg?

7.17 A gas in a 10.0 L steel cylinder is under a pressure of 4.68 atm at 22.0 °C. If the temperature is raised to 600 °C, what will be the pressure on the gas?

Note: From the combined gas law formula,

$$\frac{V_1P_1}{T_1} = \frac{V_2P_2}{T_2}$$

it follows that when the volume is kept constant, the pressure exerted by a given mass of gas is directly proportional to the absolute temperature.

7.18 A mass of helium gas contained in a 700-cm^3 vessel at 607 mmHg pressure and 22 °C is transferred to a 1000-cm^3 vessel at 110 °C. What is the pressure in the 1000-cm^3 vessel?

7.19* A cylinder contains helium gas at a pressure of 1470 $lb/in.^2$. When a quantity of helium gas that occupies a volume of 4.00 L at a pressure of 14.7 $lb/in.^2$ is withdrawn, the pressure in the tank drops to 1400 $lb/in.^2$. (Temperature remains constant.) Calculate the volume of the tank.

Hint: Let *X* equal the volume, in liters, of the cylinder. Since the pressure drops from 1470 $lb/in.^2$ to 1400 $lb/in.^2$ when the quantity of gas is removed, this quantity must have exerted a pressure of 70 $lb/in.^2$ when confined to a volume of *X* liters at the given temperature. Since we know that this quantity of gas exerted a pressure of 14.7 $lb/in.^2$ when confined to a volume of 4.00 L we can solve for *X* by applying Boyle's law. Note that we are concerned only with the pressure exerted by the gas that was *removed*, since *it* is the gas that is confined in the 4.00-L vessel.

7.20 Predict whether the volume of a gas increases, decreases, or stays the same when the following changes are made:

a. The pressure is increased from 200 mmHg to 300 mmHg; the temperature is increased from 200 °C to 300 °C.

b. The pressure is decreased from 760 mmHg to 380 mmHg; the temperature is decreased from 237 °C to −23 °C.

c. The pressure is decreased from 2 atm to 1 atm; the temperature is increased from 300 K to 600 K.

7.21 An automobile tire is inflated with air at 22 °C to a pressure of 1.8 atm. After the car is driven for several hours, the volume of the tire increases from 7.2 L to 7.8 L and the pressure increases to 1.9 atm. Find the temperature of the air inside the tire, assuming ideal gas behavior.

7.22 A balloon whose volume is 11.0 L contains gas at a temperature of 297 K and a pressure of 760 mmHg. When the balloon reaches an altitude of 20.0 km, the pressure is 63.0 mmHg and the temperature is 223 K. Find the volume of the balloon.

7.23 A sample of a gas at a pressure of 1.0 atm and a temperature of 301 K formed in a chemical reaction is transferred from a container of volume 0.98 L to one of volume 24 cm^3. The temperature of the gas is 195 K. Find the new pressure of the gas.

7.24 A gas occupies a volume of 24.5 L at 298 K and 2 atm. Find its volume at STP.

AVOGADRO'S LAW

In addition to the simple relationships between the pressure, volume, and temperature of a gas, there is also a simple relationship between the amount of a gas and these other variables. This relationship was first proposed in 1811 by Amadeo Avogadro. In modern language he stated that *equal volumes of any gases at the same temperature and pressure contain equal numbers of molecules or moles*. Thus as the quantity of a gas is increased at constant temperature and pressure, its volume increases in direct proportion. We can write this direct proportionality as

$$\frac{V_1}{n_1} = \frac{V_2}{n_2}$$

where n is the amount of gas. This law has been verified many times by experiment.

One volume of particular interest for gases is called the *standard molar volume*, the volume of 1 mol of gas at STP. It has a value of 22.4 L, which means that the volume of 1 mol of any ideal gas at 1 atm and 0 °C is 22.4 L.

We can also reason from Avogadro's law that *for different volumes of gases at the same temperature and pressure, the amount of a gas is directly proportional to the volume of that gas*. Thus if, at constant temperature and pressure, gas A occupies a volume of 100 L and gas B occupies a volume of 50 L, the number of moles of A is twice the number of moles of B.

By combining Avogadro's law with the other gas laws, we can also conclude that *at a given temperature and in a given volume of gas, the pressure is directly proportional to the number of moles of gas*. Since the number of moles of gas in a given volume is the *molar concentration* of that gas, we can also state that *at a given temperature, the pressure exerted by a gas is directly proportional to its molar concentration.*

The fact that 1 mol of any gas occupies a volume of 22.4 L at STP enables us, with the aid of Boyle's and Charles's laws, to calculate the mass of any volume of a gas and the volume of any mass of that gas under any conditions of temperature and pressure, provided, of course, that its actual chemical formula and hence its true molecular weight are known.

You should note that He, Ar, Ne, Kr, and Xe are monatomic gases; their molecular weights equal their atomic weights. The other elementary gases, namely H_2, O_2, N_2, F_2, Cl_2, Br_2, and I_2, form diatomic molecules.

PROBLEMS

7.25 What volume in liters will 2.71 mol of He gas occupy at STP?

Solution: The molar volume of a gas at STP is 22.4 L. That means that 1 mol of He will occupy a volume of 22.4 L at STP.

$$2.71 \text{ mol} \times \frac{22.4 \text{ L}}{1 \text{ mol}} = 60.7 \text{ L}$$

Note: Moles cancel, leaving the answer in liters.

7.26 What volume in liters will 0.424 mol of CO gas occupy at STP?

7.27 A volume of 50.0 L of H_2S gas, measured at STP, is what amount of H_2S?

Solution:

$$50.0 \text{ L} \times \frac{1 \text{ mol}}{22.4 \text{ L}} = 2.23 \text{ mol}$$

7.28 What amount is 18.3 L of HF gas at STP?

7.29 What volume in liters will 100 g of CH_4 gas occupy at STP?

Solution: In solving this problem our first question is: How many *moles* of CH_4 are there? We know that 1 mol of a gas occupies a volume of 22.4 L at STP. Therefore, if we know how many moles of CH_4 gas there are in 100 g of CH_4, we can multiply this number of moles by 22.4 L/mol.

The molecular weight of CH_4 is 16.0.

The complete calculation, in one operation, is

$$100 \text{ g } CH_4 \times \frac{1 \text{ mol } CH_4}{16.0 \text{ g } CH_4} \times \frac{22.4 \text{ L } CH_4}{1 \text{ mol } CH_4} = 140 \text{ L } CH_4$$

Note: The g CH_4 and mol CH_4 cancel.

7.30 What volume in liters will 40.0 g of HCl gas occupy at STP?

7.31 Calculate the mass in grams of 40.0 L of NO gas at standard conditions.

Solution: In solving this problem our first question is: What amount of NO do we have? If we know the amount in moles, we can multiply this number of moles by the number of grams of NO in 1 mol. We know that 22.4 L of NO at STP is 1 mol. Therefore, 40.0 L at STP is 40.0/22.4 mol of NO. The molecular weight of NO is 30.0, so the mass of the gas is $30.0 \times 40.0/22.4 = 53.6$ g NO.

7.32 Calculate the mass of 120 L of CO_2 gas at STP.

7.33 What mass of nitrogen is contained in 100 L of NO measured at STP?

Solution: In solving this problem we first ask: What amount of NO is there? Knowing the number of moles of NO, we can easily calculate the mass of N, since 1 mol of NO contains 14.0 g of N. The formula NO tells us that there is 1 mol N in 1 mol of NO. One mole of N has a mass of 14.0 g, 100 L of NO is 100/22.4 mol of NO, and 100/22.4 mol of NO contains 100/22.4 mol of N.

The solution, in one operation, is

$$100 \text{ L NO} \times \frac{1 \text{ mol NO}}{22.4 \text{ L NO}} \times \frac{1 \text{ mol N}}{1 \text{ mol NO}} \times \frac{14.0 \text{ g N}}{1 \text{ mol N}} = 62.5 \text{ g N}$$

Note: All units except g N cancel.

7.34 A volume of 76.1 L of H_2S, measured at STP, contains what mass of S?

7.35 What volume of SO_2, when measured at STP, will contain 50.0 g of S?

7.36 A volume of 2.00 L of a gas, measured at STP, has a mass of 5.71 g. Calculate the approximate molecular weight of the gas.

Solution: The molecular weight of a gas is the mass in grams of 1 mol of the gas. One mole of a gas occupies a volume of 22.4 L at STP. Therefore, the molecular weight of a gas is the mass in grams of 22.4 L at STP.

We can reason that since 2.00 L has a mass of 5.71 g, the mass of 1 L will be 5.71/2.00 g and the mass of 22.4 L will be $22.4 \times 5.71/2.00$ g. The calculation, in one operation, is:

$$\frac{5.71 \text{ g}}{2.00 \text{ L}} \times \frac{22.4 \text{ L}}{1 \text{ mol}} = 64.0 \text{ g/mol}$$

Since the gas laws are not exact, the molecular weight calculated in this manner is not exact.

7.37 A volume of 6.82 L of a gas, measured at STP, has a mass of 9.15 g. Calculate the approximate molecular weight of the gas.

7.38 A 3.2 L bulb contains O_2 and a 6.4 L bulb contains N_2. Both bulbs are at the same temperature and pressure. Find the ratio of the amount of O_2 to the amount of N_2. Find the ratio by mass of O_2 to N_2.

7.39 A compressed gas cylinder contains 106 g of acetylene (C_2H_2) at a pressure of 5.1 atm. After some use of the acetylene by a welder, the pressure in the cylinder falls to 4.2 atm. Find the mass of acetylene consumed, assuming the volume and the temperature of the cylinder do not change.

7.40* Flask A contains 20.0 L of CH_4 gas. Flask B contains 30.0 L of CO gas. Each volume is measured at the same temperature and pressure. If flask A contains 6.18 mol of CH_4, what amount of CO is there in flask B?

Hint: At the same temperature and pressure, the amount of gas is directly proportional to the volume of the gas. Since the volume of B is 1.50 times the volume of A, the amount of CO in B will be 1.50 times the amount of CH_4 in A.

7.41* If 25.0 g of CH_4 gas occupies a volume of 30.0 L at a certain temperature and pressure, what volume in liters will 50.0 g of CO_2 gas occupy at the same temperature and pressure?

7.42* A volume of 40.0 L of pure O_2 gas, measured at a certain temperature and pressure, was found to contain 40.0 g of oxygen. A volume of 60.0 L of CH_4 gas, measured at the same temperature and pressure, will contain what mass of CH_4?

THE IDEAL GAS LAW EQUATION *PV* = *nRT*

The equation that combines Boyle's law and Charles's law,

$$\frac{P_1V_1}{T_1} = \frac{P_2V_2}{T_2}$$

applies for a given amount of gas. This relationship can be stated in another way. For a given amount of gas the quantity PV/T is a constant. The value of this constant is proportional to the amount of gas. We can therefore write

$$\frac{PV}{T} = Rn \qquad \textbf{(7-3)}$$

where n is the amount in moles of gas and R is the symbol that we usually use to represent the proportionality constant relating the amount of gas to the PV/T term.

The constant R is called the *universal gas constant* and its value can be measured. For example, we know that at STP 1 mol of an ideal gas occupies a volume of 22.4 L. Substituting these data into Eq. (7-3), we have

$$R = \frac{1.00 \text{ atm} \times 22.4 \text{ L}}{273 \text{ K} \times 1 \text{ mol}} = 0.0821 \text{ L-atm/mol-K}$$

If pressure is expressed in millimeters of mercury, then $R = 62.4$ L-mmHg/mol-K. The exact numerical value of R depends on its units. Equation (7-3) is generally written in the form

$$PV = nRT \tag{7-4}$$

Equation (7-4) is the *ideal gas law equation*.

An alternate form of this equation can be written by substituting in the expression that defines a mole,

$$n = \frac{g}{\text{MW}} \tag{7-5}$$

where g is the mass of the sample in grams and MW is the molecular weight, giving

$$PV = \frac{g}{\text{MW}} RT \tag{7-6}$$

When employing ideal gas equations to solve problems, it is essential that you use the correct units. Temperature must always be in K and the units of pressure and volume must match the units of R. Thus if $R = 0.082$ L-atm/mol-K, then pressure must be in atmospheres and volume must be in liters.

PROBLEMS

7.43 What volume will 62.6 g of CH_4 gas occupy at 27 °C and 1.03 atm?

Solution: Unlike most of the gas law problems we have been doing up to this point, there is no change of conditions in this problem. When we are given information about a gas under a certain set of conditions and are asked for some additional information about this same set of conditions, we generally use the ideal gas equation. We solve problems of this type by substituting the data in appropriate units into $PV = nRT$:

$$n = 62.6 \text{ g } CH_4 \times \frac{1 \text{ mol } CH_4}{16.0 \text{ g } CH_4} = 3.91 \text{ mol } CH_4$$

$$P = 1.03 \text{ atm}$$

$$T = 27\,^\circ\text{C} + 273 = 300\text{ K}$$

$$V = \frac{nRT}{P} = \frac{3.91\text{ mol} \times 0.0821\dfrac{\text{L-atm}}{\text{mol-K}} \times 300\text{ K}}{1.03\text{ atm}} = 93.6\text{ L}$$

Note: The mol, K, and atm cancel.

7.44 The hydrogen gas in a 2.00-L steel cylinder at 25 °C is under a pressure of 4.00 atm. What amount of H_2 is in the cylinder?

Solution:

$$PV = nRT$$

$$n = \frac{PV}{RT} = \frac{4.00\text{ atm} \times 2.00\text{ L}}{0.0821\dfrac{\text{L-atm}}{\text{mol-K}} \times 298\text{ K}} = 0.327\text{ mol H}_2$$

Note: The L, atm, and K cancel.

7.45 What pressure, in atmospheres, will 26.0 g of He gas exert when placed in a 3.24-L steel cylinder at 200 °C?

Solution:

$$PV = \frac{g}{\text{MW}}RT$$

$$P = \frac{gRT}{\text{MW} \times V} = \frac{26.0\text{ g}}{\dfrac{4.00\text{ g}}{1\text{ mol}}} \times 0.0821\frac{\text{L-atm}}{\text{mol-K}} \times \frac{473\text{ K}}{3.24\text{ L}}$$

$$= 77.9\text{ atm}$$

Note: The g, L, K, and mol cancel.

7.46 Calculate the mass in grams of the pure H_2S gas contained in a 60.0-L cylinder at 20 °C under a pressure of 2.00 atm.

Solution:

$$PV = \frac{gRT}{\text{MW}}$$

$$g = \frac{PV \times \text{MW}}{RT} = \frac{2.00\text{ atm} \times 60.0\text{ L} \times 34.1\text{ g/mol}}{0.0821\dfrac{\text{L-atm}}{\text{mol-K}} \times 293\text{ K}}$$

$$= 170\text{ g H}_2\text{S}$$

Note: The mol, atm, L, and K cancel.

7.47 A 32.4-g sample of an ideal gas in a 12.0-L container at 25 °C exerted a pressure of 1.50 atm. Calculate its molecular weight.

Solution:

$$\mathrm{MW} = \frac{gRT}{PV} = \frac{32.4\ \mathrm{g} \times 0.0821 \frac{\mathrm{L\text{-}atm}}{\mathrm{mol\text{-}K}} \times 298\ \mathrm{K}}{1.50\ \mathrm{atm} \times 12.0\ \mathrm{L}}$$

$$= 44.0\ \mathrm{g/mol}$$

7.48 Calculate the mass of 200 L of CO_2 gas at 20 °C and 746 mmHg.

7.49 Calculate the volume in liters occupied by 234 g of CO gas at 22 °C and 740 mmHg.

7.50 A 25.0-L cylinder contains 14.2 mol of helium gas at 40 °C. What is the pressure of the helium gas?

7.51 A cylinder containing 85 g of steam at 200 °C shows a pressure of 4.0 atm. What is the volume of the cylinder in liters?

7.52 Find the mass of carbon contained in 2.34 L of carbon dioxide at STP.

7.53 A 1.00-g sample of water is placed in a sealed evacuated container of volume 1.00 L and heated to 500 K. Assume all the water has evaporated and find the pressure in the container.

7.54 A neon sign is made from 5.2 m of glass tubing of diameter 2.0 cm. The pressure required for the proper operation of the sign at 310 K is 1.5 mmHg. Find the mass of neon required to fill the sign.

7.55 How many molecules of H_2S are in a 22.40-L volume at 273.0 °C and 4.000 atm?

7.56 A volume of 120 cm^3 of a dry gaseous compound, measured at 22 °C and 742 mmHg, has a mass of 0.820 g. Calculate the approximate molecular weight of the gas.

7.57 What volume of C_2H_2 gas, measured at 25 °C and 745 mmHg, will contain 10.0 g of carbon?

7.58 Twelve liters of dry nitrogen gas, measured at 22 °C and 741 mmHg, has a mass of 13.55 g. Calculate the formula for a molecule of nitrogen gas under these conditions.

7.59 A steel tank has a volume of 4.7 L and can withstand a pressure of 75 atm before it explodes. Find the mass of O_2 that can be stored in the tank at 298 K.

7.60 A person exhales 750 g of CO_2 per day. Suppose the person is in a sealed room with the dimensions 3.0 m by 3.0 m by 2.4 m at a temperature of 290 K. Find the pressure of CO_2 in the room after one day.

Hint: The relationship 1 m^3 = 1000 L can be used for the appropriate unit conversion.

7.61 A sample of an unknown liquid is introduced into a bulb and vaporized. We assume any other gases are swept out. The mass of the bulb increases by 1.46 g. The pressure is 762 mmHg and the temperature is 332 K. The volume of the bulb is 272 cm^3. Find the molecular weight of the liquid.

7.62 A 1.07-g sample of a liquid is evaporated completely at 341 K so that its vapors fill a 0.252 L container at a pressure of 1.00 atm. Find the molecular weight of the liquid.

7.63* A 2.0-L sample of helium gas measured at 27 °C is under 2.0 times the pressure and contains 3.0 times as many molecules of gas as a sample of H_2 gas measured at 227 °C. Calculate the volume in liters of the sample of H_2 gas.

7.64* If 0.200 g of H_2 is needed to inflate a balloon to a certain size at 20 °C, what mass will be needed to inflate it to the same size at 30 °C? Assume the elasticity of the balloon is the same at 20 °C and 30 °C.

DENSITIES OF GASES

The mass of a definite volume of gas is referred to as the *density* of the gas. The definite volume that is commonly used in expressing the density of a gas is 1 L. We can show that *the densities of two gases at the same temperature and pressure are to each other as their molecular weights*.

The ideal gas equation can be written to include the density of the gas under any conditions. If density is expressed in units of grams per liter, then we can start with

$$PV = \frac{gRT}{\text{MW}}$$

which can be rearranged to

$$\frac{g}{V} = \frac{P(\text{MW})}{RT} \tag{7-7}$$

and since g/V is the mass in grams divided by the volume in liters, we can let d be the density and write

$$d = \frac{P(\text{MW})}{RT} \tag{7-8}$$

If the density is expressed in units of moles per liter, the ideal gas equation can simply be written as

$$\frac{n}{V} = \frac{P}{RT} \tag{7-9}$$

and the term P/RT is equal to the density in moles per liter.

PROBLEMS

7.65 What is the density of C_2H_2 gas at STP?

Solution: One mole of C_2H_2 has a mass of 26.0 g and occupies 22.4 L at STP. Therefore, the density is 26.0 g per 22.4 L or 1.16 g/L.

7.66 Calculate the density of CO gas at STP in grams per liter.

7.67 Calculate the density of SO_2 gas at 40 °C and 730 mmHg.

Solution: The molecular weight of SO_2 is 64.0 g/mol. Substituting in Eq. (7-8), we obtain

$$d = \frac{P(\text{MW})}{RT} = \frac{730\ \text{mmHg} \times 64.0\ \text{g/mol}}{62.4\ \dfrac{\text{L-mmHg}}{\text{mol-K}} \times 313\ \text{K}} = 2.39\ \text{g/L}$$

7.68 Calculate the density of CH_4 gas at 120 °C and 0.600 atm.

7.69 At a given temperature and pressure which is heavier, 20.0 L of C_2H_6 or 20.0 L of O_2? How many times as heavy?

Hint: At the same temperature and pressure the densities of two gases are to each other as their molecular weights. At the same temperature and pressure equal volumes of two gases contain the same amount of gas.

7.70 If the density of He gas is 0.026 g/L at a certain temperature, what is the density of Ne gas at the same temperature and pressure?

7.71 What is the temperature of a gas that has a density of 2.00 mol/L at a pressure of 44.8 atm?

7.72 The density of CO is 1.4 g/L at certain conditions of temperature and pressure. Find the density of CO_2 at the same conditions.

7.73* A quantity of N_2 occupies a volume of 1.0 L at 300 K and 1.0 atm. The gas expands to a volume of 3.0 L as the result of a change in both temperature and pressure. Find the density of the gas under these new conditions.

7.74* A given mass of helium gas, which occupied a volume of 2.0 L at STP, was allowed to expand to a volume of 4.0-L by changing the temperature and pressure. What was its density at the new temperature and pressure?

PARTIAL PRESSURE: DALTON'S LAW OF PARTIAL PRESSURES

Ideal gas behavior does not depend on the specific chemical identity of the gas. One ideal gas is the same as any other. Therefore if a container has a mixture of two different gases, each behaving ideally, the situation is the same as if the container held the same total amount of just one gas.

Suppose we place $\frac{1}{2}$ mol of CH_4 and $\frac{1}{2}$ mol of CO_2 gas together in a 22.4-L container at 0 °C. (CO_2 and CH_4 do not react with each other.) The $\frac{1}{2}$ mol (3.011×10^{23} molecules) of CH_4 will exert a pressure of 380 mmHg, and the $\frac{1}{2}$ mol of CO_2 gas will also exert a pressure of 380 mmHg. The two gases together will exert a pressure of 760 mmHg.

The pressure of 380 mmHg exerted by the $\frac{1}{2}$ mol of CH_4 is referred to as the *partial pressure* of the CH_4, and the pressure of 380 mmHg exerted by the CO_2 is the *partial pressure* of the CO_2. One-half of the molecules are CH_4 molecules and one-half are CO_2 molecules. CH_4 provides half of the molecules and exerts half of the pressure; CO_2 provides the other half of the molecules and exerts the other half of the pressure. We can conclude from this that *in a mixture of gases, the total pressure is the sum of the partial pressures of the gases in the mixture.*

In a mixture of CH_4 and CO_2 in which half of the molecules are CH_4 and half are CO_2, the *mole fraction* of the CH_4 is 0.5 and the mole fraction of the CO_2 is also 0.5. *Mole fraction* is defined as follows:

$$\text{mole fraction of } A = \frac{\text{moles of } A}{\text{total moles}}$$

Mole percent is mole fraction $\times$ 100%.

In a mixture of 2 mol of CH_4 and 3 mol of CO_2 the mole fraction of CH_4 is 0.4 $\left(\frac{2}{5}\right)$ and the mole fraction of CO_2 is 0.6 $\left(\frac{3}{5}\right)$. If the total pressure of a mixture of 2 mol of CH_4 and 3 mol of CO_2 is 800 mmHg, the partial pressure of the $CH_4(P_{CH_4})$ is found to be 320 mmHg and the partial pressure of the $CO_2(P_{CO_2})$ is 480 mmHg. But 320 mmHg is 0.4×800 mmHg and 480 mmHg is 0.6×800 mmHg. Thus *in a mixture of gases, the partial pressure of a gas is equal to its mole fraction times the total pressure.*

$$P_A = \text{mole fraction of } A \times P_{\text{total}}$$

where P_A is the partial pressure of A.

Another statement of this relationship is that *the partial pressure of a gas is directly proportional to its mole fraction.* This is one common statement of what is called *Dalton's law of partial pressures.* The direct proportionality can also be written as

$$\frac{P_A}{P_B} = \frac{\text{mole fraction of } A}{\text{mole fraction of } B}$$

We can conclude that in a mixture of nonreacting gases, for any one of the gases,

$$\text{mole fraction} = \text{pressure fraction}$$

and

$$\text{mole percent} = \text{pressure percent}$$

This means that the relative amount of a gas in a mixture is equal to its relative partial pressure. Thus if a mixture of 0.2 mol of He gas and 0.4 mol of Ne gas exerts a total pressure of 1500 mmHg, one-third of the pressure (500 mmHg) is exerted by the He and two-thirds of the pressure (1000 mmHg) is exerted by the Ne. Likewise, in a mixture of 0.2 mol of CH_4 and X mol of CO_2, if the partial pressure of the CH_4 is 18 mmHg and that of the CO_2 is 45 mmHg, there must be $45/18 \times 0.2$ mol or 0.5 mol of CO_2.

It should be emphasized that *in a mixture of nonreacting ideal gases, each gas behaves exactly as it would if it alone were present. The other gases have no effect on its behavior. Its partial pressure is the pressure that it would exert if it alone occupied the volume. Note also that the volume of every gas in a mixture of ideal gases is the same. Its volume is the volume of the container.*

PROBLEMS

7.75 In a gaseous mixture of CH_4 and C_2H_6 the amount of CH_4 is twice that of C_2H_6. The partial pressure of the CH_4 is 40 mmHg. What is the partial pressure of the C_2H_6?

Hint: The partial pressure of C_2H_6 is directly proportional to the amount of C_2H_6.

7.76 In a gaseous mixture of CH_4, C_2H_6, and CO_2, the partial pressure of the CH_4 is 50% of the total pressure. What is the mole fraction of the CH_4 in the mixture?

Hint: Mole fraction equals pressure fraction.

7.77 In a mixture of gases A, B, and C, the mole fraction of A is 0.25. What fraction of the total pressure is exerted by gas A?

7.78 A mixture of 2.0×10^{23} molecules of N_2 and 8.0×10^{23} molecules of CH_4 exerts a total pressure of 740 mmHg. What is the partial pressure of the N_2?

7.79 In a mixture of 0.200 mol of CO, 0.300 mol of CH_4, and 0.400 mol of CO_2 at 800 mmHg, what is the partial pressure of the CO?

7.80 If the 200 cm^3 of H_2 gas contained in a cylinder under pressure of 1200 mmHg is forced into a cylinder whose volume is 400 cm^3 and which already contains 400 cm^3 of CH_4 gas under a pressure of 800 mmHg, what will be the total pressure exerted by the mixture of H_2 and CH_4 gases in the 400 cm^3 cylinder if the temperature is constant?

Hint: The total pressure will be the sum of the partial pressures.

7.81 The partial pressures of the four gases contained in a 6.0-L cylinder at 1007 °C were: CO_2, 63.1 atm; H_2, 21.1 atm; CO, 84.2 atm; N_2O, 31.6 atm. What mass of CO_2 gas was in the cylinder?

Hint: Since in a mixture of gases each gas exerts the pressure that it would exert if it alone occupied the volume, we can ignore the other gases and calculate, by use of the equation $PV = nRT$, the mass of CO2 gas that would exert a pressure of 63.1 atm when placed in a 6.0-L cylinder at 1007 °C.

7.82 A mixture of 3.5×10^{23} molecules of O_2 and 6.5×10^{23} atoms of Ar has a total pressure of 0.82 atm. Find the partial pressure of Ar.

7.83 A mixture of 8.0 g CH_4 and 8.0 g Xe is placed in a container and the total pressure is found to be 0.44 atm. Find the partial pressure of CH_4.

7.84* A reaction vessel contained 5.0 L of a mixture of N_2 and O_2 gases at 25 °C and 2.0 atm pressure. The oxygen in the mixture was completely removed by causing it to oxidize an excess of electrically heated zinc wire contained in the vessel to solid ZnO. (Solid ZnO is nonvolatile.) The pressure of the nitrogen gas that remained in the vessel, measured at 25 °C, was 1.5 atm. What was the mole percent of oxygen in the original mixture? The mass percent?

Solution: $P_T = P_{O_2} + P_{N_2}$, $2\text{ atm} = P_{O_2} + 1.5\text{ atm}$, $P_{O_2} = 0.5\text{ atm}$. When V and T are constant, the number of moles of a gas is directly proportional to its partial pressure. Therefore,

$$\frac{\text{moles of } O_2}{\text{moles of } N_2} = \frac{0.50\text{ atm}}{1.5\text{ atm}}$$

$$\text{mole \% of } O_2 = \frac{0.50\text{ atm}}{2.0\text{ atm}} \times 100\% = 25\%$$

To convert mole percent to mass percent convert moles of O_2 and N_2 to grams.

Mass % of O_2

$$= \frac{0.50 \times 32 \text{ g } O_2/\text{mol } O_2}{(0.50 \times 32 \text{ g } O_2/\text{mol } O_2) + (1.5 \times 28 \text{ g } N_2/\text{mol } N_2)} \times 100\% = 28\%$$

That we do not know the exact amounts of O_2 and N_2 is not important. The significant fact is that the amount of N_2 is 3 times the amount of O_2. Once we know the mole ratios in which substances are present, the mass ratios and mass percents can then be calculated.

7.85* A mixture of 50.0 g of oxygen gas and 50.0 g of CH_4 gas is placed in a container under a pressure of 600 mmHg. What is the partial pressure of the oxygen gas in the mixture?

Solution: The partial pressure of a gas is proportional to its mole fraction. Therefore, we must first find the number of moles of O_2 and CH_4 in the mixture.

$$\text{Moles of } O_2 = 50.0 \text{ g} \times \frac{1 \text{ mol } O_2}{32.0 \text{ g } O_2} = 1.56 \text{ mol of } O_2$$

$$\text{Moles of } CH_4 = 50.0 \text{ g} \times \frac{1 \text{ mol } CH_4}{16.0 \text{ g } CH_4} = 3.12 \text{ mol of } CH_4$$

$$\text{Total moles} = 1.56 + 3.12 = 4.68 \text{ mol}$$

$$\text{Mole fraction of } O_2 = \frac{1.56}{4.68}$$

$$\text{Partial pressure of } O_2 = \text{mole fraction of } O_2 \times \text{total pressure}$$

$$= \frac{1.56}{4.68} \times 600 \text{ mmHg} = 200 \text{ mmHg}$$

7.86* In a gaseous mixture of equal masses of C_2H_6 and CO_2 the partial pressure of the C_2H_6 is 22 mmHg. What is the partial pressure of the CO_2?

7.87* What mass of pure CO gas would have to be mixed with 40 g of pure CH_4 gas to form a mixture in which the partial pressure of the CO is equal to the partial pressure of the CH_4?

7.88* Exactly 1.100 g of carbon dioxide was introduced into a 1.00-L flask that contained some pure oxygen. The flask was warmed to 100 °C and the pressure was found to be 815 mmHg. No chemical reaction occurred. Calculate the mass of oxygen in the flask.

7.89* For the gaseous compound whose chemical formula is C_3H_8, calculate

a. volume of 3.00 mol at STP

b. volume of 32.0 g at 18 °C and 752 mmHg

c. amount in 140 L of the gas at STP

d. mass of 100 L of the gas at 20 °C and 700 mmHg

e. density of the gas in g/L at STP

f. percent by mass of carbon

g. partial pressure of C_3H_8 in a gaseous mixture of equal masses of C_3H_8 and CH_4 at 750 mmHg

h. density of the gas in g/L at 80 °C and 500 mmHg

STOICHIOMETRY OF GASES

Stoichiometric calculations on reactions in the gas phase are basically the same as the ones we did in Chapters 4, 5, and 6. The differences are primarily related to the different ways in which we measure the quantities of gases and liquids and solids. We generally do not weigh gases directly, but rather we measure their volumes, pressures, and temperature. When doing stoichiometric calculations on gases, therefore, we must convert information about pressure, volume, and temperature to information about mass and amount.

PROBLEMS

7.90 What volume of O_2 gas will be required to burn 50 L of H_2 gas? The volumes of both gases are measured at STP. The equation for the reaction is

$$2H_2 + O_2 \rightarrow 2H_2O.$$

Solution: The equation tells us that 1 mol of O_2 reacts with 2 mol of H_2. At STP 1 mol of H_2 is 22.4 L. Therefore, 50 L is 50/22.4 mol of H_2.

$$\frac{1}{2} \text{ of } \frac{50}{22.4} \text{ mol or } \frac{50}{2 \times 22.4} \text{ mol } O_2 \text{ will be required}$$

$$\frac{50}{2 \times 22.4} \text{ mol } O_2 \times \frac{22.4 \text{ L } O_2}{1 \text{ mol } O_2} = 25 \text{ L } O_2$$

We see that the volume of O_2 required is one-half the volume of H_2 just as the amount of O_2 is one-half the amount of H_2. In other words, at STP *the volumes of the two gases are to each other as their amounts*. Since both gases will respond equally to changes of temperature and pressure, it follows that *when measured at the same temperature and pressure, the volumes of gases involved in a reaction are to each other as the amounts of the gases in the equation for the reaction.* This very important generalization is an obvious consequence of the fact that *at constant temperature and pressure, the volume occupied by a gas is directly proportional to the amount of that gas.*

7.91 What volume of CO_2 gas, measured at 200 °C and 1.20 atm pressure, will be formed when 40.0 g of carbon is burned? The equation is $C + O_2 \rightarrow CO_2$.

Solution: Moles of CO_2 formed = moles of C burned.

$$\frac{40.0 \text{ g C}}{12.0 \text{ g/mol}} = \frac{40.0}{12.0} \text{ mol C burned}$$

Therefore, 40.0/12.0 mol of CO_2 will be formed.

To find the volume occupied by 40.0/12.0 mol of CO_2 at 200 °C and 1.2 atm we use the ideal gas equation, $PV = nRT$.

$$V = \frac{nRT}{P} = \frac{40.0}{12.0} \text{ mol} \times 0.0821 \frac{\text{L-atm}}{\text{mol-K}} \times \frac{473 \text{ K}}{1.20 \text{ atm}} = 108 \text{ L}$$

The entire calculation, in one operation, is

$$\frac{40.0 \text{ g C}}{12.0 \text{ g C/mol C}} \times \frac{1 \text{ mol } CO_2}{1 \text{ mol C}} \times 0.0821 \frac{\text{L-atm}}{\text{mol-K}} \times \frac{473 \text{ K}}{1.20 \text{ atm}} = 108 \text{ L}$$

7.92 What mass of tin would be formed if an excess of pure SnO were reduced with 1500 cm^3 of dry hydrogen gas measured at 300 °C and 740 mmHg?

Hint: $SnO + H_2 \rightarrow Sn + H_2O$

Using $PV = nRT$, first calculate the number of *moles* of H_2.

7.93 Ammonia gas is oxidized by oxygen gas in the presence of a catalyst as follows:

$$4NH_3 + 5O_2 \rightarrow 6H_2O + 4NO$$

What volume of oxygen will be necessary to oxidize 500 L of NH_3 gas? What volumes of NO and steam will be formed? All gases are measured under the same conditions of temperature and pressure.

Solution: At constant temperature and pressure the *volumes* of gases are directly proportional to the amount of each gas in the equation for the reaction. Therefore, in this problem the volumes of NH_3, O_2, H_2O (steam), and NO will be in the ratio 4:5:6:4.

Accordingly, $\frac{5}{4} \times 500$ or 625 L of O_2 will be needed, and $\frac{6}{4} \times 500$ or 750 L of H_2O and $\frac{4}{4} \times 500$ or 500 L of NO will be formed.

7.94 What volume, in cubic feet, of oxygen gas will be required for the oxidation of 6000 ft^3 of SO_2 gas in the contact process for manufacturing sulfuric acid? What volume of SO_3 gas will be formed? All gases are measured under the same conditions of temperature and pressure.

Hint: At constant temperature and pressure the *volumes* of gases are directly proportional to the amount of each gas in the equation for the reaction. It makes no difference in what unit volumes are expressed as long as the same unit (liters, cubic feet, quarts, cubic centimeters) is used for each gas.

7.95* When 1.82 g of zirconium metal (atomic weight 91.22) reacts with excess HCl at 27 °C and 720 mmHg, 1.04 L of dry H_2 gas, measured at 27 °C and 720 mmHg, is liberated. Write a balanced equation for the reaction that occurs when Zr is treated with HCl.

7.96* Equal volumes of hydrogen and oxygen, both at room temperature and atmospheric pressure, were introduced into a completely evacuated reaction bomb. The bomb was sealed and heated to 120 °C; the pressure of the mixture of gases in the bomb was found to be 100 mmHg at 120 °C. An electric arc inside the container was turned on, which caused the reaction $2H_2 + O_2 \rightarrow 2H_2O$ to take place. When the reaction was over, the bomb was cooled until the temperature was again 120 °C. An examination revealed that there was no liquid water in the bomb. What was the new pressure inside the bomb?

Solution: At the same temperature and pressure equal volumes of O_2 and H_2 contain equal numbers of moles of O_2 and H_2. Since, from the equation $2H_2 + O_2 \rightarrow 2H_2O$, we see that H_2 and O_2 combine in the ratio of 2 mol of H_2 to 1 mol of O_2 to form 2 mol of H_2O and since, at constant volume and temperature, the pressure is directly proportional to the amount, 50 mmHg worth of H_2 will combine with 25 mmHg worth of O_2 to form 50 mmHg worth of H_2O. There will be an excess of 25 mmHg worth of O_2, so the final pressure will be 75 mmHg.

7.97* A quantity of C_2H_4, when burned to CO_2 and H_2O, yielded 120 L of CO_2 measured at a certain temperature and pressure. Measured at the same temperature and pressure, a quantity of C_5H_{12}, when burned to CO_2 and H_2O, yielded 50.0 L of CO_2. Calculate the mass ratio of the quantities of C_2H_4 and C_5H_{12} that were burned.

Hint: Since, at constant temperature and pressure, the amounts of gases are directly proportional to their volumes, the moles of CO_2 from C_2H_4 and the moles of CO_2 from C_5H_{12} are in the ratio of 120 to 50. Since 1 mol of C_2H_4 yields 2 mol of CO_2 and 1 mol of C_5H_{12} yields 5 mol of CO_2, the C_2H_4 and

C_5H_{12} must have been present in what mole ratio to yield this 120/50 ratio of CO_2? Knowing the mole ratios you can calculate the mass ratios.

7.98* At a very high temperature 2 volumes of H_2S gas decomposed completely to give 3 volumes of a mixture of H_2 gas and sulfur vapor. Write the equation for the reaction.

7.99 A sample of $N_2O_3(g)$ has a pressure of 0.034 atm. It then undergoes complete decomposition to $NO_2(g)$ and $NO(g)$. Find the total pressure of the mixture of gases, assuming the volume and temperature do not change.

7.100* A mixture of $CO(g)$ and $O_2(g)$ in a 1.0 L container at 1000 K has a total pressure of 2.2 atm. After some time the pressure falls to 1.9 atm as the result of formation of CO_2. Find the amount of CO_2 that forms.

7.101 A sample of $C_2H_2(g)$ has a pressure of 0.78 atm. After some time a portion of it reacts to form $C_6H_6(g)$. The total pressure of the mixture of gases is then 0.39 atm. Assume the volume and temperature do not change. Find the fraction of C_2H_2 that has undergone reaction.

7.102 A 10-L container is filled with 0.10 mol of $H_2(g)$ and heated to 3000 K. The pressure is found to be 3.0 atm. Find the partial pressure of $H(g)$ that forms from the H_2 at this temperature.

7.103* When CO_2 gas, in a steel bomb at 427 °C and a pressure of 10 atm, is heated to 1127 °C, the pressure rises to 22.5 atm. The following reaction occurs: $2CO_2 \rightarrow 2CO + O_2$. Calculate the mole percent of CO_2 decomposed.

Hint: In Probs. 7.103–7.105 we observe reactions at some point before they have gone to completion and calculate the mole percent or fraction of reactant that has reacted at that point. In making the calculation we can assume that we have, in effect, stopped the reaction momentarily.

In order to calculate the mole percent of CO_2 decomposed we must know how many moles of CO_2 we had at the start and how many we have at the moment the reaction is observed. Using $PV = nRT$ we can, from the facts given, calculate how many moles of CO_2 we have *per liter* at the start and how many moles of $CO_2 + CO + O_2$ we have *per liter* at observation time. Although we do not know the volume of the reaction vessel, we are justified in selecting 1 L of it for our calculations because the temperature, pressure, and concentrations of gases will be completely uniform throughout the vessel.

Having calculated moles per liter of CO_2 at the start and of the gas mixture at observation time, and knowing that the reaction that takes place is $2CO_2 \rightarrow 2CO + O_2$, we can then set up an algebraic equation involving X (the number of moles of CO_2 decomposed) that will enable us eventually to calculate the mole percent of CO_2 decomposed.

7.104* When a sample of acetylene C_2H_2 is treated with a catalyst, some of it is converted to benzene C_6H_6 according to the equation $3C_2H_2 \rightarrow C_6H_6$. The density of the gaseous mixture of C_2H_2 and C_6H_6 at 27 °C and a pressure of 0.44 atm is 0.760 g/L. What fraction of the C_2H_2 originally present has changed to C_6H_6?

Hint: See Prob. 7.103. The law of conservation of matter dictates that the mass per volume of C_2H_2 at the start must equal the mass per volume of the mixture of gases at observation time. This relationship will enable us to calculate the number of moles of C_2H_2 per liter at the start.

7.105* A mixture of equal masses of C_4H_{10} and O_2 gases, when heated in a reaction vessel at 400 °C, reacted slowly to form CO_2 gas and steam (H_2O). At the end of 1 min the total pressure of the mixture of gases (C_4H_{10}, O_2, CO_2, and H_2O) was 2.96 atm and the density of the mixture was 2.000 g/L. What mole percent of the C_4H_{10} had been oxidized at the end of the 1-min interval?

7.106* A certain hydrocarbon gas was mixed in a steel reaction bomb with the exact amount of O_2 gas required to burn it completely to CO_2 and H_2O (steam). The mixture was ignited by a spark, which caused all the hydrocarbon to react with all the O_2 to form CO_2 and H_2O. The pressures before and after ignition, measured at the same temperature, were the same. The partial pressures of the CO_2 and steam were the same. Calculate the molecular weight of the hydrocarbon.

7.107 A mixture of CS_2 and CH_4 is burned to a mixture of CO_2, SO_2, and H_2O at 450 °C. The partial pressures of the CO_2, SO_2, and H_2O in the mixture are 175 mmHg, 200 mmHg, and 150 mmHg, respectively. What was the mass percent of CH_4 in the original mixture?

Hint: Since the reaction takes place in one reaction vessel, the three gases, CO_2, H_2O, and SO_2, are at the same temperature and occupy the same volume. Therefore, the numbers of moles of the three gases are in the same ratio as their partial pressures.

$$CS_2 + 3O_2 \rightarrow CO_2 + 2SO_2$$
$$CH_4 + 2O_2 \rightarrow CO_2 + 2H_2O$$

Since the partial pressures of the SO_2 and H_2O are, respectively, 200 mmHg and 150 mmHg, they must be present in the *ratio* of 2.00 mol of SO_2 to 1.50 mol of H_2O. Therefore, as the equations testify, the CS_2 and CH_4 must have been present in the *ratio* of 2.00 mol of CS_2 to 1.50 mol of CH_4. From the known mole ratio the mass percent of CH_4 can be calculated.

7.108 A mixture contains solid carbon, gaseous CS_2, and gaseous CH_4. The mixture is oxidized completely to give a gaseous mixture of CO_2, SO_2, and steam (H_2O). The ratio of the partial pressures of the CO_2, H_2O, and SO_2 in this gaseous mixture is 5.00 for the CO_2 to 2.00 for the H_2O to 1.00 for the SO_2. What was the percent by mass of solid carbon in the original mixture?

7.109 A 10.0-g sample of a mixture of $CuSO_4{\cdot}5H_2O$ and $CaCO_3$ was heated, resulting in decomposition of the carbonate and dehydration of the hydrate. If 5.00 cm^3 of water vapor (steam) was produced for every 2.00 cm^3 of CO_2, both volumes being measured at the same temperature and pressure, calculate the mass percent of $CaCO_3$ in the original mixture.

7.110* A mixture of C_2H_6 gas and CS_2 gas was placed in a reaction vessel at 200 °C. An excess of O_2 gas was then added to the mixture. When the resulting mixture was ignited by a spark, all the C_2H_6 and CS_2 was oxidized to CO_2, SO_2, and H_2O. When the reaction was complete, the partial pressures of the CO_2 and SO_2, measured at 200 °C, were 108 mmHg and 120 mmHg, respectively. Calculate the mass percent of C_2H_6 in the original mixture of C_2H_6 and CS_2.

Hint: See Prob. 7.107.

7.111* A gaseous mixture of equal masses of CH_4 and C_2H_6 plus excess O_2 was contained in a reaction vessel at 300 °C. The partial pressure of the CH_4 in this mixture was 15 mmHg. The mixture was ignited, which resulted in combustion of all the CH_4 and C_2H_6 to yield a gaseous mixture of CO_2, H_2O (all steam, no liquid water), and O_2. The mole fraction of the CO_2 in this gaseous mixture was 0.20. What was the partial pressure of the oxygen gas in the original mixture?

7.112* A gaseous mixture contained in a 1.0-L reaction vessel at 127 °C and 10 atm pressure consisted of CS_2 and CH_4 and an excess of oxygen. The mixture was ignited by a spark and complete oxidation to CO_2, SO_2, and H_2O (steam) occurred. After the reaction was complete, the pressure in the 1.0-L vessel, measured at 527 °C, was 17.1 atm. What mass of CS_2 was in the mixture?

Hint: See Prob. 5.53.

7.113* A gaseous mixture of CH_4 and CS_2 was added to an excess of oxygen gas. The resulting mixture was placed in an evacuated, constant-volume reaction vessel. The temperature of the mixture was 320 °C. The mixture was ignited, resulting in the complete oxidation of CH_4 and CS_2 to CO_2, H_2O, and SO_2. The temperature of the reaction vessel was brought back to the original value,

320 °C. The total pressure of the mixture of gases (no liquid) in the vessel at 320 °C was 79 mmHg, and the partial pressure of the SO_2 was 8.0 mmHg. Calculate the total pressure of the original mixture of CH_4, CS_2, and O_2.

7.114* A mixture of $CH_4(g)$ and $C_2H_6(g)$ has a total pressure of 0.53 atm. Just enough $O_2(g)$ is added to the mixture to bring about its complete combustion to $CO_2(g)$ and $H_2O(g)$. The total pressure of these two gases is found to be 2.2 atm. Assuming constant volume and temperature, find the fraction of CH_4 in the mixture.

VAPOR PRESSURE

When a dry gas, hydrogen for example, is collected over water at a given temperature, water evaporates, so molecules of water vapor, $H_2O(g)$, mix with the hydrogen. This evaporation continues until the rate at which water molecules evaporate equals the rate at which they condense. The hydrogen gas is then said to be *saturated* with water vapor. The pressure of this water vapor is called the *vapor pressure* of water and its value depends only on the temperature.

The vessel in which the hydrogen is collected contains a mixture of hydrogen and water vapor, and each gas (hydrogen and water vapor) exerts its own partial pressure. The total pressure is the sum of the two partial pressures. It follows, therefore, that when a gas is collected over water only a part of the total pressure is exerted by the gas itself; some of the pressure is exerted by the water vapor. Therefore, when the gas laws are used, the partial pressure of the water vapor must be subtracted from the total pressure under which the gas is collected to give the actual pressure on the gas itself. Table A.1 in the appendix gives the vapor pressure of water at various temperatures.

PROBLEMS

7.115 The volume of a dry gas is 600 cm^3 at 25 °C and 750 mmHg. What volume would this gas occupy if collected over water at a pressure of 746 mmHg and 32 °C?

Solution: Since the gas is to be collected over water, the vapor pressure of water at 32 °C must be subtracted from the pressure of 746 mmHg.

$$\begin{aligned} \text{Vapor pressure of water at 32 °C} &= 35.4\ \text{mmHg} \\ &\quad \text{(see appendix Table A.1)} \\ \text{Actual new pressure} &= 746\ \text{mmHg} - 35.4\ \text{mmHg} \\ &= 710.6\ \text{mmHg} \end{aligned}$$

Using this pressure of 710.6 mmHg as the new pressure in place of the 746 mmHg, we can then solve in the manner shown in Prob. 7.14.

$$\text{New volume} = 600\ \text{cm}^3 \times \frac{750\ \text{mmHg}}{710.6\ \text{mmHg}} \times \frac{305\ \text{K}}{298\ \text{K}} = 648\ \text{cm}^3$$

7.116 The volume of a dry gas at 758 mmHg and 12 °C is 100 ft^3. What volume will this gas occupy if stored over water at 22 °C and a pressure of 740 mmHg?

7.117 A sample of 100 cm^3 of dry gas, measured at 20 °C and 750 mmHg, occupied a volume of 105 cm^3 when collected over water at 25 °C and 750 mmHg. Calculate the vapor pressure of water at 25 °C.

7.118 A 1.20-L volume of a gas was collected over water at 26 °C and 1.0 atm. Calculate the amount of gas that was collected.

7.119 What volume in liters will 300 g of oxygen occupy when collected over water at 20 °C and 735 mmHg?

7.120 The vapor pressure of water at 121 °C is 2 atm. Some liquid water is injected into a sealed vessel containing air at 1 atm pressure and 121 °C and is allowed to come to equilibrium with its vapor. What is the total pressure in the sealed container?

Solution: As long as there is some liquid water present and the temperature remains at 121 °C the partial pressure of the water vapor in equilibrium with the liquid water will remain constant at 2 atm; it cannot rise above 2 atm or fall below 2 atm. If we neglect the small solubility of air in water, the partial pressure of the air will remain at 1 atm. The total pressure will then be 3 atm.

7.121 Binary compounds of alkali metals and hydrogen react with water to liberate $H_2(g)$. The H_2 from the reaction of a sample of NaH with an excess of water fills a volume of 0.49 L above the water. The temperature of the gas is 305 K and the total pressure is 1.0 atm. Find the amount of H_2 liberated. Find the mass of NaH that reacted.

7.122* Three 1-L flasks, all at 27 °C, are interconnected with stopcocks, which are initially closed. The first flask contains 1 g of H_2O. The second flask contains O_2 at a pressure of 1 atm. The third flask contains 1 g of N_2.

a. The temperature is kept constant at 27 °C. At this temperature the vapor pressure of water is 0.0380 atm. The stopcocks are all opened. When equilibrium is reached, what is the pressure in the flasks?

b. If the temperature of the whole system is raised to 100 °C, what is the pressure?

Hint: Will there be any liquid H_2O present in the final system at 27 °C? At 100 °C?

7.123* At 37 °C liquid A has a vapor pressure of 58.4 mmHg and liquid B a vapor pressure of 73.6 mmHg. To a sealed, evacuated container of volume 100 L maintained at 37 °C are added 0.20 mol of gaseous A and 0.50 mol of gaseous B.

What is the total pressure in the container when equilibrium is reached? (Each vapor is completely insoluble in the other liquid.)

RELATIVE HUMIDITY

Air that contains a high concentration of water vapor is said to be *humid. Humidity* refers to the water vapor content of a gas. If air is saturated with water vapor at a given temperature its *relative humidity* is 100%; that is, it contains all (100%) the water vapor it can hold. If it contains only half the water vapor it can hold (50% saturated), its relative humidity is 50%. *Relative humidity* is, therefore, the ratio between the concentration of water vapor actually present and the concentration that would be present if the gas were saturated with water vapor. The simplest way to express the moisture content of a gas is in terms of the water vapor pressure; the relative humidity is then the ratio between the partial pressure of the water vapor and the equilibrium (saturated) vapor pressure at that temperature.

PROBLEMS

7.124 The partial pressure of the water vapor in the air in a room is 11.6 mmHg at 22 °C. What is the relative humidity of the air in the room?

Solution: Relative humidity is the ratio between the partial pressure of the water vapor and the equilibrium (saturated) vapor pressure at that temperature. From Table A.1 of the appendix we learn that the vapor pressure of water at 22 °C is 19.8 mmHg.

$$\text{Relative humidity} = \frac{11.6\ \text{mmHg}}{19.8\ \text{mmHg}} \times 100\% = 58.6\%$$

7.125* A weather report gives the temperature as 22.5 °C, barometric reading 750 mmHg, and relative humidity 75%. What is the mole fraction of water vapor in the atmosphere? (See Table A.1.)

7.126* At 24 °C the vapor pressure of H_2O is 22.2 mmHg. If the relative humidity in a sealed container is 30% at 24 °C and 57% when the temperature is lowered to 13 °C, what is the vapor pressure of H_2O at 13 °C?

DIFFUSION AND EFFUSION

If a gas is introduced into a container that already contains a gas, the new gas will move through the other gas to fill the container uniformly. This process is called *diffusion*. If a container filled with gas has a small hole in one wall and a vacuum on the other side of the wall, the gas will pass through the hole. This process is called *effusion*. The rates of both these processes depend on the particular gases involved. For two gases the rates of diffusion (or effusion) are inversely proportional to the square roots of their densities. This generalization is known as *Graham's law*.

$$\frac{\text{Rate of diffusion of gas A}}{\text{Rate of diffusion of gas B}} = \frac{\sqrt{\text{density of B}}}{\sqrt{\text{density of A}}}$$

PROBLEMS

7.127 Gas A is 9 times as dense as gas B. In a given diffusion apparatus and at a certain temperature and pressure, gas B diffuses 15 cm in 10 s. In the same apparatus and at the same temperature and pressure, how fast will A diffuse?

Solution:

$$\frac{\text{Rate of A}}{\text{Rate of B}} = \frac{\sqrt{\text{density of B}}}{\sqrt{\text{density of A}}}$$

Substituting in this formula, we have

$$\frac{\text{rate of A}}{15\ \text{cm}/10\ \text{s}} = \frac{\sqrt{1}}{\sqrt{9}} = \frac{1}{3}$$

$$\text{rate of A} = \frac{15\ \text{cm}/10\ \text{s}}{3} = 5\ \text{cm in } 10\ \text{s}$$

7.128 The density of CH_4 is 16.0 g per 22.4 L. The density of HBr is 81.0 g per 22.4 L. If CH_4 diffuses 2.30 ft in 1 min in a certain diffusion apparatus, how fast will HBr diffuse in the same apparatus at the same temperature and pressure?

7.129 How do the rates of diffusion of HBr and SO_2 compare?

Hint:

$$\frac{\text{Rate of A}}{\text{Rate of B}} = \frac{\sqrt{\text{density of B}}}{\sqrt{\text{density of A}}}$$

Since the density of a gas is directly proportional to its molecular weight, we can write

$$\frac{\text{rate of A}}{\text{rate of B}} = \frac{\sqrt{\text{molecular weight of B}}}{\sqrt{\text{molecular weight of A}}}$$

7.130 In a given effusion apparatus 15.0 cm^3 of HBr gas was found to effuse in 1 min. What volume (in cm^3) of CH_4 gas would effuse in 1 min in the same apparatus at the same temperature?

7.131 An unknown gas diffuses at the rate of 8 cm^3/s in a piece of apparatus in which CH_4 gas diffuses at 12 cm^3/s. Calculate the approximate molecular weight of the gas.

7.132* The average kinetic energy of gas molecules is calculated from the formula $KE = \frac{1}{2}mv^2$ and is directly proportional to the absolute temperature. If H_2 gas molecules move at an average speed of 1.2 km/s at 300 K, what will their average speed, in kilometers per second, be at 1200 K?

7.133 An unknown gas was found to diffuse 2.2 ft in the same time that methane gas (CH_4) diffused 3.0 ft. The unknown gas was found to contain 80% C and 20% H. Calculate the exact molecular weight of the gas.

7.134 In a given diffusion apparatus 15.0 cm^3 of HBr gas diffused in 1 min. In the same apparatus and under the same conditions 33.7 cm^3 of an unknown gas diffused in 1 min. The unknown gas contained 75% carbon and 25% hydrogen. Calculate its exact molecular weight.

7.135 Find the ratio of the average speed of N_2 at 300 K to that of Ar at 400 K.

Chapter 8

Thermochemistry

Almost every physical or chemical change is accompanied by the evolution or absorption of heat. Processes that evolve heat are called *exothermic* and those that absorb heat are called *endothermic*. Heat evolved is expressed as a negative quantity, heat absorbed as a positive quantity. The magnitude of the quantity of heat that accompanies a change depends on many factors, including the conditions under which the process takes place. Most commonly we carry out chemical and physical changes in containers that are open to the atmosphere—in effect, at constant pressure. The heat that accompanies a process at constant pressure corresponds to a change in a property of the system called its *enthalpy* or *heat content*. Enthalpy is represented by the symbol H, and the change in the enthalpy as the result of a physical or chemical change is ΔH. By definition ΔH is the enthalpy of the products minus the enthalpy of the reactants.

We can systematize and correlate a great deal of chemical information by the use of ΔH. For convenience we give names to certain common kinds of enthalpy changes. The ΔH of any chemical reaction is called the *heat of reaction*. The ΔH of the rapid reaction of a substance with O_2 is called the *heat of combustion* of the substance. The ΔH of the reaction in which a substance is formed from its elements is called the *heat of formation* of the substance. The ΔH values of various phase changes are also given names. We refer to the *heat of vaporization* for the ΔH of the conversion of liquid to gas, the *heat of fusion* for the conversion of solid to liquid, and the *heat of sublimation* for the conversion of solid to gas.

The value of ΔH is also a function of the conditions of the process. Accordingly we wish to be very precise about the conditions of a change when giving its ΔH so that we can make meaningful comparisons between different processes. By convention we define a standard state. In the standard state the pressure of every component in the system is exactly 1 atm. We designate the enthalpy change under these conditions as

$\Delta H°$, the *standard enthalpy*. We will usually be interested in processes that take place at 298 K (25 °C) unless otherwise indicated. We also assume that all substances are in their most stable form at 1 atm and 298 K.

The *standard heat of formation* of a substance, $\Delta H_f°$, is an important quantity. It is the heat liberated or absorbed when 1 mol of the substance forms from its elements, each of which must be in its most stable form at 1 atm and 298 K. Thus the $\Delta H_f°$ of H_2O refers to the formation of liquid water from $H_2(g)$ and $O_2(g)$. A tabulation of $\Delta H_f°$ values can be found in Table A.7 in the appendix. There is a simple relationship between the values of $\Delta H_f°$ of the components of a reaction and the $\Delta H°$ of the reaction: The heat of the reaction is equal to the sum of the heats of formation of the products minus the sum of the heats of formation of the reactants.

$$\Delta H° = \sum \Delta H_f°(\text{products}) - \sum \Delta H_f°(\text{reactants}) \quad \textbf{(8-1)}$$

Another quantity related to the heat of reaction is the *bond energy*. When chemical bonds form, heat is given off; therefore, heat is required to break a bond. The bond energy is the average heat required to break a given type of chemical bond in a compound in the gas phase into its constituent atoms also in the gas phase. Bond energies are usually expressed in units of kJ/mol or kcal/mol and are always positive numbers. A table of bond energies can be found in Table A.8 in the appendix. These average bond energy values can be used to estimate many heats of reaction. The heat of reaction is related to the bond energies by the relationship

$$\Delta H = \sum \text{bonds broken} - \sum \text{bonds made} \quad \textbf{(8-2)}$$

You should remember that bond energies are defined for substances in the vapor phase, so this method can be used only when all the components of the reaction are gases.

A quantity of heat can be expressed in a number of different units. The SI unit is the *joule* (J). A common unit is the *calorie* (cal).

$$1 \text{ cal} = 4.184 \text{ J}$$

The calorie is the quantity of heat required to raise the temperature of 1 g of water by 1 °C (actually from 14.5 °C to 15.5 °C).

For large quantities we can use the *kilojoule* (kJ) or the *kilocalorie* (kcal).

$$1 \text{ kJ} = 1000 \text{ J} \text{ and } 1 \text{ kcal} = 1000 \text{ cal}$$

Calorimetry is the branch of science that deals with the measurement of heat. A device for carrying out such a measurement is called a calorimeter. In a calorimeter we measure temperature change and use known relationships between heat and temperature to find the heat.

The heat capacity of a system or a substance is defined in the following way:

$$\text{heat capacity} = \frac{\text{heat}}{\Delta T} = \frac{q}{\Delta T}$$

The heat capacity of a calorimeter can be measured by measuring the temperature change caused by a fixed quantity of electrical energy. The temperature is usually measured in kelvins (or Celsius, which is equivalent for a ΔT). Once we know the heat capacity of the calorimeter and we measure the temperature change, it is easy to calculate q. If the calorimeter operates at constant pressure, then $q = \Delta H$.

Heat capacity is the heat (q) required to cause a temperature change (ΔT) in a substance. The value depends not only on the nature of the substance, but also on the quantity of substance. So we can define the molar heat capacity as the quantity of heat required to change the temperature of 1 mol of the substance by 1 K.

$$\text{C(molar)} = \frac{q}{n\Delta T}, \text{ where } n \text{ is the amount in moles}$$

The quantity of heat required to change the temperature of 1 g of the substance by 1 K is called the specific heat of the substance.

$$\text{Specific heat capacity} = \frac{q}{m\Delta T}, \text{ where } m \text{ is the mass in grams.}$$

Notice the relationship between molar heat capacity and specific heat capacity:

$$\text{specific heat capacity} \times \text{molar mass} = \text{molar heat capacity}$$

Thus the specific heat of water is 4.184 J/g·K (1.000 cal/g-°C) and its heat capacity is 75.38 J/mol-K (18.02 cal/mol-°C) at 15 °C.

Calculations involving these quantities depend on the relationship between four parameters. Given any three, we can find the fourth. Usually we will be finding q.

At constant pressure the relationship between the heat capacity, the enthalpy, and the temperature change is

$$C = \Delta H/\Delta T \qquad \textbf{(8-3)}$$

PROBLEMS

8.1 Given that the molar heat capacity of $H_2O(l)$ is 75.4 J/mol·K and of $Hg(l)$ is 27.9 J/mol·K, calculate the heat required to raise the temperature of 1.0 kg of $H_2O(l)$ from 298 K to 308 K, and then calculate the heat required to bring about the same temperature change in the same mass of $Hg(l)$.

Solution: Since we are given the molar heat capacities, we first convert the mass of H_2O to amount. Then we use the relationship that defines molar heat capacity:

$$1000 \text{ g } H_2O \times \frac{1 \text{ mol } H_2O}{18.0 \text{ g } H_2O} \times 75.4 \text{ J/mol·K} \times (308 \text{ K} - 298 \text{ K}) = 42{,}000 \text{ J}$$

and

$$1000 \text{ g Hg} \times \frac{1 \text{ mol Hg}}{200.6 \text{ g Hg}} \times 27.9 \text{ J/mol·K} \times (308 \text{ K} - 298 \text{ K}) = 1400 \text{ J}$$

Note how much more heat is required for H_2O compared to Hg. This difference reflects the unusually high heat capacity of water.

8.2 What mass of water can be heated from 20 °C to 60 °C by 20,500 J of heat?

8.3 The heat capacity of methyl alcohol (CH_4O) is 80.3 J/mol·K. What quantity of heat will be evolved when the temperature of 2610 g of methyl alcohol falls from 22 °C to 2 °C?

8.4 Find the quantity of heat that must be liberated by a reaction to raise the temperature of a calorimeter whose heat capacity is 75.4 kJ/K by 1.00 K.

8.5 The heat of combustion of methane to carbon dioxide and liquid water is −885.4 kJ/mol. It is found that when a 5.347-g sample of methane is burned in a calorimeter, the temperature of the calorimeter increases from 296.14 K to 299.88 K. Find the heat capacity of the calorimeter.

8.6 Find ΔH for the combustion of ethanol, C_2H_5OH, from the following data: The heat capacity of the calorimeter is 34.65 kJ/K. The combustion of 1.765 g of ethanol raises the temperature of the calorimeter from 294.33 K to 295.84 K.

8.7 One tablespoon of peanut butter has a mass of 16 g. It is combusted in a calorimeter whose heat capacity is 120 kJ/K. The temperature of the calorimeter rises from 295.2 K to 298.4 K. Find the caloric content of the peanut butter. (*Hint:* 1 food calorie is 4.184 kJ.)

8.8 A 51.42-g sample of ice is placed in a calorimeter whose heat capacity is 2.348 kJ/K. The temperature of the calorimeter drops from 282.47 K to 275.16 K, as all the ice melts. Calculate the heat of fusion of ice.

8.9 The heat of combustion of a sample of coal is 25.1 kJ/g. What mass of this coal would have to be burned in order to generate enough heat to raise the temperature of 1530 g of water from 283.0 K to 337.0 K?

Solution:

$$\frac{\text{Joules required to heat the water}}{\text{Joules evolved per gram of coal burned}} = \text{mass of coal burned}$$

$$\frac{1530 \text{ g } H_2O \times 54.0 \text{ K} \times \dfrac{4.184 \text{ J}}{1 \text{ g } H_2O \times 1\text{K}}}{\dfrac{25{,}100 \text{ J}}{1 \text{ g coal}}} = 13.8 \text{ g coal}$$

8.10 A sample of 12 g of a certain grade of coal gave off enough heat to raise the temperature of 222 mol of water from 285 K to 303 K. Calculate the heat of combustion of the coal.

8.11 A sample of 2.500 g of sulfur, when burned to SO_2, raised the temperature of 1080 g of water from 22.5 °C to 27.5 °C. Calculate the heat of formation of SO_2.
Hint: To form 1 mol of SO_2 one must burn 1 mol of S. Therefore, the heat of formation of SO_2 is the same as the heat of combustion of S.

8.12 When burned in oxygen, 10.00 g of phosphorus generated enough heat to raise the temperature of 2950 g of water from 18.0 °C to 38.0 °C. Calculate the heat of formation of P_4O_{10}.
Hint: Note that 4 mol of P yields 1 mol of P_4O_{10}.

8.13 The heat of fusion of ice is 333 J/g. The heat of combustion of methyl alcohol is 715 J/mol. What mass of methyl alcohol, CH_3OH, must be burned in order to generate enough heat to melt a block of ice of mass 6010 g?

8.14 The heat of combustion of gasoline is about 3.0×10^4 kJ/L. The energy of complete fission of 100 kg of uranium is 1.0 megaton, and 1.0 megaton $= 1.4 \times 10^{14}$ kJ. What volume of gasoline must be burned to yield the same amount of energy possible from fission of 1.0 kg of uranium?

8.15 The heat of combustion of methane (CH_4), the chief constituent of natural gas, is 891 kJ/mol. The heat capacity of iron is 26.4 J/mol·K; its melting point is 1803 K and its heat of fusion is 14,900 J/mol. Assuming no heat loss, calculate the mass of CH_4 that must be burned to cause 76.5 mol of iron, initially at 303 K, to melt.

Hint: Divide the problem into two parts. First calculate the heat required to raise the temperature to the melting point and then the quantity of heat that must be absorbed for melting to occur. Be sure to keep the units consistent, since both J and kJ are used.

8.16 Calculate the heat required to raise the temperature of a 5.46-g block of aluminum from 298.0 K to 317.6 K (heat capacity of Al = 24.3 J/mol·K).

8.17 Calculate the heat evolved when the temperature of 1.0 mol of graphite falls 100 K (heat capacity of graphite = 8.54 J/mol·K).

8.18 The heat of combustion of methane is 890.5 kJ/mol. A tank of water is heated by combustion of methane. Calculate the temperature increase in 20.56 kg of water as the result of the combustion of 2.00 mol of methane, assuming no heat loss to the surroundings.

8.19 The heat of combustion of octane (C_8H_{18}) is 1303 kJ/mol. Find the mass of octane that must be burned in order to raise the temperature of a 1.00-kg block of iron by 200 K, assuming no heat loss (heat capacity of Fe = 24.8 J/mol·K).

8.20* An ice cube of mass 9.0 g is added to a cup of coffee at 363 K that contains 120 g of liquid. Assume the heat capacity of coffee is the same as that of water. The heat of fusion of ice is 6.0 kJ/mol. Find the temperature of the coffee after the ice melts.

8.21 Calculate $\Delta H°$, the standard heat of reaction, for

$$CaO(s) + CO_2(g) \rightarrow CaCO_3(s)$$

using the values of $\Delta H_f°$ in Table A.7 in the appendix.

Solution: Substituting the values from Table A.7 of the appendix into Eq. (8-1), we have

$$\Delta H° = \Delta H_f°(\text{products}) - \Delta H_f°(\text{reactants})$$

$$\Delta H° = \left[\left(-1207\,\frac{\text{kJ}}{\text{mol}} \times 1\text{ mol}\right)\right] - \left[\left(-636\,\frac{\text{kJ}}{\text{mol}} \times 1\text{ mol}\right) + \left(-393.5\,\frac{\text{kJ}}{\text{mol}} \times 1\text{ mol}\right)\right]$$

$$= -178\text{ kJ}$$

8.22 Calculate $\Delta H°$ for the following reaction at 25 °C:

$$2C_2H_6(g) + 7O_2(g) \rightarrow 4CO_2(g) + 6H_2O(l)$$

Solution: Applying the method of Prob. 8.21, we have

$$\Delta H° = \left[\left(-393.5\,\frac{\text{kJ}}{\text{mol}} \times 4\text{ mol}\right) + \left(-286\,\frac{\text{kJ}}{\text{mol}} \times 6\text{ mol}\right)\right] - \left[\left(-84.7\,\frac{\text{kJ}}{\text{mol}} \times 2\text{ mol}\right) + \left(0\,\frac{\text{kJ}}{\text{mol}} \times 7\text{ mol}\right)\right] = -3120\text{ kJ}$$

You should note that ΔH_f° of $O_2(g) = 0$. *The heat of formation of any element in its standard state is equal to zero* by definition. Thus the ΔH_f° of $Br_2(l)$ at 25 °C and 1 atm is equal to zero, but ΔH_f° of $Br_2(g)$ or of Br is not equal to zero. The element must be in its most stable state at 1 atm for ΔH_f° to be equal to zero.

8.23 The heat of vaporization of $H_2O(l)$ at 298 K is 44.1 kJ/mol. Using the result from Prob. 8.22, calculate ΔH for the reaction

$$C_2H_6(g) + \tfrac{7}{2}O_2(g) \rightarrow 2CO_2(g) + 3H_2O(g)$$

Solution: This equation differs in two ways from that in Prob. 8.22. All the molar quantities are cut in half; therefore, we can begin by dividing the calculated ΔH° by 2, to give 3120/2 = 1560 kJ/mol. Furthermore, one of the products is $3H_2O(g)$ rather than $3H_2O(l)$, so the heats of reaction will differ by ΔH_{vap} of $H_2O \times 3$ mol of H_2O = 44.1 kJ/mol × 3 mol = 132 kJ. Several approaches can be used to work out the direction of the difference. We can reason that since heat must be supplied to evaporate water, the reaction that forms $H_2O(g)$ will be less exothermic (less negative) than the reaction that forms $H_2O(l)$; some of the heat produced will be needed for the evaporation of the water. $\Delta H = -1560\text{ kJ} + 132\text{ kJ} = -1428\text{ kJ}$.

Alternatively we can use *Hess's law of heat summation*, which states that the energy liberated in a reaction is independent of the path that is followed and is the algebraic sum of the heat changes in the steps of the reaction. We can bring about the reaction in this problem by a path involving two steps of known ΔH.

$$C_2H_6 + \tfrac{7}{2}O_2(g) \rightarrow 2CO_2(g) + 3H_2O(l) \quad \Delta H = -1560\text{ kJ} \quad \textbf{(a)}$$

$$3H_2O(l) \rightarrow 3H_2O(g) \quad \Delta H = 132\text{ kJ} \quad \textbf{(b)}$$

Algebraic combination of Eqs. (a) and (b) gives

$$C_2H_6(g) + \tfrac{7}{2}O_2(g) \rightarrow 2CO_2(g) + 3H_2O(g)$$
$$\Delta H_1 + \Delta H_2 = -1428\text{ kJ}$$

8.24 Use the data given in Table A.7 to calculate ΔH° for the following reactions at 298 K:

a. $HCl(g) + NH_3(g) \rightarrow NH_4Cl(s)$

b. $2NO_2(g) \rightarrow N_2O_4(g)$

c. $C_2H_4(g) + 3O_2(g) \rightarrow 2CO_2(g) + 2H_2O(l)$

d. $C_2H_2(g) + 2H_2(g) \rightarrow C_2H_6(g)$

8.25 Use the data given in Table A.7 to calculate the heat of combustion of $C_3H_8(g)$ to $CO_2(g)$ and $H_2O(l)$.

8.26 Given that $\Delta H°$ for the reaction $2SO_2(g) + O_2(g) \rightarrow 2SO_3(g)$ is -196 kJ and that $\Delta H_f°$ of $SO_2(g)$ is -297 kJ/mol, calculate $\Delta H_f°$ of $SO_3(g)$ at 298 K. *Hint:* Let $x = \Delta H_f°$ of $SO_3(g)$ and substitute into the equation of Prob. 8.21.

8.27 Use the data given in Table A.7 to calculate ΔH for the following reactions at 298 K:

a. $2HBr(g) \rightarrow H_2(g) + Br_2(g)$, given ΔH_{vap} of $Br_2(l) = 31$ kJ/mol

b. $2Fe(s) + 3H_2O(g) \rightarrow Fe_2O_3(s) + 3H_2(g)$, given ΔH_{vap} of $H_2O(l) =$ 44.1 kJ/mol

8.28 The $\Delta H_f°$ of $TiI_3(s)$ is -335 kJ/mol and $\Delta H°$ for the reaction $2Ti(s) + 3I_2(g) \rightarrow 2TiI_3(s)$ is -856 kJ. Calculate the ΔH of sublimation of $I_2(s)$. Note that I_2 is a solid at 298 K.

8.29 Given

$$NO(g) + NO_2(g) \rightarrow N_2O_3(g) \quad \Delta H_1° = -40 \text{ kJ} \qquad \textbf{(a)}$$

$$N_2O_4(g) \rightarrow 2NO_2(g) \quad \Delta H_2° = 58 \text{ kJ} \qquad \textbf{(b)}$$

calculate $\Delta H°$ for

$$2N_2O_3(g) \rightarrow 2NO(g) + N_2O_4(g) \qquad \textbf{(c)}$$

Solution: If some algebraic combination of Eqs. (a) and (b) gives Eq. (c), Hess's law of heat summation can be applied. We can see that such a combination can be found since all the substances in Eq. (c) appear in Eq. (a) or Eq. (b) and the substance not in Eq. (c), $NO_2(g)$, appears in both Eqs. (a) and (b) and can be made to cancel. The appropriate combination can be deduced by inspection or by the following reasoning. Equation (c) has 2 mol of $N_2O_3(g)$ on the left, and $N_2O_3(g)$ appears in Eq. (a) on the right. Reverse Eq. (a) and multiply it by a factor of 2, giving Eq. (d): $2N_2O_3(g) \rightarrow 2NO(g) + 2NO_2(g)$. Reversing an equation has the effect of changing the sign of its ΔH. Therefore, $\Delta H_4° = -2(\Delta H_1°)$. Next, Eq. (c) has $N_2O_4(g)$ on the right; $N_2O_4(g)$ appears in Eq. (b) on the left. Reverse Eq. (b) and change the sign of $\Delta H_2°$ to give Eq. (e): $2NO_2(g) \rightarrow N_2O_4(g)$, $\Delta H_5° = -\Delta H_2°$. Combine Eqs. (d) and (e) to give $2N_2O_3(g) + 2NO_2(g) \rightarrow 2NO(g) + 2NO_2(g) + N_2O_4(g)$. Cancel like terms ($2NO_2(g)$) from both sides to give Eq. (c). Since Eq. (d) plus Eq. (e) = Eq. (c), $\Delta H_4° + \Delta H_5° = \Delta H_3° = -2(-40 \text{ kJ}) + (-58 \text{ kJ}) = 22$ kJ.

8.30 Given the equations

$$C(s) + O_2(g) \rightarrow CO_2(g) \quad \Delta H = -395 \text{ kJ}$$

$$C(s) + \tfrac{1}{2}O_2(g) \rightarrow CO(g) \quad \Delta H = -118 \text{ kJ}$$

calculate ΔH for the reaction:

$$CO_2(g) \rightarrow CO(g) + \tfrac{1}{2}O_2(g)$$

8.31* Given the equations

$$SO_2(g) + \tfrac{1}{2}O_2(g) \rightarrow SO_3(g) \quad \Delta H = 89.5 \text{ kJ}$$
$$S(s) + \tfrac{3}{2}O_2(g) \rightarrow SO_3(g) \quad \Delta H = -204.2 \text{ kJ}$$

calculate the heat of formation of $SO_2(g)$ under these nonstandard state conditions.

8.32* Calculate ΔH for the reaction $H_2(g) + \tfrac{1}{2}O_2(g) \rightarrow H_2O(l)$, which is not carried out in the standard state, from the following data for processes carried out at the same conditions:

$$C(s) + 2H_2O(g) \rightarrow CO_2(g) + 2H_2(g) \quad \Delta H = 163 \text{ kJ}$$
$$C(s) + \tfrac{1}{2}O_2(g) \rightarrow CO(g) \quad \Delta H = -121 \text{ kJ}$$
$$H_2O(g) \rightarrow H_2O(l) \quad \Delta H = -40.6 \text{ kJ}$$
$$CO(g) + \tfrac{1}{2}O_2(g) \rightarrow CO_2(g) \quad \Delta H = -283 \text{ kJ}$$

8.33* Use Hess's law to prove that

$$\Delta H^\circ = \sum \Delta H_f^\circ(\text{products}) - \sum \Delta H_f^\circ(\text{reactants})$$

Hint: Pick a general reaction like $AB + CD \rightarrow AC + BD$, where A, B, C, and D are different elements.

8.34 Using the table of bond energies, Table A.8 in the appendix, calculate ΔH° for the reaction

$$H_2(g) + I_2(g) \rightarrow 2HI(g)$$

Solution: Since the bond energy is the heat that must be absorbed in order to break a bond, it is always a positive quantity and is usually given in kJ/mol. It is defined when all species are in the gas phase. The heat of reaction can be calculated by using Eq. (8-2):

$$\Delta H = \sum \text{bonds broken} - \sum \text{bonds made}$$

The reaction of interest must therefore be analyzed in order to determine the bonds that break and the bonds that form. In the above reaction, the H—H bond in the reactant $H_2(g)$ is broken, the I—I bond in the reactant $I_2(g)$ is broken, and two H—I bonds are formed, one in each $HI(g)$. Substituting, we have

$$\Delta H = [436\text{ kJ} + 151\text{ kJ}] - [2(299\text{ kJ})] = -11\text{ kJ}$$

8.35 Using the bond energies given in Table A.8, calculate ΔH for the following reactions.

a. $2HBr(g) \rightarrow H_2(g) + Br_2(g)$
b. $H_2(g) \rightarrow 2H(g)$
c. $CH_4(g) + Cl_2(g) \rightarrow CH_3Cl(g) + HCl(g)$
d. $2NH_3(g) \rightarrow 3H_2(g) + N_2(g)$
e. $4NH_3(g) + 5O_2(g) \rightarrow 4NO(g) + 6H_2O(g)$

8.36 Using the bond energies given in Table A.8, calculate ΔH for the reaction

$$H_2C{=}CH_2 + Cl_2 \rightarrow Cl{-}CH_2{-}CH_2{-}Cl$$

8.37* The heat of formation of $HF(g)$ is -269 kJ/mol; use the bond energy values in Table A.8 to calculate the bond energy of the H—F bond.

Hint: Let x = the bond energy and substitute it into the relationship of Prob. 8.34 applied to the reaction for forming 1 mol of HF from its elements.

8.38* Given that ΔH_f° for $CH_4(g)$ is -74.9 kJ/mol, use the bond energy values in Table A.8 to calculate the heat of vaporization of graphite.

Hint: The heat of formation refers to the process $C(s) + 2H_2(g) \rightarrow CH_4(g)$; the bond energy can be applied to the process $CH_4(g) \rightarrow 4H(g) + C(g)$. The heat of vaporization refers to the process $C(s) \rightarrow C(g)$. Use Hess's law.

8.39 The ClCl and FF bond energies are 243 kJ/mol and 158 kJ/mol respectively. The ΔH_f° of $ClF_3(g)$ is -159 kJ/mol. Calculate the average bond energy of the ClF bond in ClF_3.

8.40 Repeat the calculation of Prob. 8.39 on $ClF_5(g)$. $\Delta H_f^\circ = -240$ kJ/mol. Account for the difference with the answer in Prob. 8.39.

8.41 The heat of atomization is the heat required to convert a molecule in the gas phase into its constituent atoms in the gas phase. It is used to calculate average bond energies. Without using any tabulated bond energies, calculate the average CCl bond energy from the following data: The heat of atomization of CH_4 is 1660 kJ/mol and the heat of atomization of CH_2Cl_2 is 1486 kJ/mol.

Chapter 9

Quantum Theory and the Structure of Atoms

The study of the interaction between electromagnetic radiation and matter has played a prominent role in the development of our current ideas on the structure of matter. All electromagnetic radiation can be described as waves of electric and magnetic fields moving with a velocity $c = 2.998 \times 10^8$ m/s. A *wave* is a regular succession of maxima and minima or peaks and valleys. The distance between two consecutive maxima is called the *wavelength* λ. The *frequency* ν is the number of maxima that pass a fixed point in unit time, usually one second. From these definitions it follows that $c = \lambda\nu$, that is, wavelength and frequency are inversely proportional.

But electromagnetic radiation also has properties that we associate with particles rather than with waves. The energy of electromagnetic radiation of a given frequency is not infinitely divisible. The smallest value of the energy is called a *quantum*, and the energy of a given amount of this radiation must be an integral multiple of the quantum. The idea of a quantum of energy is more consistent with the concept of a particle than with that of a wave. The energy of the quantum of electromagnetic radiation is proportional to its frequency, $E = h\nu$, where E is the energy in joules, ν is the frequency in reciprocal seconds (s^{-1}), and h is a proportionality constant called *Planck's constant*, which has the value 6.626×10^{-34} J-s.

A classic phenomenon that illustrates this idea is the *photoelectric effect*. When light impinges on a metal, electrons may be emitted. But only when the frequency of light is equal to or greater than a value characteristic for each metal will electrons be emitted. This value is called the *threshold frequency* ν_0 and the energy of a single quantum of this light is given by $E = h\nu_0$. This energy must also be the binding energy ε with which a single electron is held by the metal. When the frequency of light impinging on the metal is greater than the threshold frequency, only a part of the available energy is needed to overcome the attraction between the metal and the electron. The excess energy results *not* in the emission of more electrons but in the electrons being emitted with excess kinetic energy. The number of electrons emitted depends rather on the intensity, that is, the number of quanta, of the incident light.

These relationships are given by

$$\mathrm{h}\nu = \varepsilon + \frac{1}{2}mv^2 \tag{9-1}$$

where ν is the frequency of the incident light and $\frac{1}{2}mv^2$ is the excess kinetic energy.

When a pure element is heated—as, for example, hydrogen atoms in stars—radiation is emitted. It is found that this emitted radiation consists of only a relatively small number of wavelengths, so when the radiation is passed through a prism it appears as a number of lines, called the *atomic spectrum*, characteristic for each element. The measured wavelengths of these lines for hydrogen obey a surprisingly simple numerical relationship:

$$\overline{\nu} = \frac{1}{\lambda} = R_{\mathrm{H}}\left(\frac{1}{n_{\mathrm{L}}^2} - \frac{1}{n_{\mathrm{H}}^2}\right) \tag{9-2}$$

where λ is the wavelength of the line, R_{H} is the Rydberg constant, with a value of $1.10 \times 10^5\ \mathrm{cm}^{-1}$, and n_{L} and n_{H} are arbitrarily assigned small positive integers such that $n_{\mathrm{H}} > n_{\mathrm{L}}$. The symbol $\overline{\nu}$ is called the wave number and is defined as $1/\lambda$. Every time two small integers are substituted for n_{L} and n_{H} into this equation, a wavelength can be calculated that corresponds to that of a line observed in the atomic spectrum of hydrogen.

Observations such as these eventually led to the quantum description of the atom in which only certain energies are possible for the electrons. The emission of a quantum of radiation $h\nu$ thus corresponds to the transition of an electron from a high energy level to a low energy level. The difference in energy between the two levels, ΔE, is equal to $h\nu$. Similarly when a quantum of radiation is absorbed, an electron goes from a low energy level to a higher one. Again, $\Delta E = h\nu$.

In order to understand the behavior of electromagnetic radiation we must think in terms of a wave-particle duality. That is, radiation has some wave characteristics and some particle characteristics. The properties of matter also require a wave-particle duality. Associated with a moving particle is a wavelength defined by

$$\lambda = \frac{h}{mv} \tag{9-3}$$

where λ is the wavelength, h is Planck's constant, m is the mass of the particle, and v is its velocity. The quantity mv is sometimes written as p, the momentum. For macroscopic particles such as a baseball, m is so large that λ is inconsequentially small. However, for submicroscopic particles such as an electron, m is so small that λ has an appreciable value.

PROBLEMS

9.1 What is the frequency of green light whose wavelength is 520 nm?
Hint: Substitute into $\nu = c/\lambda$ using the conversion 1 nm $= 10^{-9}$ m.

9.2 What is the wavelength of a radio wave whose frequency is 6×10^{11} s^{-1}?

9.3 What is the velocity of a sound wave whose frequency is 1000 s^{-1} and whose wavelength is 33 cm?

9.4 What is the energy of a quantum of gamma radiation whose frequency is 4.0×10^{19} s^{-1}?

9.5 What is the energy of a quantum of ultraviolet radiation whose wavelength is 300 nm?

9.6 How many quanta of radiation whose frequency is 1×10^{15} s^{-1} will be required for an energy of 1 J?

9.7* We require 4.184 J to raise the temperature of 1 g of water 1 K. Water is exposed to infrared radiation of $\lambda = 2.8 \times 10^{-4}$ cm. Assume that all the radiation is absorbed and converted to heat. How many quanta will be required to raise the temperature of 1.0 g of water 2.0 K?

9.8 What is the binding energy of an electron in a metal whose threshold frequency for the photoelectric effect is 5.0×10^{14} s^{-1}? What is it for 1 mol of electrons?

9.9 What will be the kinetic energy of an electron emitted from the metal in Prob. 9.8 when it is exposed to radiation of frequency 2.0×10^{5} s^{-1}?

9.10 Calculate the threshold frequency necessary for the photoelectric effect to occur in a metal that is found to emit electrons with kinetic energy of 2.2×10^{-29} J per electron when exposed to radiation of frequency 1.0×10^{5} s^{-1}.

9.11 A certain metal is found to have a threshold frequency of 4.5×10^{14} s^{-1} for the photoelectric effect. The mass of an electron is 9.1×10^{-28} g. Calculate the velocity of an electron emitted by this metal when it is exposed to radiation of wavelength 550 nm.

9.12 A series of four lines in the atomic spectrum of hydrogen corresponds to the transition of an electron from the fifth energy level to each of the lower energy levels. Calculate the wavelength of each line.

Hint: In Eq. (9-2) n_H represents the higher energy level and n_L the lower energy level involved in the emission of radiation. Thus $n_H = 5$ and $n_L = 1, 2, 3$, and 4. Substitution of these values into Eq. (9-2) will give the appropriate wavelengths.

9.13 When hydrogen is at relatively low temperatures, its electron is in the first energy level, called the *ground state*. What is the longest wavelength of radiation that can be absorbed by hydrogen?

Hint: Equation (9-2) can also be used to correlate the absorption of radiation by an electron when it goes from a low to a high energy level.

9.14 What is the difference in energy between the second and third levels of hydrogen?

Hint: Calculate the wavelength of the radiation corresponding to this transition and then the frequency and corresponding energy of this radiation.

9.15* What is the energy required to remove an electron from hydrogen in its ground state?

9.16 An electron at 300 K is found to have a wavelength of 6.0 nm. Its mass is 9.1×10^{-28}g. What is its velocity?

9.17 A xenon atom travels with a velocity of 2.4×10^4 cm/s at 300 K. What is its wavelength?

9.18 A baseball of mass 100 g is thrown with a velocity of 90 mph. What is its wavelength? (1 mile = 1.61×10^5 cm.)

ELECTRONIC CONFIGURATIONS OF ATOMS

Only certain energy levels are allowed for the electrons in multielectron atoms. The *ground-state electronic configuration* of an atom, the configuration of lowest energy, is the one in which the electrons occupy the lowest energy levels available, subject to certain restrictions. In order to describe an electronic configuration, we first describe the allowed energy levels and then specify the ones that contain electrons.

The description of energy levels is based on the mathematical solutions of the Schrödinger wave equation for hydrogen, which relates the allowed energies to the mass and wave motion of a particle and whose solutions exactly describe the hydrogen atom. The approximation is made that the energy levels for each electron in a multielectron atom are like those of hydrogen. An allowed energy level can be described as a region of space that we enclose with an imaginary surface whose shape and location are obtained from the solution to the wave equation. This surface is a convenient way for us to represent an *orbital*, which is the total picture of the allowed energy level. Three integers, called *quantum numbers*, specify the energy, shape, and direction of the orbital.

The first of these integers is called the *principal quantum number* and is represented by n. The value of n specifies almost completely the energy associated with the orbital. The principal quantum number may have any integral value ≥ 1. The lower its value, the lower the energy. The second of these integers is called the *angular momentum quantum number* l. The value of l is specified *after* the value of n and is restricted by the value of n. The l may have any integral value from 0 to $n - 1$. The value of l further specifies the energy of the orbital and also gives information about its shape. The third of these integers is called the magnetic quantum number m_l. Its values are restricted by the values of l. It can equal $-l, -l + 1, \ldots, 0, \ldots, l - 1, l$. It serves to specify the directional properties of the orbital but is not related to its energy.

By specifying, within the given restrictions, values for the three quantum numbers, an orbital, or an allowed orbital, is described. However to complete the description of an electron in this orbital, it is also necessary to have information about its spin. The *spin quantum number* m_s, which can have a value of $+\frac{1}{2}$ or $-\frac{1}{2}$, indicates relative direction of spin. Thus if two electrons both have a spin of $+\frac{1}{2}$ or of $-\frac{1}{2}$, they are spinning the same way and are said to be *unpaired* or *parallel*; if one is $+\frac{1}{2}$ and one is $-\frac{1}{2}$, they are spinning in opposite ways and are said to be *paired*.

By convention certain letters are used to replace the numerical values of the l and m_l quantum numbers. Thus $l = 0$ corresponds to a spherical orbital called an *s orbital*. When $l = 1$ we have a two-lobed orbital called a *p orbital*. When $l = 1$, m_l can have the values $-1, 0, +1$, which correspond to three p orbitals, differing in their direction but not their energy, represented by p_x, p_y, and p_z. When $l = 2$ we have a *d orbital*. When $l = 2$, m_l can have the values $-2, -1, 0, +1, +2$, which correspond to five d orbitals differing in their direction but not their energy, represented by d_{xy}, d_{xz}, d_{yz}, d_{z^2}, and $d_{x^2-y^2}$. When $l = 3$ we have an f orbital of which there are seven, differing in direction but not energy. The value of the principal quantum number precedes the letter designations. Thus a description of the hydrogenlike set of allowed orbitals for a multielectron atom begins:

$$1s, 2s, 2p_x, 2p_y, 2p_z, 3s, 3p_x, 3p_y, 3p_z, 3d_{xy}, 3d_{xz}, 3d_{yz}, 3d_{z^2}, 3d_{x^2-y^2}, \ldots, \text{etc.}$$

The ground-state electronic configuration of an atom is worked out from these orbitals by the *Aufbau* (German for "buildup") method, in which the electrons are placed into the available orbitals in order of increasing energy, subject to certain restrictions. The *Pauli exclusion principle* states that in a given species no two electrons can have the same four quantum numbers. Thus an orbital can hold a maximum of two electrons and they must spin in opposite directions. In a situation where more than one orbital of equal energy is available to be filled, *Hund's rules* specify the manner in which the electrons will occupy these orbitals.

A lower-energy configuration is obtained with one electron in each of two different orbitals of the *same* energy than with two electrons in the same orbital. When there is only one electron in each of two orbitals, it is better to have them spinning the same way.

We write electronic configurations by indicating the number of electrons in each orbital with a superscript. Thus the ground-state electronic configurations of the first few elements are: H, $1s^1$; He, $1s^2$; Li, $1s^2 2s^1$; Be, $1s^2 2s^2$; and B, $1s^2 2s^2 2p^1$. We can indicate relative spin with arrows pointing up or down. Thus C is $1s^2 2s^2 2p_x\uparrow 2p_y\uparrow$. The unpaired electrons in the $2p$ orbital spin the same way. The electron configuration for N is $1s^2 2s^2 2p_x\uparrow 2p_y\uparrow 2p_z\uparrow$, and O is $1s^2 2s^2 2p_x^2 2p_y\uparrow 2p_z\uparrow$. When a group of orbitals of the same energy are filled or almost filled, it is common to write them together. Thus F is $1s^2 2s^2 2p^5$ and Ne is $1s^2 2s^2 2p^6$.

You should note that with neon all the orbitals of principal quantum number 2 are filled and the next element, Na, begins with $3s$. It is common to abbreviate the electronic configuration of Na as $[\text{Ne}]3s^1$, of Mg as $[\text{Ne}]3s^2$, and of Al as $[\text{Ne}]3s^2 3p^1$. When argon is reached the $3s$ and $3p$ orbitals are filled, giving $[\text{Ne}]3s^2 3p^6$, and the electronic configurations of elements after argon are written beginning with [Ar]. Similarly the $4s$ and $4p$ orbitals are filled in krypton, $[\text{Ar}]3d^{10} 4s^2 4p^6$, so electronic configurations of elements after Kr are written beginning with [Kr].

The order in which the orbitals fill follows the regular sequence described above until element 19, potassium. At this point there are so many electrons present that the hydrogenlike approximation is not always a good one, and a number of exceptions must be committed to memory. In general the $4s$ fills before the $3d$; the $5s$ before the $4d$; and the $6s$ before the $4f$ before the $5d$. The following table (p. 111) will help you to remember this generalization, the diagonal lines indicating the order of filling.

There are numbers of exceptions, however. One type of exception seems to reflect stability associated with a half-filled or filled set of orbitals of the same energy. For example Cr has the configuration $[\text{Ar}]4s^1 3d^5$, and Cu has $[\text{Ar}]4s^1 3d^{10}$.

Another difficulty after element 19 appears in the relative ease of removal of electrons from an atom. For example, it is found that in the series of elements from Sc to Zn the $4s$ electron is more easily removed than the $3d$ electron.

Understanding ground-state electronic configurations is extremely important for understanding the periodic table. Columns consist of elements with the same electronic arrangement of their outermost electrons, called the *valence electrons*. For example, column IA of the periodic table is composed of elements whose valence-electron configuration can be represented by ns^1. Thus Na is $[\text{Ne}]3s^1$, K is $[\text{Ar}]4s^1$,

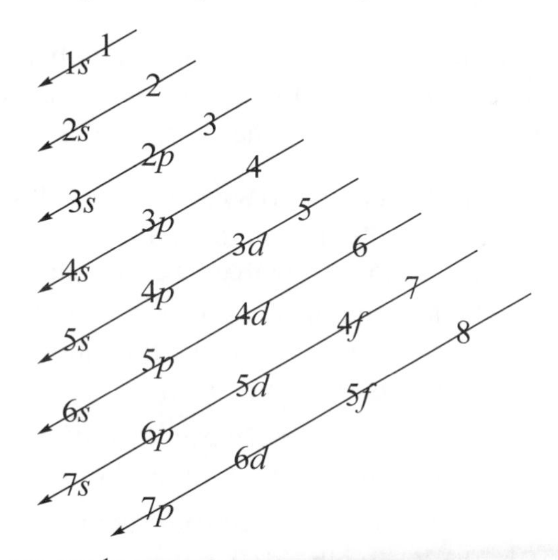

Rb is [Kr]$5s^1$, and Cs is [Xe]$6s^1$. Also, it is the valence-electron configuration that determines most of the chemical behavior of an element.

PROBLEMS

9.19 Which of the following orbital designations are not possible? 6*s*, 2*d*, 8*p*, 4*f*, 1*p*, 3*f*.

9.20 Write the ground-state electronic configuration of the following: (a) Sr, (b) Co, (c) Se, (d) Ba.

Solution:

a. The atomic number of strontium is 38. The atomic number of krypton, the noble gas that most closely precedes strontium in the periodic table, is 36. The difference between these numbers is the number of electrons outside the krypton core (two in this case) that must be shown to specify the electronic configuration of strontium. The orbital that begins filling after the krypton core is completed is the 5s orbital. Since the two electrons can be accommodated in the 5s orbital, the ground-state electronic configuration of strontium is [Kr]$5s^2$.

b. Cobalt, atomic number 27, has $27 - 18 = 9$ electrons outside the argon core in the neutral atom. These electrons fill the 4*s* orbital and partially fill the 3*d* orbital. The ground-state electronic configuration of Co is therefore [Ar]$4s^2 3d^7$.

9.21 Write the ground-state electronic configuration of each of the following rare gases: (a) Kr, (b) Xe, (c) Rn.

9.22 Write the ground-state electronic configuration of each of the following ions: (a) P^{3+}, (b) Se^{2+}, (c) Br^-, (d) Sb^+.

Solution: (a) Since phosphorus is atomic number 15, the P^{3+} ion has $15 - 3 = 12$ electrons. There are $12 - 10 = 2$ electrons outside the neon core. The electronic configuration of the P^{3+} ion is therefore [Ne]$3s^2$.

9.23 Name the element that fits each of the following descriptions:

a. The lightest atom containing a half-filled *p* subshell

b. The heaviest rare gas that has no electrons in *d* orbitals

c. An element that has two electrons with quantum numbers $n = 4$, $l = 0$ and eight electrons with quantum numbers $n = 3$, $l = 2$

d. Two elements with two unpaired $3p$ electrons

9.24 Write the ground-state electronic configurations of the following elements and identify their relationship: (a) As, (b) Sb, (c) Bi.

9.25 Write the ground-state electronic configurations of the following elements: (a) Rb, (b) Y, (c) Nb, (d) Tc.

9.26 Write the ground-state electronic configurations of the following species: (a) S^{2-}, (b) Cl^-, (c) Ti^{4+}.

9.27 Name four ions that have the ground-state electronic configuration [Xe].

9.28 Without consulting a periodic table predict the atomic numbers of the next three elements in the column under F in the periodic table.

9.29 Without consulting a periodic table predict which element between 32 and 35 will most closely resemble element 52.

9.30 Name an element to fit each of the following.

a. An element whose dipositive cation has the same electronic configuration as He

b. An element whose tetrapositive cation has the same electronic configuration as the monoanion of Cl

c. An element with a half-filled $4s$ shell and a half-filled *d* shell

d. An element whose tripositive cation has a half-filled $3d$ shell

9.31 How many unpaired electrons would be found in the following atoms in their ground states? (a) Mn, (b) Sc, (c) Fe, (d) Zn.

9.32 Write the ground-state electronic configurations of the cations that form from sulfur by the loss of one to six electrons. Indicate which ones have one or more unpaired electrons.

9.33 Without consulting a periodic table predict the electronic configuration of the element under Ag, atomic number 47.

9.34 The first excited state refers to the electronic configuration with an energy closest to but higher than that of the ground state. Write electronic configurations of the first excited state of the following: (a) C, (b) Ne, (c) Li.

9.35 The elements with atomic numbers 58 through 70, called the *rare earths*, all have very similar properties. Suggest a reason for this behavior based on their electronic configurations.

9.36* Imagine a universe in which the m_l quantum number is allowed only positive values ($m = 0, 1, 2, \ldots, l$). Assuming that all other quantum numbers are unaffected, that the Pauli exclusion principle still holds, and that the name corresponding to an element of a given atomic number is the same, give the following:

a. The new electronic configuration for fluorine

b. The electronic configuration for the element below fluorine in the new periodic table

c. The numbers and names of the first three rare gases in the new periodic table

9.37* From its electronic configuration predict which of the first 10 elements would be most similar to the as yet unknown element of atomic number 116.

9.38* Imagine a universe in which the value of the m_s quantum number can be $+\frac{1}{2}$, 0, and $-\frac{1}{2}$ instead of just $\pm\frac{1}{2}$. Assuming that all the other quantum numbers can take only the values possible in our world and that the Pauli exclusion principle applies, give the following:

a. The new electronic configuration for nitrogen

b. The electronic configuration for the element below nitrogen in the new periodic table

c. The numbers of the first three rare gases in the new periodic table

Chapter 10

The Structure of Molecules

The atoms in a molecule are held to each other by *chemical bonds*. Bonds are most commonly classified as *ionic* or *covalent*, depending on the extent to which the electrons in the bond are associated with each of the atoms bonded together. For example, in the molecule H_2 each H atom contributes its 1*s* electron to the bond and the electron is equally associated with, or shared between, the two H atoms. This bond is a covalent bond. On the other hand, in a molecule such as CsF the Cs contributes a 6*s* electron and the fluorine a 2*p* electron to the bond, but the electron pair of the bond is primarily associated with the fluorine. This bond is an ionic bond. Most chemical bonds fall between these two extremes and are classified as covalent or ionic based on the nature of the atoms of the bonds and on observations of chemical properties.

The vast majority of chemical bonds we encounter are covalent bonds. For convenience we can divide covalent bonds into two types, polar and nonpolar. In nonpolar bonds both atoms have an equal or almost equal association with the electrons in the bond. For example, homonuclear diatomic molecules such as Cl_2, F_2, and N_2 have nonpolar bonds, and bonds between C and H or C and C are nonpolar. In many bonds that are covalent one of the two atoms in the bond clearly has a greater association with the shared electrons than the other. These bonds are called *polar* bonds; common examples are H—O, H—N, and H—halogen bonds.

We can often predict the type of bond by use of a devised property called *electronegativity*, which is a rough measure of the relative attraction of a bonded atom for the shared electrons. The greater the difference in electronegativity between the two atoms of the bond, the more ionic the bond; the smaller the difference in electronegativity, the more nonpolar the bond. The electronegativity of an atom can be correlated with its position in the periodic table; electronegativity increases in going from left to right and from bottom to top of the table. Thus fluorine is the most electronegative element, followed by oxygen, nitrogen, and chlorine, while cesium is the least electronegative element, followed by rubidium, potassium, and barium. A rough quantitative rule is: When the electronegativity difference between the two atoms of the bond is greater than 1.7, the bond is ionic; when the difference is less than 1.7, it is covalent and polar; and when the difference is less than 0.5, it is covalent and nonpolar. Table A.9 in the appendix lists some electronegativity values.

We can understand many aspects of bond formation by focusing on the *valence electrons* of the atoms of the bond. These are the electrons outside the completed noble gas shell with the exception of electrons in completed *d* or *f* shells. The valence electrons generally are those of highest or sometimes second-highest principal quantum number.

We can often understand bond formation as a reflection of the stability that seems to be associated with the electronic configuration ns^2np^6 of noble gases. That is, if an atom can achieve this electronic configuration by forming bonds in which it loses, gains, or shares electrons, it will do so. This observation is known as the *octet rule* and works well for the lighter elements and groups IA, IIA, VA, VIA, and VIIA of the periodic table, with the exception of hydrogen and helium, which can accommodate only two valence electrons. Despite the fact that a great deal of bond formation violates the octet rule, this rule is often a very useful starting point for working out the bonding in molecules.

The most common notation chemists use to represent the bonding in covalent molecules is a slight modification of the electron dot picture originally devised by G. N. Lewis. It is based on the following rules:

1. Only the valence electrons are shown.
2. A *pair* of shared electrons is represented by a dash between the two atoms sharing them. Thus when two atoms are sharing one pair of electrons, they are connected by a single dash called a *single bond*; when they are sharing two pairs of electrons, they are connected by two dashes called a *double bond*; and when they are sharing three pairs of electrons, they are connected by three dashes called a *triple bond*.
3. An unshared or nonbonding pair of electrons is represented by a pair of dots written next to the atom with which the electrons are associated.
4. All the valence electrons of each atom must be accounted for by either dots or dashes.

Some of the electronic properties of the elements that help in writing Lewis structures are

1. Hydrogen can accommodate no more than two electrons.
2. In stable compounds C, N, O, and F are *never* surrounded by more than eight electrons (the octet rule) and only very rarely, in the case of N, by fewer.
3. Compounds of many other elements often have octets around all the atoms but there are many exceptions among the heavier elements beginning with silicon.
4. Elements of group IIIA often have incomplete octets.
5. Compounds with an odd number of valence electrons must have at least one atom that violates the octet rule.

In many cases you can follow a series of steps to arrive at a correct Lewis structure for a compound. We illustrate the method for HNO_3, nitric acid.

Step 1 Determine the total number of valence electrons by adding the number of valence electrons of each atom of the molecule. The number of valence electrons of an atom in an A column of the periodic table is the number of the column. Thus elements in column IA have one valence electron, in column IIA have two valence electrons, and so on to column VIIA with elements that have seven valence electrons. (No such simple correlation can be used for elements in the B columns of the periodic table.) There are $1 + 5 + (6 \times 3) = 24$ valence electrons in HNO_3.

Step 2 Connect all the atoms in the molecule using no more than one dash (one electron pair) between two atoms. You may need certain information to connect the atoms correctly.

a. Hydrogen can have only one electron pair and therefore only one bond.

b. Normally, we do not find oxygen–oxygen bonds. The exceptions are the peroxides, in which two oxygens are joined to each other.

c. Sometimes an experimental observation is necessary. For example, in N_2O, the attachment is found to be N—N—O, not N—O—N.

d. Most strong oxyacids have the H on oxygen. Thus for HNO_3 we write

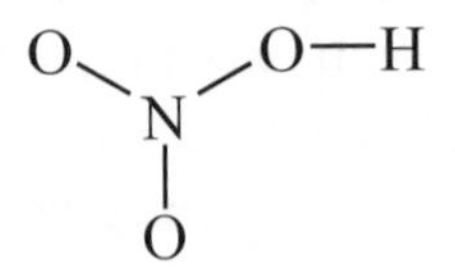

because we know (or are told) that it is a strong acid.

Note: The octet rule prohibits more than four atoms joined to a central one. Connecting the atoms is the most difficult part of writing structures, but as we get more facts, it becomes easier.

Step 3 Count the number of electrons used for the dashes (two per dash) and subtract this number from the total available to see how many electrons are left over. For HNO_3 we have used 4 dashes $\times$ 2 electrons = 8 electrons and have $24 - 8 = 16$ electrons left over.

Step 4 Locate those atoms that do not yet have completed octets. Calculate how many more electrons are needed to complete each octet and add these numbers.

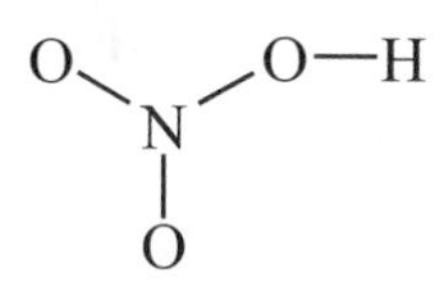

Two of the O's need six, one of the O's needs four, and the N needs two more electrons to complete their octets. H is complete.

The total number of electrons needed to complete each octet is $(2 \times 6) + (1 \times 4) + (1 \times 2) = 18$.

Step 5 Compare the numbers from steps 3 and 4. If they are the same, fill out the octets with pairs of dots so that each atom is surrounded by eight electrons. If they are not the same, the result of step 3 may be less than that of step 4, almost always by an even number of electrons. Divide this number by 2. The result gives the number of additional dashes or bonds required. Put in these dashes between two atoms with incomplete octets and then add dots to complete all the octets. Generally, one or two dashes are added where a dash is already found. That is, cyclic structures are not used unless some specific information is available indicating they should be. Thus,

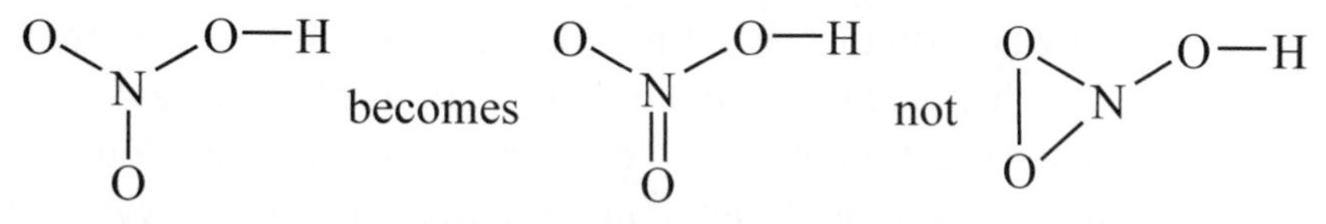

Two or three dashes (two or three pairs of electrons) between two atoms are common and are called a double bond and triple bond, respectively.

Step 6 Use the remaining electrons to complete the remaining octets with dots.

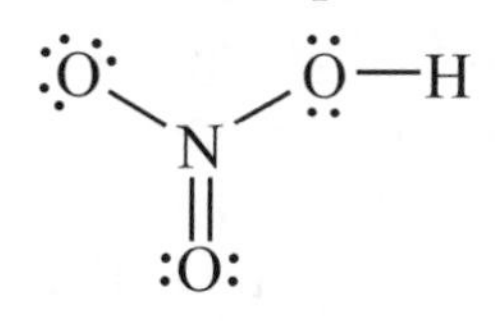

Step 7 Check to see that the correct number of electrons has been used.

Step 8 If the number from step 3 is less than the number from step 4 and the atoms are such that no more dashes may be added, then one or more atoms will have an incomplete octet. For example, in BF_3 the boron, a group IIIA element, has an incomplete octet because fluorine can form only one bond.

Step 9 If the number from step 3 is greater than the number from step 4 then one or more of the atoms will be surrounded by more than eight electrons. Such an atom is said to have an expanded octet. In many molecules only one atom will be permitted to expand its octet. For example in ClF_3, this atom must be the Cl, since F may not expand its octet. The structure is

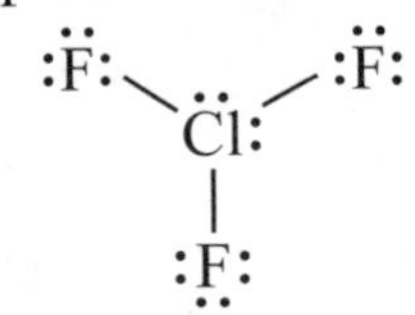

In other molecules it will be possible to expand the octet of more than one of the atoms. Often it will be the least electronegative atom which is expanded. For example in PCl_5 it is the P which has the expanded octet. The structure is

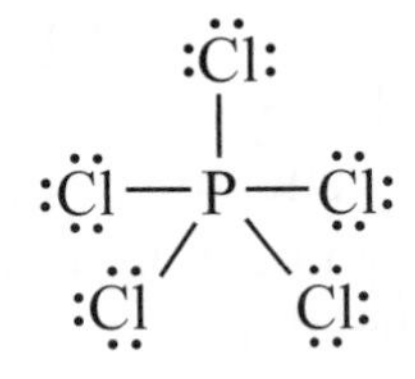

Step 10 The principle of electroneutrality is often helpful in arriving at the best Lewis structure, especially for compounds of C, N, and O. Normally C forms four bonds, N forms three bonds and has one nonbonding electron pair, and O forms two bonds and has two nonbonding electron pairs. Often in writing a Lewis structure we find that this normal complement of bonds is not obtained. An unusual number of bonds may cause destabilization and result in the creation of *formal charges*. Thus when an atom with a complete octet forms one more bond than normal, for example oxygen with three bonds or nitrogen with four bonds, it is thought of as bearing a formal charge of +1. When an atom with a complete octet forms one less bond than normal, for example carbon with three bonds or oxygen with one bond, it is thought of as bearing a formal charge of −1. In neutral compounds the sum of any formal charges that exist must be zero. The observation that the existence of formal charges is destabilizing is often helpful in arriving at the best Lewis structure for a compound. For example, two choices are apparent for HCN: $H{-}C{\equiv}N:$ and $:\overset{(-)}{C}{\equiv}\overset{(+)}{N}{-}H$. The first structure is correct and consistent with avoiding formal charges.

PROBLEMS

10.1 List the following compounds in order of increasing ionic bond character:
a. NH_3 **b.** H_2O **c.** HF **d.** CH_4 **e.** H_2
Hint: The greater the electronegativity difference between two atoms, the more ionic the bond. Use a periodic table or Table A.9 in the appendix.

10.2 List the following compounds in order of increasing ionic bond character:
a. HI **b.** I_2 **c.** CsI **d.** CsF **e.** NaI **f.** HF **g.** NCl_3.

10.3 Give the number of an element that fits each of the following descriptions without consulting a periodic table:
a. The most electronegative element in the first row (Li to Ne).
b. The least electronegative element in the third row.
c. The most electronegative element among those with atomic numbers 3, 11, 12, 13, 35.
d. Two third-row elements with four valence electrons.
e. The element closest in atomic number to Te (52) that has the same valence-shell electronic configuration with a different principal quantum number.
f. The element among those with atomic numbers 7, 8, 15, 17 that forms the least polar bond with H.
g. The two lowest-numbered elements that often expand their octets.
h. The element among those with atomic numbers 16, 17, 33, 34, 35 that forms the most ionic compound with lithium.

10.4 What is the correlation between valence-electron configuration and the normal number of bonds formed in the series of elements having atomic numbers 3 through 9?

10.5 Which of the following molecules cannot completely satisfy the octet rule?
a. N_2O **b.** NO **c.** N_2O_3 **d.** NO_2 **e.** N_2O_4 **f.** Cl_2O **g.** N_2O_5

10.6 Which of the following molecules cannot have an expanded octet?
a. H_2SO_4 **b.** H_3PO_4 **c.** OF_2 **d.** $HClO_4$ **e.** HNO_3

10.7 Draw Lewis structures for the following molecules:
a. NCl_3 **b.** NF_3 **c.** OF_2 **d.** N_2

Solution: (a) Since there are 5 valence electrons on the nitrogen atom and 7 valence electrons on the chlorine atom, the NCl_3 molecule has $5 + (3 \times 7) = 26$ valence electrons. The correct structure must show all these electrons. The atoms in this molecule can be connected in only one way since the Cl cannot form more than one bond to N:

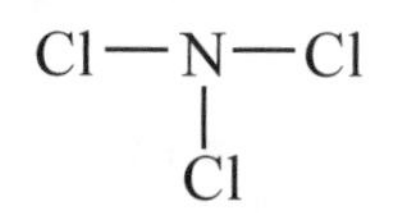

This skeleton structure uses $3 \times 2 = 6$ valence electrons. There are $26 - 6 = 20$ electrons remaining to complete octets. Each Cl atom requires 6 electrons to complete its octet and the N atom requires two electrons. The total number of required electrons is $(3 \times 6) + 2 = 20$. Since the total required and the number remaining are the same, we just complete the octet about each atom, using pairs of dots for the nonbonding electron pairs:

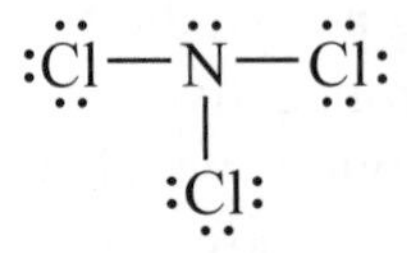

A final count of the electrons in the completed structure provides an additional check on the structure.

10.8 Draw Lewis structures for (a) HNO_2 (b) N_2O_3 (no N—N bond) (c) C_2H_2 (d) C_2H_4.

Solution: (a) There are $1 + 5 + (2 \times 6) = 18$ valence electrons in HNO_2. Since HNO_2 is nitrous acid, an oxyacid, the most probable skeleton structure is

H—O—N—O

which has an O—H bond and no O—O bond.

This structure uses $3 \times 2 = 6$ electrons. There are $18 - 6 = 12$ electrons remaining. The first O requires 4 electrons, the N requires 4 electrons, and the second O requires 6 electrons for completion of their octets. A total of 14 electrons is thus required, while only 12 electrons remain. Another electron pair must be shared. This pair is most reasonably shared between N and the second O, giving the structure H—O—N=O. This structure requires 10 electrons to complete all octets. Since 10 electrons remain, we can fill in the required number of nonbonding electron pairs to give:

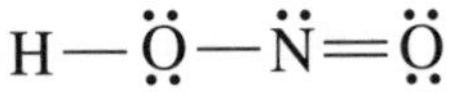

A final count of the electrons in the completed structure provides an additional check on the structure.

10.9 (a) Draw the Lewis structure for CH_4O, methyl alcohol. (b) Ethyl alcohol, C_2H_6O, is very similar in behavior to methyl alcohol. Draw the Lewis structure of ethyl alcohol based on the structure of methyl alcohol. (c) Dimethyl ether is a compound that also has the composition C_2H_6O. Draw its Lewis structure.

10.10 Draw Lewis structures for the following molecules with incomplete octets: (a) BF_3 (b) $AlCl_3$ (c) NO

10.11 Draw Lewis structures for the following molecules using expanded octets as needed: (a) H_3PO_4 (b) $HClO_3$ (c) ICl_5

10.12 Draw Lewis structures for the following ions: (a) CH_3^+ (b) NH_4^+ (c) ClO^- (d) I_3^-

10.13 Use the principle of electroneutrality to draw Lewis structures for (a) CNBr (b) HClO

10.14* Draw the Lewis structure of P_4 based on the observations that no expanded octets or double bonds are present and that all the P atoms are equivalent.

10.15 Draw two possible Lewis structures for NH_3O and predict which one represents the more stable compound.

10.16 Draw Lewis structures for (a) $SiCl_4$ (b) PCl_5 (c) SCl_4 (d) $BrCl_5$. Which ones have the same electron configurations as the corresponding chlorides of the first-row elements?

RESONANCE

In drawing Lewis structures, you will often find that there is a choice for the location of a multiple bond. For example the skeleton structure of HNO_3 requires the addition of one more shared electron pair. Starting from

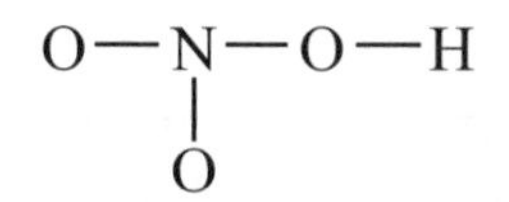

we can write

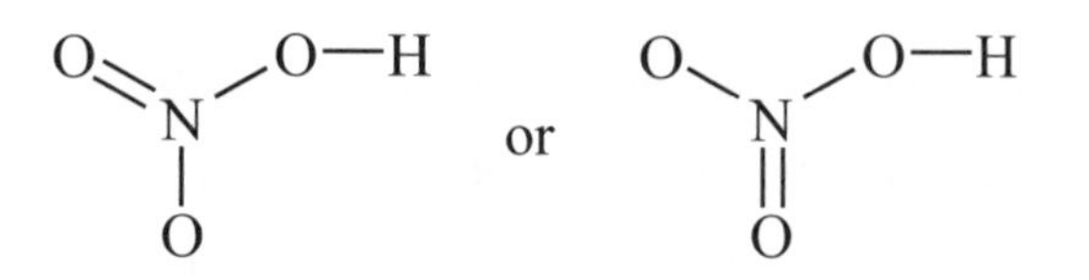

Similarly when we write the Lewis structure of N_2O, we must place two multiple bonds in N—N—O. We can write N=N=O, N≡N—O, or N—N≡O.

Whenever such a situation arises, the correct structure of the molecule cannot be represented by a single Lewis structure; and any single Lewis structure that can be written does not correspond to a real species. The correct structure of the molecule is a blend of two or more of these "imaginary" Lewis structures. They are called *contributing structures* and the molecule is called a *resonance hybrid* and is said to be *resonance-stabilized.* It is important to understand that none of the contributing structures ever exist; but they are a convenient way of describing a real structure as a *blend* of them. For example, experimental evidence indicates that NO_3^-, the nitrate ion, has only one type of oxygen and one type of nitrogen–oxygen bond. This evidence is consistent with a description of nitrate as an equal blend of the three imaginary contributing structures.

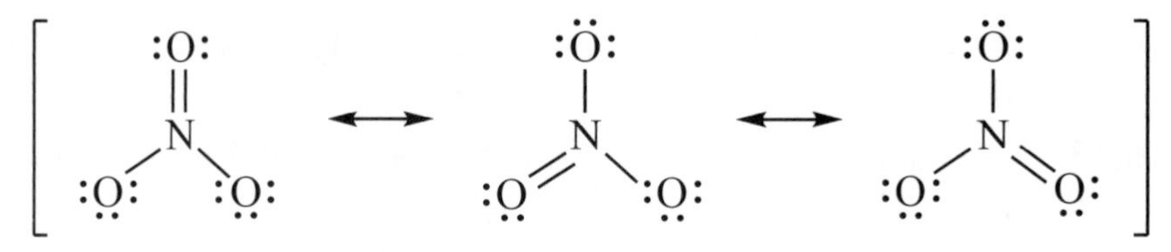

Contributing structures are separated by the double-headed arrow, which should not be confused with the arrows used in chemical equations. Contributing structures never involve changes in the original attachments of the atoms; they only change the way the extra electron pairs required for multiple bonding are distributed.

PROBLEMS

10.17 Draw the important contributing structures for the following compounds:
a. N_2O_4 **b.** O_3 **c.** CH_3NO_2 (C—N bond)

10.18 Draw the important contributing structures for the following ions:
a. CO_3^{2-} **b.** SO_4^{2-} **c.** NO_2^-

10.19 Draw the important contributing structures for the following odd-electron molecules:
a. NO **b.** NO_2

Hint: Extra contributing structures arise here because of the possibility of having the odd electron and incomplete octet on N or O.

10.20 Draw two contributing structures for a molecule with the following properties: formula C_6H_6; one type of C, one type of H, one type of carbon–carbon bond intermediate between a single and double bond; and cyclic structure.

10.21 Account for the observation that the allyl cation

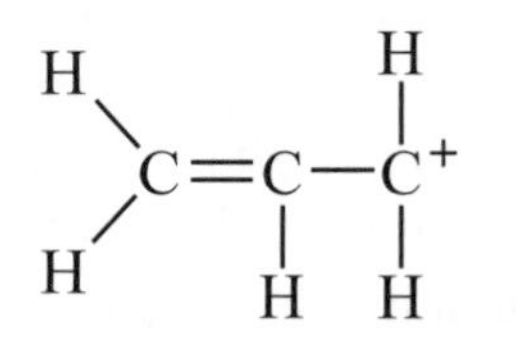

is more stable than the corresponding cation without the double bond.

HYBRIDIZATION

Another approach to the description of molecular structure is to think of bonds as arising by combinations of atomic orbitals. For example, the bond in H_2 can be pictured as a molecular orbital, which holds the two electrons of the bond, formed by combining the 1*s* orbital of each hydrogen atom. A useful starting point is thus to consider the origin of each of the two electrons in a bond and specify the bond orbital as a combination of the two orbitals, one on each atom, that originally held the electron.

Experiments readily demonstrate that this simple picture requires modification. For example, the two O—H bonds in water should be formed from a $2p_x$ and a $2p_y$ orbital each overlapping with the 1*s* orbital of hydrogen. Since the $2p_x$ and $2p_y$ orbitals are perpendicular, the bond angle in H–O–H (drawn as O bonded to two H atoms) is predicted to be 90°. It is found to be 104.5°.

A great many such observations can be accounted for by postulating the existence of *hybridization* or mixing of atomic orbitals. A *hybrid orbital* can be described as a blend of atomic orbitals. Thus if the 2*s* and a 2*p* orbital of an atom are mixed, two new hybrid orbitals are formed, each a blend of one part 2*s* and one part 2*p*. Such orbitals are called 2*sp*. If the 2*s* and two 2*p* orbitals of an atom are mixed, three hybrid orbitals are formed, each a blend of one part 2*s* and two parts 2*p*. Such orbitals are called $2sp^2$. Similarly, the 2*s* and the three 2*p* orbitals form four $2sp^3$ hybrid orbitals.

A hybridization picture of a molecule is often a useful aid for understanding molecular structure and properties because we can formulate hybrid orbitals to fit observed bond angles. We can propose a hybridization for each individual atom in the molecule. To determine the hybridization of an atom we count the number of other atoms to which it is bonded and the number of nonbonding valence-electron pairs that surround it. For example, the C of CH_4 is bonded to four atoms; the N of $:NH_3$ is bonded to three atoms and has one nonbonding pair, a total of four; the O of $H_2\ddot{O}:$ is bonded to two atoms and has two nonbonding pairs, a total of four. Since four $2sp^3$ hybrid orbitals are formed from four atomic orbitals, the hybridization of C, N, and O in these compounds is conveniently designated $2sp^3$. In general, when the total of atoms and nonbonding valence-electron pairs is four, we will designate the atom as sp^3-hybridized.

In the compounds BF_3 and $H_2C{=}CH_2$, we propose $2sp^2$ hybridization for B and C, respectively, since the total of atoms and nonbonding valence-electron pairs around these atoms is three. Compounds such as H—C≡C—H and H—C≡N have a total of two atoms and no nonbonding valence-electron pairs around C. The C is 2*sp*-hybridized. Atoms that are bonded to only one other atom, such as the F atoms in BF_3 or two of the O atoms in HNO_3 need not be considered as hybridized.

In atoms with expanded octets, mixing of *d* orbitals with *s* and *p* orbitals can lead to more complex hybridizations. When a total of five atoms and nonbonding valence-electron pairs surrounds a single atom, the atom is sp^3d-hybridized. When a total of six atoms and nonbonding valence-electron pairs surrounds a single atom, the atom is sp^3d^2-hybridized.

Hybridization helps us to understand the formation of double and triple bonds. When an atom forms three sp^2 hybrid orbitals, there remains one unhybridized *p* orbital. This orbital can overlap with the unhybridized *p* orbital of another sp^2-hybridized atom to form a second bonding orbital and accommodate the second pair of electrons for the double bond.

In single bonds, the two atomic orbitals that form the bond point toward each other along the internuclear line, while in the second bond of the double bond the unhybridized *p* orbitals are perpendicular to the internuclear line and parallel to each other. These two types of bonds are called σ and π bonds, respectively. They can be diagrammed as shown for $H_2C{=}CH_2$:

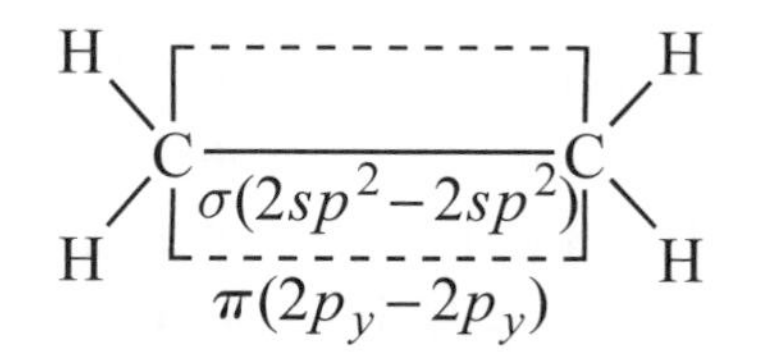

A triple bond can be thought of as consisting of a σ bond from overlap of *sp* orbitals and two π bonds from overlap of two pairs of mutually perpendicular unhybridized *p* orbitals.

PROBLEMS

10.22 What is the hybridization of the asterisked atoms in each of the following:
a. $^*CH_2Cl_2$ **b.** $^*AlCl_3$ **c.** $Br{-}C^*{\equiv}N$ **d.** $^*CO_2$
e. HN^*O_3 **f.** $H_2O_2^*$

10.23 Fill in the blanks in the statement below for each of the following compounds:
a. $H{-}C{\equiv}N{:}$ **b.** $H_2C{=}\ddot{N}{-}H$ **c.** $H_3C{-}NO_2$
d. $H_3C{-}N{\equiv}C^*{:}$
1. The hybridization of the (asterisked) C is _____.
2. The hybridization of the N is _____.

10.24 Indicate for each of the bonds in CH_2O the bond type and the atomic orbitals that overlap to form the bond.

Solution: Before you consider the details of the bonding of a molecule, you should draw the Lewis structure:

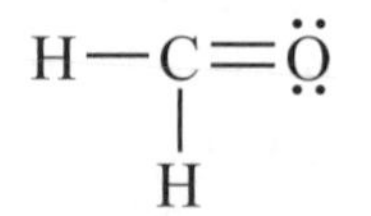

The hybridization of the C atom is $2sp^2$ since it is bonded to three atoms and has no nonbonding valence-electron pairs. Each of the C—H single bonds is a σ bond. The orbitals that overlap to form this bond are a 1*s* of hydrogen and a $2sp^2$ of carbon. The

C=O consists of a σ bond and a π bond. The σ bond is formed from overlap of the $2sp^2$ orbital of C with one of the $2p$ orbitals of O, say the $2p_x$ orbital. The π bond is formed from the overlap of a $2p$ orbital of C with a $2p$ orbital of O. Since p orbitals must be parallel to form π bonds, the directional subscripts of these two orbitals must be the same and must be different from the subscript of the $2p$ orbital that forms the σ bond. We can say that the π bond is formed from $2p_y$ of C and $2p_y$ of O.

10.25 Indicate for each of the bonds in the following molecules the bond type (σ or π) and the atomic orbitals that overlap to form the bond: (a) CH_3Cl (b) C_2H_2 (c) HCN (d) NH_2OH

10.26 Indicate for each of the bonds in CO_2 the bond type and the orbitals that overlap to form the bond.

10.27 Indicate for each of the bonds in the two important contributing structures of N_2O the bond type and the orbitals that overlap to form the bond.

10.28 If the hybridization in NO_3^- is imagined to be $2sp^2$ on the N, draw one overlap picture consistent with the description of NO_3^- by three contributing Lewis structures.

MOLECULAR GEOMETRY

If we imagine bonds as straight lines between atoms, then any two bonds from the same atom intersect to form a bond angle. We can describe molecular geometry by designating these bond angles. In simple cases we can also describe molecular geometry by indicating an overall arrangement of some or all of the atoms. For example, a molecule is *linear* if its atoms lie on a straight line; it is *planar* if its atoms lie in a plane; it is *tetrahedral* if it has four atoms attached to a central atom and these four atoms lie at the corners of a solid figure called a *tetrahedron*.

We can understand and predict molecular geometry using a number of different *models* or simplified pictures of molecular structure. In the *hybridization model* we can calculate the ideal geometry associated with each type of hybridization, as shown in the following table.

Hybridization	Ideal Geometry	Bond Angles
sp	Linear	180°
sp^2	Planar	120°
sp^3	Tetrahedral	109°28'
sp^3d	Trigonal bipyramidal	120° and 90°
sp^3d^2	Octahedral	90°

Another model that we can use is based on the idea that the electrons in the valence shell of each atom of a molecule will arrange themselves in such a way as to minimize repulsions between them. This model is called the *valence-shell-electron-pair-repulsion* model or the *VSEPR* model. To apply the method we total the number of electron pairs in σ bonds and the nonbonding valence-shell electron pairs about an atom. (We do not count electrons in π bonds.) Associated with a given total are ideal bond angles and geometry that minimize the repulsions between the valence-shell electron pairs, as shown in the following table.

Electron Pairs	Ideal Geometry	Bond Angles
2	Linear	180°
3	Planar	120°
4	Tetrahedral	109°28'
5	Trigonal bipyramidal	120° and 90°
6	Octahedral	90°

The geometry of a molecule can be described, as we have seen, either by specifying bond angles or by giving the overall geometry or shape of the molecule. In molecules with nonbonding valence-electron pairs, these pairs influence the bond angles but they are not included when we give the overall shape of the molecule. For example, CH_4, $:NH_3$, and $H_2\ddot{O}:$ all have bond angles close to 109°28'. But we describe only CH_4 as tetrahedral, with the four H atoms lying at the corners of a tetrahedron. The NH_3 molecule is pyramidal, the N and the three H atoms lying at four corners of a pyramid, while the H_2O molecule is bent; the three atoms do not lie on a straight line.

In most cases the actual geometry of a molecule is only approximated by the value predicted by its hybridization or VSEPR model. Superimposed are a number of other effects that can generally be related to the minimization of interelectronic repulsions. Thus the H—N—H bond angles in NH_3 are somewhat less than 109° because a nonbonding electron pair occupies more space than a bonded pair. The nonbonding pair tends to push the bonded pairs and therefore the bond angles closer together. In water the H—O—H angle is still smaller because two nonbonding pairs are pushing the bonds together. A number of other factors of this type operate. In $H_2C{=}O$, the H—C—H angle is predicted to be 120°. It is found to be 125°. The electronegative oxygen holds the electrons in the C═O relatively closely, so repulsions between these electrons and the ones in the C—H bonds are not too serious. This allows the C—H bond electrons to move farther from each other, spreading the angle. A similar effect is seen in CH_3Cl, where the H–C–Cl angle is slightly smaller than 109° and the H–C–H angles are slightly larger.

PROBLEMS

10.29 Predict the most likely geometry for the following species:
a. N_2O **b.** AlF_3 **c.** $SiCl_4$ **d.** H_2S

10.30 Predict the most likely geometry for the following:
a. SiF_4 **b.** PCl_3 **c.** H_2Se **c.** NH_2^-

10.31 Predict the most likely geometry for the following:
a. CH_3^+ **b.** NO_3^- **c.** NOCl **d.** NO_2^-

10.32 The compound C_3O_2 is found to have linear geometry and no formal charges. Write a reasonable Lewis structure based on the observed geometry.

10.33 Predict which of the following molecules are likely to have linear structures:
a. $BeBr_2$ **b.** HOBr **c.** $HgBr_2$ **d.** $PbBr_2$ **e.** Cl_2O

10.34 What is the most likely geometry for the following ions?
a. NO_3^- **b.** CO_3^{2-} **c.** CH_3^- **d.** NH_4^+

10.35 What is the most likely geometry for the following species?
a. PCl_5 **b.** SF_6 **c.** H_5IO_6 **d.** SO_4^{2-}

10.36* Considering only the positions of the atoms predict the geometry of the following species, which have expanded octets and which include nonbonded pairs of electrons:
a. $BrCl_3$ **b.** $BrCl_5$ **c.** I_3^- **d.** SCl_4

Solution: Work out the expected geometry in the usual way and then consider the most advantageous placement of the atoms and the nonbonded pairs in the available positions. In $BrCl_3$ the Br is surrounded by three atoms and two nonbonded pairs, a total of five, corresponding to a trigonal bipyramid. The two nonbonded pairs are found to be in the equatorial plane, leaving the three Cl atoms at the corners of a distorted T.

Chapter 11

Concentration of Solutions

The composition of a solution is expressed as its concentration. A concentration is always a fraction. The numerator of the fraction is the quantity of solute. The denominator is the quantity of solvent or of solution.

Let's look at some **concentration expressions**. We want to be able to calculate concentrations from data given about the quantities of the components of the solution and we want to be able to convert from one concentration expression to another.

We can classify the various concentration expressions in a systematic way. To do so we need to indicate the quantity and its units expressed in the numerator and the quantity and its units expressed in the denominator and also whether the denominator refers to the solvent or the solution. Note that in most expressions the denominator refers to solution, not solvent.

MASS–MASS EXPRESSIONS

Both the numerator and the denominator are mass. In mass–mass expressions the denominator almost always refers to the solution. The simplest mass–mass expression is the mass fraction:

$$\text{mass fraction} = \frac{\text{mass of solute}}{\text{mass of solution}}$$

In this expression mass can be expressed in any unit we choose, provided that the unit is the same in both the numerator and the denominator.

If solutions are relatively dilute then mass fraction becomes somewhat inconvenient because the numbers will be small. Instead we use closely related units simply for the purpose of working with more manageable numbers.

Thus the mass percent is mass fraction $\times$ 100. Sometimes concentrations are more conveniently expressed as mass percent.

For very dilute solutions we can use parts per million (ppm), which is defined as mass fraction $\times 10^6$ and for even more dilute solutions we can use parts per billion (ppb), which is defined as mass fraction $\times 10^9$.

This procedure of adjusting the numbers for convenience can be used for many expressions of concentration.

VOLUME–VOLUME EXPRESSIONS

In General Chemistry these units will be less common because they are convenient mainly for solutions of two or more liquids. The concentrations of such solutions can be expressed in exactly the same way as for mass–mass expressions. Put the volume of the solute (or the component of interest) in the numerator and the volume of the solution in the denominator, taking care that the units of volume are the same in both. (Generally we will assume that the volume of the solution is the sum of the volumes of the liquid components.) You can use the same numerical operations (percent, ppm, ppb) to make the numbers more convenient. In addition there is one special volume–volume expression of concentration that is used for solutions of alcohol and water. It is called proof and is defined as

$$\text{proof} = 2 \times \text{volume percent}$$

AMOUNT–AMOUNT EXPRESSIONS

These may seem a little different than the others but they are not. In the numerator is the amount of solute, measured in moles, as usual. In the denominator is the amount of solution measured in moles. But how do we measure the amount of solution when it contains at least two things of different molecular weights? The amount of solution is the amount of solvent plus the amount of each solute.

Suppose we have a solution containing two components 1 and 2. Let's call the amount of 1 $\mathbf{n_1}$ and the amount of 2 $\mathbf{n_2}$. The mole fraction, **X**, of each component is then

$$X_1 = \frac{n_1}{n_1 + n_2} \text{ and } X_2 = \frac{n_2}{n_1 + n_2}$$

We can generalize this expression for however many components there are in the solution.

Sometimes it is more convenient to use mole percent:

$$\text{mole percent} = \text{mole fraction} \times 100\%$$

MIXED EXPRESSIONS OF CONCENTRATION

A number of the most commonly used expressions of concentration are ones in which the quantity of solute in the numerator and the quantity of solution or solvent in the denominator are measured differently.

Molarity

The most common expression of this type is *molarity* (M). It is an *amount–volume* unit. Recall that in this unit the numerator contains the amount of solute (in moles of course) and the denominator contains the volume of *solution* in liters (L).

Molality

A second common mixed expression is called *molality* (m). The numerator of this expression contains the amount of solute (in moles). In the denominator is the mass of *solvent* (in kilograms).

Mass–Volume and Amount–Volume

Sometimes it is convenient to express the quantity of solution as a volume when the quantity of solute is expressed as mass or amount. There are many variations possible: mass of solute in kg or g or mg or μg/volume of solution in m^3 or L or mL or μL etc.

Similarly we can express the amount of solute in mol or mmol or μmol and the volume of solution in any of these units of volume. Which of these units we may choose to use is governed by convenience.

CONVERSION OF CONCENTRATION EXPRESSIONS

There are two basic types of conversions. One is straightforward and is merely an extension of the unit conversions we did in Chapters 2–6. What we require are the appropriate conversion factors. So for example if we want to convert from a mass–mass concentration expression to an amount–amount concentration expression, we need the relationship between mass and amount for the substances of interest. Conversions among expressions of concentration that are based on just mass or amount simply require the use of the usual conversion factors that we obtain from the Table of Atomic Weights.

Another type of conversion between expressions of concentration is the conversion between two concentration expressions, one of which includes a volume, generally the volume of the solution.

In order to do such conversions we need to know the relationship between the mass and the volume of a solution. This relationship is the density, $d = \text{mass/volume}$. We generally start by finding the mass of a given volume of solution.

PROBLEMS

11.1 An alloy of Cu and Al is prepared from 65.6 g of Cu and 423.1 g of Al. Find the mass fraction of each component of this solid solution.

Solution: First find the mass of the solution. It is the sum of the masses of the components: 65.6 g + 423.1 g = 488.7 g of solution. The mass fraction of each component is its mass divided by the mass of the solution:

$$\text{fraction Cu} = \frac{65.6 \text{ g Cu}}{488.7 \text{ g solution}} = 0.134$$

$$\text{fraction Al} = \frac{423.1 \text{ g Al}}{488.7 \text{ g solution}} = 0.866$$

Note that the sum of the mass fractions of all the components must equal 1.

11.2 Calculate the mass fraction of bromine in a solution prepared from 22.7 g of bromine and 98.7 g of carbon tetrachloride.

11.3 Calculate the mass of sugar in 65.6 g of a solution that is 0.153 sugar by mass.

11.4 Calculate the mass of gold required to prepare 38.5 g of an alloy that is 18/24 parts gold by mass.

11.5 The density of ethyl alcohol at 293 K is 0.79 g/mL and the density of chloroform at this temperature is 1.5 g/mL. Calculate the mass fraction of chloroform in a solution prepared from equal volumes of these two liquids.

11.6 Twelve grams of NaCl is dissolved in 68 g of water. Calculate the mass percent of NaCl in the solution.

Solution: Calculation of the mass percent requires the mass of solution that is equal to the mass of solute plus the mass of solvent. Twelve grams of NaCl in 68 g of H_2O gives 12 + 68 = 80 g of solution. Substituting into the expression for mass percent gives

$$\text{mass percent} = \frac{12\text{ g}}{80\text{ g}} \times 100\% = 15\%$$

Note: In the calculation of a percent or a fraction any units can be used provided the numerator and the denominator have the same units.

11.7 What mass of NaCl is there in 60 g of a 15% solution of NaCl in water?

Solution: 15% means that 0.15 of the mass of the solution is NaCl.

$$0.15 \times 60\text{ g NaCl} = 9.0\text{ g NaCl}$$

11.8 What mass of sugar would have to be dissolved in 60 g of water to yield a 25% solution?

Solution: We want a solution in which the mass of the sugar is 0.25 of the mass of the solution. The mass of the solution is the mass of the sugar plus the mass of the water. Let X = mass of the sugar.

$$X + 60\text{ g} = \text{mass of the solution}$$
$$X = 0.25\,(X + 60\text{ g})$$
$$X = 20\text{ g sugar}$$

11.9 What mass of water and what mass of salt would you use to prepare 80 g of a 5.0% salt solution?

Solution: In a 5.0% salt solution, the mass of the salt is 0.050 of the mass of the solution.

$$0.050 \times 80\text{ g} = 4.0\text{ g salt}$$
$$\text{mass of the solution} = \text{mass of salt} + \text{mass of water.}$$
$$80\text{ g solution} - 4.0\text{ g salt} = 76\text{ g water}$$

11.10 What is the mass percent of NaF in a solution prepared from 1.22 mol of NaF and 41.2 mol of H_2O?

11.11 What mass of water is required to dissolve 1.7 g of KNO_3 so that the resulting solution is 7.9% KNO_3 by mass?

11.12 A 10-g sample of NH_4Cl is dissolved in 100 g of a 10% solution of NH_4Cl in water. Calculate the mass percent of the resulting solution.

11.13 You are given 100 g of a 10.0% solution of $NaNO_3$ in water. How many more grams of $NaNO_3$ would you have to dissolve in the 100 g of 10.0% solution to change it to a 20.0% solution?

11.14 A 60-g sample of a 12% solution of NaCl in water was mixed with 40 g of a 7.0% solution of NaCl in water. What was the mass percent of the resulting solution?

11.15 The mass of 15 cm^3 of a solution is 12 g. Calculate the density of the solution.

Solution: Density means grams per cubic centimeter. Therefore, to find the density we divide the mass in grams by the volume in cm^3. That is,

$$\text{density} = \frac{12\ \text{g}}{15\ \text{cm}^3} = 0.80\ \text{g/cm}^3$$

11.16 The density of a solution is 1.80 g/cm^3. What volume will 360 g of the solution occupy?

Solution: To find the volume in cm^3 occupied by 360 g of a solution we use the density in the same way as any other conversion factor:

$$360\ \text{g} \times \frac{1\ \text{cm}^3}{1.80\ \text{g}} = 200\ \text{cm}^3$$

11.17 A 44.00% solution of H_2SO_4 has a density of 1.343 g/cm^3. What mass of H_2SO_4 is there in exactly 60 cm^3 of this solution?

Solution: The mass in grams of 60 cm^3 of this solution will be

$$60\ \text{cm}^3\ \text{solution} \times \frac{1.343\ \text{g}}{1\ \text{cm}^3} = 80.58\ \text{g solution}$$

$$\frac{0.4400\ \text{g}\ H_2SO_4}{1\ \text{g solution}} \times 80.58\ \text{g solution} = 35.46\ \text{g}\ H_2SO_4$$

The entire calculation, in one operation, is

$$60\ \text{cm}^3\ \text{solution} \times \frac{1.343\ \text{g}}{1\ \text{cm}^3} \times \frac{0.4400\ \text{g}\ H_2SO_4}{1\ \text{g solution}} = 35.46\ \text{g}\ H_2SO_4$$

11.18 A solution is prepared by mixing 10 g of CH_3OH and 100 g of H_2O. Calculate the mole fraction of each component.

Solution: To calculate the mole fraction of a component of a solution we must know the number of moles of that component and the total number of moles of all the components in the solution. The moles of CH_3OH

$$n_{CH_3OH} = 10\text{ g } CH_3OH \times \frac{1\text{ mol } CH_3OH}{32\text{ g } CH_3OH} = 0.31$$

The moles of H_2O

$$n_{H_2O} = 100\text{ g } H_2O \times \frac{1\text{ mol } H_2O}{18\text{ g } H_2O} = 5.5$$

Then

$$f_{CH_3OH} = \frac{n_{CH_3OH}}{n_{CH_3OH} + n_{H_2O}} = \frac{0.31}{0.31 + 5.5} = 0.053$$

Once the fractions of all components of a solution but one are known, the remaining fraction is most easily calculated by remembering that the sum of the fractions is 1. In this case $f_{CH_3OH} + f_{H_2O} = 1$ or $f_{H_2O} = 1 - 0.053 = 0.947$.

11.19 Calculate the mole fraction of copper in an alloy made from 124 g of copper and 124 g of aluminum.

11.20 Calculate the mole fraction of CCl_4 in a solution prepared from 22.7 g Br_2 and 98.7 g CCl_4.

11.21 Calculate the mole fraction of ethylene glycol, $C_2H_6O_2$, in a solution that is 74.5% by mass ethylene glycol.

11.22 The density of mercury is 13.6 g/mL at 273 K. A solution of silver in mercury, called an amalgam, is prepared from 30.0 mL of mercury and 2.34 g of silver. Calculate the mole fraction of silver in the silver amalgam.

11.23 Find the mass of glucose $C_6H_{12}O_6$ that must be dissolved in 250 g of water to prepare a solution in which the mole fraction of glucose is 0.10.

11.24 Find the molality of ascorbic acid (MW = 176) in a solution prepared from 1.94 g of ascorbic acid and 50.1 g H_2O.

Solution: In order to find the molality (m) we need to express the amount of solute in moles and the mass of solvent in kg.

$$1.94\text{ g AA} \times 1\text{ mol AA}/176\text{ g AA} = 0.0110\text{ mol AA}$$

$$50.1\text{ g } H_2O \times 1\text{ kg}/1000\text{ g} = 0.0501\text{ kg } H_2O$$

$$\text{molality} = 0.0110\text{ mol AA}/0.0501\text{ kg } H_2O = 0.220\ m\text{ in AA}$$

11.25 Calculate the molality of Br_2 in a solution prepared from 22.7 g Br_2 and 98.7 g CCl_4.

11.26 A solution is prepared at 25 °C by mixing 15 g of Na_2SO_4 with 125 cm^3 of H_2O. The density of H_2O at this temperature is 1.0 g/cm^3. What is the molality of Na_2SO_4 in the solution?

Solution: Molality is defined as the amount of solute dissolved in 1 kg of solvent. There is 15 g Na_2SO_4 × (1 mol Na_2SO_4/142 g Na_2SO_4) = 0.106 mol of Na_2SO_4 and there is 125 cm^3 × 1.0 g/cm^3 = 125 g of H_2O. Substituting into the expression for molality, we have

$$\text{molality} = \frac{0.106 \text{ mol}}{125 \text{ g}} \times \frac{1000 \text{ g}}{1 \text{ kg}} = 0.85\ m$$

11.27 Calculate the mole fraction and the molality of NaBr in a solution prepared at 25 °C from 17.2 g of NaBr and 196 cm^3 of H_2O.

11.28 What mass of glucose $C_6H_{12}O_6$ must be dissolved in 150 cm^3 of water to prepare a solution that is 0.34 m in glucose?

11.29 Calculate the molality of glycerin $C_3H_8O_3$, in a solution prepared from equimolar quantities (the same amounts) of glycerin and water.

11.30 Calculate the molality of water in a solution that is 74.5% ethylene glycol by mass.

11.31 Calculate the masses of water and acetone, C_3H_6O, necessary for preparation of 120 g of a solution that is 2.2 m in acetone.

11.32 What is the mole fraction of glucose in the solution in Prob. 11.28?

11.33* A solution of C_2H_6O in water has a mole fraction of $H_2O = 0.97$. Calculate the molality of the C_2H_6O.

11.34* To 223 cm^3 of a solution of density 1.12 g/cm^3 containing 9.20% $C_2H_6O_2$ (ethylene glycol) by weight in water is added 21.4 g of $C_3H_8O_3$ (glycerol). Calculate the mole fraction of glycerol in the resultant solution.

11.35 Calculate the molality and mole fraction of ethanol in 25.4 proof wine at 293 K. The density of ethanol is 0.789 g/mL and the density of water is 0.998 g/mL at this temperature.

11.36 What mass of NaOH is required to prepare 1.0 L of 1.0 M NaOH?

Solution: 1.0 M NaOH will contain 1.0 mol of NaOH dissolved in enough water to make 1 L of solution. One mole of NaOH is 40 g of NaOH. Therefore 40 g is required.

11.37 What mass of K_2SO_4 is required to prepare 1.00 L of 0.500 *M* K_2SO_4?

Solution: One liter of 0.500 *M* K_2SO_4 will contain one-half mole of K_2SO_4 dissolved in enough water to make a liter of solution. One-half mole of K_2SO_4 is 87.1 g. Therefore, 87.1 g of K_2SO_4 is required.

11.38 What mass of $Al_2(SO_4)_3$ is required to prepare 300 cm^3 of 0.200 *M* $Al_2(SO_4)_3$?

Solution: One liter of 0.200 *M* $Al_2(SO_4)_3$ contains 0.200 mol of $Al_2(SO_4)_3$. 300 cm^3 is 0.300 L; 0.300 L of 0.200 *M* $Al_2(SO_4)_3$ contains 0.3×0.2 or 0.0600 mol of $Al_2(SO_4)_3$. One mole of $Al_2(SO_4)_3$ is 342 g. Therefore, $0.0600 \times 342\text{ g} = 20.5\text{ g}$ is required. In one step:

$$0.300\text{ L} \times \frac{0.200\text{ mol } Al_2(SO_4)_3}{\text{L}} \times \frac{342\text{ g } Al_2(SO_4)_3}{\text{mol } Al_2(SO_4)_3} = 20.5\text{ g } Al_2(SO_4)_3$$

Note: Liters and moles cancel, leaving the answer in grams.

11.39 A solution is prepared from 12 g of NaOH dissolved in enough water to give 500 cm^3 of solution; calculate the molarity of the solution.

Solution: To find the molarity means to find the number of moles of solute that are present in 1000 cm^3 (1 L) of solution.

$$1\text{ mol NaOH} = 40\text{ g NaOH}$$

Therefore,

$$12\text{ g NaOH} \times \frac{1\text{ mol NaOH}}{40\text{ g NaOH}} = 0.30\text{ mol NaOH}$$

The 0.30 mol of NaOH is present in 500 cm^3 of solution. Since the molarity is the number of moles per 1000 cm^3, then,

$$1000\text{ cm}^3 \times \frac{0.30\text{ mol}}{500\text{ cm}^3} = 0.60\text{ mol}$$

Therefore, the solution is 0.60 *M*.

11.40 A solution of $Cu(NO_3)_2$ contains 100 mg of the salt per cm^3. Calculate the molarity of the solution.

Solution: 100 mg = 0.1 g. The 0.1 g in 1 cm^3 is the same concentration as 100 g in 1000 cm^3. Therefore, this is a solution containing 100 g of $Cu(NO_3)_2$ per liter. To find the molarity we must find the number of moles per liter. There are 187.5 g of $Cu(NO_3)_2$ in a mole. Therefore,

$$100 \text{ g } Cu(NO_3)_2 \times \frac{1 \text{ mol } Cu(NO_3)_2}{188 \text{ g } Cu(NO_3)_2} = 0.53 \text{ mol } Cu(NO_3)_2$$

Therefore, the solution is 0.53 *M*.

11.41 What mass of KOH is required to prepare 400 cm^3 of 0.18 *M* KOH?

11.42 What volume of 0.306 *M* Na_2CO_3 solution can be prepared from 164 g of Na_2CO_3?

11.43 A solution of NaCl contains 14 g of NaCl in 750 cm^3 of solution. What is the molarity of the solution?

11.44 If 230 cm^3 of 0.30 *M* Na_2SO_4 is evaporated to dryness, what mass of dry Na_2SO_4 will be obtained?

11.45 The solubility of magnesium carbonate is 0.011 g/100 mL of water. Calculate the molarity of magnesium carbonate in a saturated solution, assuming no volume change of the water on formation of the solution.

11.46 A saturated solution of potassium permanganate at 293 K is 0.40 *M*. Calculate the solubility of potassium permanganate measured in grams/100 mL of water, assuming no volume change of the water on formation of the solution.

11.47 Calculate the molarity of a solution that contains 1.22 g of hemoglobin (MW = 68,300) in 165 mL of solution.

11.48 The density of a 22.0% solution of methanol in water by mass is 0.963 g/mL. Calculate the molarity of the methanol in the solution.

11.49 The density of a solution of 18.0% lead(II) nitrate by mass in water is 1.18 g/mL. Calculate the molarity of the lead nitrate.

11.50* A solution of 80.00% sulfuric acid by mass has a density of 1.727 g/mL. A volume of 50.00 mL of this solution is diluted with water until the volume of the solution is 75.00 mL. Calculate the molarity of sulfuric acid in the new solution.

11.51 A solution of sugar is prepared by mixing 436 mL of 2.05 *M* sugar solution with 238 mL of 3.34 *M* sugar solution. Assume the volume of the resulting solution is the sum of the volume of the two solutions. Calculate the molarity of sugar in the new solution.

11.52 A 10.0-cm^3 volume of a 70.0% solution of sulfuric acid of density 1.61g/cm^3 was dissolved in enough water to give 25.0 cm^3 of solution. What was the molarity of the final solution?

11.53 A 3.58 M H_2SO_4 solution is 29.0% H_2SO_4 by mass. Calculate the density of 3.58 M H_2SO_4.

11.54 To what volume in cm^3 must 44.20 cm^3 of a 70.00% solution of sulfuric acid whose density is 1.610 g/cm^3 be diluted to give 0.4000 M H_2SO_4?

11.55 A 1.00 M solution of CH_3OH in water has a density of 0.91 g/cm^3. Calculate the mole fraction, mass percent, and molality of the CH_3OH.

Solution: From the density, 1 L of solution has a mass of 0.910 $g/cm^3 \times 1000\ cm^3 =$ 910 g. It contains 1.00 mol $=$ 32.0 g of CH_3OH and therefore $910 - 32 = 878$ of H_2O.

$$878\text{ g of } H_2O \times \frac{1\text{ mol } H_2O}{18.0\text{ g } H_2O} = 48.8\text{ mol } H_2O$$

$$f_{CH_3OH} = \frac{n_{CH_3OH}}{n_{CH_3OH} + n_{H_2O}} = \frac{1.00}{1.00 + 48.8} = 0.0201$$

$$\text{Mass percent} = \frac{32.0}{910} \times 100 = 3.52\%$$

$$\text{Molality} = \frac{1\text{ mol}}{(910 - 32)\text{ g}} \times \frac{1000\text{ g}}{1\text{ kg}} = 1.14\ m$$

11.56 A 1.30 M solution of KBr in water has a density of 1.16 g/cm^3. Calculate the mole fraction, mass percent, and molality of the KBr.

11.57 A solution that is 5.30% LiBr by mass has a density of 1.04 g/cm^3. Calculate its molarity.

11.58 What amount of hydrogen will be liberated from 320 cm^3 of 0.50 M H_2SO_4 by an excess of magnesium?

Solution:

$$Mg + H_2SO_4 \rightarrow MgSO_4 + H_2$$

0.50 M H_2SO_4 contains 0.50 mol of H_2SO_4 per liter

$$0.320\text{ L } H_2SO_4 \times \frac{0.50\text{ mol } H_2SO_4}{1\text{ L } H_2SO_4} = 0.16\text{ mol } H_2SO_4$$

1 mol of H_2SO_4 liberates 1 mol of H_2.

Therefore, 0.16 mol of H_2 will be liberated.

11.59 What amount of hydrogen will be liberated from 400 cm^3 of 0.40 M HCl by an excess of zinc?

Solution:

$$Zn + 2HCl \rightarrow ZnCl_2 + H_2$$

2 mol of HCl yields 1 mol of H_2

1 mol of HCl yields 0.5 mol of H_2

0.40 M HCl contains 0.40 mol of HCl per liter

$$0.400 \text{ L} \times \frac{0.40 \text{ mol}}{1 \text{ L}} = 0.16 \text{ mol HCl}$$

Therefore, 0.080 mol ($\frac{1}{2}$ of 0.16 mol) of H_2 will be liberated.

11.60 What amount of hydrogen gas will be liberated when excess Mg reacts with the following?
a. 600 cm^3 of 0.80 M H_2SO_4
b. 600 cm^3 of 0.80 M HCl

11.61 A volume of 600 cm^3 of 0.40 M HCl was treated with excess Mg. The evolved H_2 gas was all used to reduce CuO to Cu. What mass of free copper was formed?

11.62 A 447-cm^3 volume of Na_2CO_3 solution was warmed with an excess of sulfuric acid until all action ceased. A 5.0-L volume of dry CO_2 gas, measured at standard conditions, was given off. Calculate the molarity of the sodium carbonate solution.

Solution: The analysis of this problem could run something like this: To find the molarity we must find the number of moles of Na_2CO_3 per liter of solution. If we know what amount of Na_2CO_3 there is in 447 cm^3, we can calculate the amount in 1000 cm^3 (1 L). From the equation

$$Na_2CO_3 + H_2SO_4 \rightarrow Na_2SO_4 + H_2O + CO_2$$

or, more simply, from the formulas Na_2CO_3 and CO_2, we can see that 1 mol of Na_2CO_3 yields 1 mol of CO_2. Therefore, if we knew what amount of CO_2 was formed we would know the amount of Na_2CO_3 present in the 447 cm^3 of solution.

We know that 5.0 L of CO_2 was evolved. Since 1 mol of CO_2 occupies a volume of 22.4 L at STP, 5.0 L of CO_2 is 5/22.4 mol. Therefore, 5/22.4 mol of Na_2CO_3 is present in the 447 cm^3 (0.447 L) of solution.

$$\frac{5}{22.4}\text{ mol} \div 0.447\text{ L} = 0.5\text{ mol/L}$$

Therefore, the solution is 0.5 *M*.

11.63 A beaker contained 130 cm^3 of hydrochloric acid. The contents were treated with excess zinc. A volume of 7.13 L of dry hydrogen gas, measured at 22 °C and 738 mmHg, was obtained. Calculate the molarity of the acid.

11.64 A 17.4-cm^3 sample of a 70.0% solution of sulfuric acid whose density is 1.61 g/cm^3 was diluted to a volume of 100 cm^3 and was then treated with a large excess of zinc. The evolved hydrogen gas was combined with chlorine gas to form HCl. This HCl gas was then dissolved in enough water to form 200 cm^3 of hydrochloric acid. There was no loss of material in the reactions. Calculate the molarity of the hydrochloric acid.

11.65 A 44.0% solution of H_2SO_4 has a density of 1.343 g/cm^3. A volume of 25.0 cm^3 of a 44.0% H_2SO_4 solution was treated with an excess of Zn. What volume did the dry hydrogen gas that was liberated occupy at STP?

Solution: $Zn + H_2SO_4 \rightarrow ZnSO_4 + H_2$. Therefore, 1 mol of H_2SO_4 will liberate 1 mol of H_2.

We will calculate what amount of H_2SO_4 is present in the solution.

$$25.0\text{ cm}^3 \times 1.343\text{ g/cm}^3 \times 0.440 = 14.8\text{ g } H_2SO_4$$

$$14.8\text{ g } H_2SO_4 \times \frac{1\text{ mol } H_2SO_4}{98.1\text{ g } H_2SO_4} = 0.151\text{ mol } H_2SO_4$$

Therefore, 0.151 mol of H_2 was evolved. One mole of H_2 occupies a volume of 22.4 L at STP. Thus

$$0.151\text{ mol} \times \frac{22.4\text{ L}}{1\text{ mol}} = 3.38\text{ L } H_2$$

11.66 What volume of dry HCl gas, measured at 25 °C and 740 mmHg, can be prepared by combining chlorine gas with the hydrogen that will be liberated when 100 cm^3 of a 20.0% solution of H_2SO_4 of density 1.14 g/cm^3 is treated with an excess of aluminum?

11.67 What volume of 0.250 *M* HCl will be required to neutralize 120 cm^3 of 0.800 *M* KOH?

Solution: One mole of HCl will neutralize 1 mol of KOH. Therefore, moles of HCl required = moles of KOH present in the solution. The 0.800 *M* KOH contains 0.800 mol of KOH per liter; 0.250 M HCl contains 0.250 mol of HCl per liter.

$$0.120\text{ L} \times \frac{0.800\text{ mol KOH}}{1\text{ L}} = \text{moles KOH present}$$

$$X\text{ L} \times \frac{0.250\text{ mol HCl}}{1\text{ L}} = \text{moles HCl required}$$

Therefore, since moles HCl required = moles KOH present,

$$X\text{ L HCl} \times \frac{0.250\text{ mol}}{1\text{ L}} = 0.120\text{ L} \times \frac{0.800\text{ mol}}{1\text{ L}}$$

$$X = 0.384\text{ L}$$

11.68 If 12.0 g of NaOH was required to neutralize 82.0 cm^3 of sulfuric acid, calculate the molarity of the acid.

11.69 A 25.0-cm^3 volume of NaOH solution exactly neutralized 40 cm^3 of 0.10 M H_2SO_4. Calculate the molarity of the NaOH.

11.70 What volume of 0.250 M $AgNO_3$ will be required to precipitate the chloride from 80.0 cm^3 of 0.400 M NaCl? What mass of AgCl will be precipitated?

11.71* An 85-g sample of an antimony sulfide ore containing 40% by weight of Sb_2S_3 and 60% inert material is oxidized until all the S in the Sb_2S_3 is converted to SO_3. This SO_3 is dissolved in enough water to give 200 cm^3 of solution. What volume of 0.400 M NaOH will be required to completely neutralize the contents of the 200 cm^3 of solution?

Hint: $Sb_2S_3 \rightarrow 3SO_3 \rightarrow 3H_2SO_4 \rightarrow 6NaOH$

$$\text{Moles NaOH required} = 6 \times \text{moles } Sb_2S_3$$

$$\text{cm}^3\text{ NaOH required} \times \frac{0.400\text{ mol NaOH}}{1000\text{ cm}^3\text{ solution}} = \text{moles NaOH required}$$

11.72* A 16-g mixture of sodium and potassium, when allowed to react with water, gave a solution that neutralized 602.5 cm^3 of 0.40 M H_2SO_4. What mass of sodium was in the mixture?

Hint: One mole of Na yields 1 mol of NaOH; 1 mol of K yields 1 mol of KOH. Moles of hydroxides = 2 × moles of H_2SO_4.

11.73* A 14.8-g mixture of Na_2CO_3 and $NaHCO_3$ was dissolved in enough water to make 400 cm^3 of solution. When the 400 cm^3 of solution was treated with excess 2.00 M H_2SO_4 and boiled to remove all dissolved gas, 3.73 L of dry CO_2 gas, measured at 740 mmHg and 22.0 °C, was obtained. Calculate the molarity of the Na_2CO_3 and of the $NaHCO_3$ in the 400 cm^3 solution.

Hint: X = molarity of Na_2CO_3 and Y = molarity of $NaHCO_3$.

Moles of Na_2CO_3 in 400 cm^3 of solution = $0.400X$.
Moles of $NaHCO_3$ in 400 cm^3 of solution = $0.400Y$.
One mole of Na_2CO_3 yields 1 mol of CO_2.
One mole of $NaHCO_3$ yields 1 mol of CO_2.

11.74* To a beaker containing 164 cm^3 of a solution of $CuSO_4$ was added 10.00 g of magnesium metal. When the reaction was complete, a mixture of Mg and Cu having a mass of 14.45 g remained in the beaker. Calculate the molarity of the original $CuSO_4$ solution.

Chapter 12

Properties of Solutions

A number of the physical properties of a solution, to a first approximation, do not depend on the identity of the solute but only on the number of solute particles that are dissolved. These properties are called *colligative properties*, and some examples are vapor pressure, boiling point, melting point, and osmotic pressure. A study of these properties often reveals important information about the solute, such as its molecular weight or its state in solution. The approach we will use is an approximation similar to the one we use in treating gases as ideal gases. Dilute solutions are treated as ideal solutions with the result that relatively simple relationships govern colligative properties.

At a given temperature the vapor pressure of a liquid solvent is always greatest when the liquid is pure and it decreases as solute is dissolved in it. For dilute solutions, *Raoult's law*

$$P_A = P_A^{\circ} f_A$$

states that the vapor pressure of solvent $A(P_A)$ above a solution is equal to the vapor pressure of the pure liquid (P_A°) at the given temperature multiplied by the mole fraction (f_A) of the solvent in the solution. The same relationship also applies to solutions with two or more volatile components, for example water and ethanol. The vapor pressure of each volatile component above the solution at a certain temperature is given by the vapor pressure of the pure solvent at that temperature multiplied by its mole fraction in the solution.

Raoult's law may also be applied to the calculation of the melting point and boiling point of the solvent in a solution that contains only nonvolatile solutes. The boiling point of the solvent in such a solution is *higher* than it is when the solvent is pure, and the freezing point of the solvent in such a solution *is lower* than when it is pure. These phenomena are referred to as *boiling-point elevation* and *freezing-point depression*. We can demonstrate for the ideal solution that the magnitude of the boiling point elevation or freezing point depression depends on the number of nonvolatile solute particles and on the nature of the solvent. An approximate relationship is $\Delta T = K_b m$ for boiling point and $\Delta T = K_f m$ for freezing point, where ΔT is the difference between the boiling or freezing point of the pure solvent and the solvent in the solution (expressed as a positive number), m is the total molality of solute particles, and K_b and K_f are the boiling-point constant and freezing-point constant, respectively. These constants are experimentally determined quantities characteristic of each solvent. For water, $K_b = 0.52$ and $K_f = 1.86$ when the temperature is measured in kelvins or °C.

Another property of solutions that depends on the number of solute particles is the osmotic pressure. It can be measured by placing the solution in a tube fitted at one end with a semipermeable membrane through which the solvent but not the solute can pass. If the tube is placed in a container of the pure solvent, the solvent will enter the tube, causing the liquid level in the tube to rise to a certain height. The difference in height between the level of the pure solvent and the level of the solution that has risen in the tube can be used to calculate the osmotic pressure π. The osmotic pressure (π) is related to the amount of solute particles by a relationship that closely resembles the ideal gas equation:

$$\pi V = nRT$$

where V is the volume of solution (in liters), n is the amount of solute particles (in moles) in that volume, T is the temperature (in K), and $R = 0.082$ L-atm/K-mol, the ideal gas constant. Measurement of osmotic pressure is a useful way of finding the molecular weight of large molecules, since very small osmotic pressures can be measured easily.

We should always recognize that colligative properties depend on the amount of solute particles in the solution, which is not necessarily the same as the amount of solute used in preparing the solution. When strong electrolytes such as salts, strong acids, and strong bases are dissolved in polar solvents such as water they dissociate into ions. Thus when 1 mol of NaCl dissolves in water it produces 2 mol of particles: 1 mol of Na^+ ions and 1 mol of Cl^- ions. Similarly each mole of $Mg(OH)_2$ produces 3 mol of particles ($1Mg^{2+}$ and $2OH^-$), and each mole of $Ca_3(PO_4)_2$ produces 5 mol of particles ($3Ca^{2+}$ and $2PO_4^{3-}$). When weak electrolytes such as weak acids and weak bases dissolve in polar solvents, they undergo partial dissociation to produce varying numbers of solute particles. Thus it is possible by measuring some of the colligative properties of a solution whose composition is known to gain information about the behavior of substances when they dissolve.

PROBLEMS

12.1 A 60.0-g quantity of a nonelectrolyte dissolved in 1000 g of H_2O lowered the freezing point 1.02 °C. Calculate the approximate molecular weight of the nonelectrolyte.

Solution: One mole of nonelectrolyte in 1000 g of water would have depressed the freezing point 1.86 °C. Since a depression of 1.02° was observed, 1.02/1.86 mol of nonelectrolyte must have been dissolved.

$$\frac{1.02}{1.86}\text{ mol} = 60\text{ g}$$

$$1\text{ mol} = \frac{1.86}{1.02} \times 60\text{ g} = 109\text{ g}$$

The molecular weight is approximately 109.

12.2 When 4.20 g of a nonelectrolyte was dissolved in 40.0 g of water, a solution that froze at −1.52 °C was obtained. Calculate the approximate molecular weight of the nonelectrolyte.

Hint: The molecular weight is the mass of solute that will depress the freezing point 1.86 °C when dissolved in 1000 g of solvent, so we first find the concentration in grams of solute per 1000 g of water.

$$1000\text{ g water} \times \frac{4.20\text{ g solute}}{40.0\text{ g water}} = 105\text{ g solute}$$

A solution containing 105 g of solute in 1000 g of water has the same *concentration* as one containing 4.20 g of solute in 40.0 g of water. Therefore, 105 g of solute will depress the freezing point of 1000 g of solvent by exactly the same number of degrees that 4.20 g of solute will depress the freezing point of 40.0 g of solvent, namely 1.52 °C. Continue as in Prob. 12.1.

12.3 A 20.0-g quantity of $C_6H_{10}O_5$, a nonelectrolyte, is dissolved in 250 g of H_2O. Calculate the boiling point of the solution at 1 atm.

Solution: First we calculate the molality of the solution. Since the molecular weight of $C_6H_{10}O_5$ is 162,

$$\text{molality} = 20.0\text{ g } C_6H_{10}O_5 \times \frac{1\text{ mol } C_6H_{10}O_5}{162\text{ g } C_6H_{10}O_5} \times \frac{1}{250\text{ g } H_2O} \times 1000\text{ g } H_2O$$

$$= 0.494\ m$$

Since pure water boils at 100 °C, the boiling point of the solution is 100 °C + ΔT. Thus

$$BP_{solution} = \Delta T + 100\ °C = 0.52 \times m = 0.52 \times 0.494$$
$$BP_{solution} = 0.52 \times 0.494 + 100$$
$$= 100.26\ °C$$

12.4 When 4.96 g of a nonelectrolyte was dissolved in 54.0 g of H_2O, a solution was obtained that boiled at 100.41 °C at 1 atm. Calculate the approximate molecular weight of the nonelectrolyte.

12.5 A 6.00-g sample of C_2H_5OH, a nonelectrolyte, was dissolved in 363 g of H_2O. Calculate the freezing point of the solution.

12.6 What mass of $C_3H_5(OH)_3$, a nonelectrolyte, must be dissolved in 458 g of H_2O to give a solution that will freeze at −4.00 °C?

12.7 When 12 g of a nonelectrolyte was dissolved in 300 g of water, a solution that froze at −1.62 °C was obtained. What was the approximate molecular weight of the nonelectrolyte?

12.8 When 5.12 g of the nonionizing solute naphthalene ($C_{10}H_8$) is dissolved in 100 g of CCl_4 (carbon tetrachloride), the boiling point of the CCl_4 is raised by 2 K. What is the boiling-point constant of CCl_4?

12.9 A compound was found, on analysis, to consist of 50.00% oxygen, 37.50% carbon, and 12.50% hydrogen. When dissolved in 100.0 cm^3 of water, 1.666 g of the compound gave a nonconducting solution that froze at −1.00 °C. Calculate the molecular weight of the compound.

12.10 Calculate the boiling point of a solution of 4.39 g of naphthalene, $C_{10}H_8$, in 99.5 g of carbon tetrachloride (BP = 349.7 K, K_b = 5.03).

12.11 Calculate the mass of sugar, $C_6H_{12}O_6$, that must be dissolved in 1 kg of ethanol (K_b = 1.22) to raise its boiling point 1.0 K.

12.12 A solution of 10.8 g of an unknown substance in 101.2 g of *n*-octane (BP = 398.8 K, K_b = 4.02) boils at 401.4 K. Find the molecular weight of the unknown substance.

12.13 Find the freezing point of a solution that is 20% by mass ethylene glycol, $C_2H_6O_2$, in water.

12.14 A solution of 0.358 g of an unknown substance in 6.45 g of camphor (MP = 453.0 K, K_f = 40.0) freezes at 446.8 K. Find the molecular weight of the unknown.

12.15 Calculate the mass of glycerol, $C_3H_8O_3$, that must be dissolved in 1000 g of water to form a solution with a freezing point of 268.2 K.

12.16 A solution is prepared by dissolving 2.37 g of NaCl in 100 g of water. What is the boiling point of the solution?

Hint: This problem is treated like Prob. 12.3 except that we must include the fact that each mole of NaCl gives 2 mol of solute particles when it dissolves in water. In order to calculate the boiling point we multiply the molality calculated from the formula weight of NaCl by 2.

$$\text{Molality} = 2 \times \frac{2.37}{58.5} \Big/ 100 \times 1000$$

12.17 A solution is prepared by dissolving 1.27 g of an electrolyte whose molecular weight is 310 in 50 g of water, and the freezing point of the solution is found to be -0.762 °C. What amount of solute particles is formed in solution from each mole of solute?

Solution: First calculate the freezing point depression assuming no dissociation

$$\Delta T = K_f m = (1.86) \frac{(1.27/310 \times 1000)}{50} = 0.152 \text{ °C}$$

Then divide this value into the observed value: $0.762/0.152 = 5.01$. Since the observed depression is 5 times the depression calculated by assuming no dissociation, we see that each mole of solute has dissociated into 5 mol of particles.

12.18 A solution of 6.27 g of K_2SO_4 in 50 g of water is prepared. What is its boiling point?

12.19 What mass of $Mg(NO_3)_2$ must be dissolved in 100 g of water to raise the boiling point to 101.21 °C?

Hint: First calculate the molality of solute particles for the boiling-point elevation. Then assume that each mole of magnesium nitrate dissociates completely in order to calculate the number of moles of solute needed. Then calculate the required mass.

12.20 A solution of a salt whose molecular weight is 142.1 is prepared from 2.84 g of salt and 500 g of water. Its freezing point is observed to be -0.223 °C. How many moles of solute particles are formed in solution from each mole of salt that is dissolved?

12.21 Calculate the freezing point of a solution prepared from 5.46 g of potassium iodide and 134 g of water.

12.22 Calculate the boiling point of a solution prepared from 15.3 g of sodium sulfate and 98.3 g of water.

12.23 Calculate the molality of sodium chloride in water at 252 K, which is the eutectic temperature—the lowest temperature that a saturated solution of NaCl in water can reach and still remain liquid.

12.24* A metal M of atomic weight 96 reacts with fluorine to form a salt that can be represented as MF_x. In order to determine x and therefore the formula of the salt, a boiling-point elevation experiment is carried out. A 9.18-g sample of the salt is dissolved in 100 g of water and the boiling point of the solution is found to be 374.38 K. Find the formula of the salt.

12.25 When 1.16 g of acetic acid ($C_2H_4O_2$) is dissolved in 50 g of benzene, the freezing point of the solution is found to be depressed by 1.024 °C. The value of K_f for benzene is 5.12. Describe the state of acetic acid in benzene solution.

12.26 When $HClO_2$, a weak acid, is dissolved in water, it undergoes partial dissociation according to the equation

$$HClO_2(aq) \rightleftarrows H^+(aq) + ClO_2^-(aq)$$

A solution is prepared that contains 1.64 g of $HClO_2$ in 100 g of water. The freezing point of this solution is found to be −0.649 °C. Calculate the fraction of $HClO_2$ that dissociates.

Solution: First proceed as in Prob. 12.17 to calculate what the freezing-point depression would be if there were no dissociation:

$$\Delta T = 1.86\frac{°C}{m} \times \frac{1.64 \text{ g } HClO_2}{68.5 \text{ g } HClO_2/1 \text{ mol } HClO_2} \times \frac{1}{0.100 \text{ kg } H_2O}$$
$$= 0.445\ °C$$

Divide this calculated ΔT into the observed ΔT: 0.649/0.445 = 1.46. This verifies that dissociation was not complete since, if it were, we would find the observed ΔT to be twice the calculated one. To calculate the fraction dissociated let X = fraction dissociated. From the equation given for the reaction, for each mole of $HClO_2$ that dissociates 2 mol of particles are formed. If X dissociates, $2X$ particles are formed and $1 - X$ particles of $HClO_2$ remain. Thus the total number of particles is $2X + (1 - X) = X + 1$. We have calculated that this number of particles produces 1.46 times the freezing-point depression as if there were no dissociation. Thus $X + 1 = 1.46$, $X = 0.46$, which is the fraction of $HClO_2$ that has dissociated.

12.27* When HNO_2 is dissolved in water, it partially dissociates according to the equation $HNO_2(aq) \rightarrow H^+(aq) + NO_2^-(aq)$. A solution is prepared that contains 7.050 g of HNO_2 in 1000 g of water. Its freezing point is found to be −0.2929 °C. Calculate the fraction of HNO_2 that has dissociated.

12.28* It is found that when a solution containing 39.2 g of H_2SO_4 in 1000 g of water is prepared, the H_2SO_4 dissociates completely into H^+ and HSO_4^- and that 17.8 mol % of the HSO_4^- further dissociates into $H^+ + SO_4^{2-}$. Calculate the freezing point of the solution.

12.29 A solution of 42.2 g of glucose ($C_6H_{12}O_6$), a nonelectrolyte, in 200 g of water is prepared at 25 °C. What is the vapor pressure of water above the solution?

Solution: Assuming ideal solution behavior we can substitute into the Raoult's law expression; using the value of $P^\circ_{H_2O}$ from Table A.1 in the appendix we obtain

$$P_{H_2O} = P^\circ_{H_2O} f_{H_2O} = (23.6\ \text{mmHg})\left(\frac{\frac{200}{18}}{\frac{200}{18} + \frac{42.2}{180}}\right) = 23.1\ \text{mmHg}$$

12.30 A solution is prepared from 27.1 g of K_2SO_4 and 1000 g of water. Calculate the vapor pressure of water above the solution at 100 °C.

Solution: Since pure water boils at 100 °C, its vapor pressure is 760 mmHg at that temperature. The vapor pressure above the solution depends on the mole fraction of water based on the number of moles of solute in the solution. Since 1 mol of K_2SO_4 gives 3 mol of solute particles, 27.1 g gives $3 \times 27.1/174 = 0.467$ mol of solute particles.

$$f_{H_2O} = \frac{\frac{1000}{18}}{\frac{1000}{18} + 0.467} = 0.992$$

and $P = 760\ \text{mmHg} \times 0.992 = 754\ \text{mmHg}$.

Note that the solution does not boil at 100 °C, since its vapor pressure is less than 760 mmHg.

12.31 A solution of 50 g of ethylene glycol ($C_2H_6O_2$), a nonelectrolyte, and 50 g of water is prepared. Calculate the vapor pressure of water above this solution at 40 °C.

12.32 A solution of 75 g of NH_4Cl (a salt) in 150 g of water is prepared. Calculate the vapor pressure of water above the solution at 35 °C.

12.33 Find the vapor pressure of water above a solution prepared from 18.22 g of lactose, $C_{12}H_{22}O_{11}$, and 81.46 g of water at 338.0 K. The vapor pressure of pure water at this temperature is 0.2467 atm.

12.34 The density of a 3.742 *M* solution of glycerol, $C_3H_8O_3$, in water at 298.0 K is 1.0770 g/mL. The vapor pressure of pure water at this temperature is 0.03126 atm. Calculate the vapor pressure of water above the solution.

12.35 The vapor pressure of pure benzene, C_6H_6, is 0.132 atm at 299 K. Calculate the mass of hexachlorobenzene, C_6Cl_6, that must be dissolved in 25 g of benzene to lower its vapor pressure to 0.126 atm.

12.36 A solution is prepared from 26.7 g of an unknown compound and 116.2 g of acetone, C_3H_6O, at 313 K. The vapor pressure of pure acetone at this temperature is 0.526 atm. The vapor pressure of acetone above the solution is 0.501 atm. Find the molecular weight of the unknown compound.

12.37 Calculate the vapor pressure in atm above a solution prepared from 24.3 g of magnesium nitrate and 249 g of water at 308 K.

12.38* The density of a 0.438 *M* solution of potassium chromate at 298 K is 1.063 g/mL. Calculate the vapor pressure of water in atm above the solution.

12.39 A solution is prepared at 70 °C from 80 g of ethanol (C_2H_6O) and 100 g of water. Calculate the vapor pressure of each component above the solution and the mole fraction of ethanol in the vapor. The vapor pressure of pure ethanol at this temperature is 543 mmHg and that of pure water is 234 mmHg.

Solution: For a solution with more than one volatile component, Raoult's law can be applied to each separately. First we calculate the mole fraction of each component.

$$X_{H_2O} = \frac{\frac{100}{18}}{\frac{100}{18} + \frac{80}{46}} = 0.762$$

$$X_{C_2H_6O} = 1 - X_{H_2O} = 0.238$$

$$P_{H_2O} = P^\circ_{H_2O} X_{H_2O}$$

$$= 234 \text{ mmHg} \times 0.762 = 178 \text{ mmHg}$$

$$P_{C_2H_6O} = P^\circ_{C_2H_6O} X_{C_2H_6O}$$

$$= 543 \text{ mmHg} \times 0.238 = 129 \text{ mmHg}$$

Since for a given set of conditions moles are directly proportional to vapor pressures, the mole fraction of ethanol in the vapor is

$$f_{C_2H_6O} = \frac{129}{129 + 178} = 0.420.$$

12.40 Calculate the vapor pressure of ethanol (C_2H_6O) and of H_2O and the mole fraction of H_2O in the vapor above a solution prepared at 30 °C by mixing 100 g of ethanol and 100 g of water. The vapor pressure of pure ethanol at this temperature is 78.0 mmHg.

12.41* The vapor above the solution in Prob. 12.40 is separated and condensed. Calculate the mole fraction of water in the vapor above this condensate. The vapor above this condensate is again separated and condensed. Calculate the mole fraction of water in the vapor above the second condensate. This problem describes a process called *fractional distillation*.

12.42* A solution prepared at 30 °C from 50.0 g benzene (C_6H_6) and 50.0 g of toluene (C_7H_8) has a total vapor pressure of 80.8 mmHg. A solution prepared at 30 °C from 75.0 g of C_7H_8 and 25.0 g of C_6H_6 has a total vapor pressure of 59.9 mmHg. Calculate the vapor pressure of pure C_6H_6 and pure C_7H_8 at 30 °C.

12.43 At 333 K the vapor pressure of benzene, C_6H_6, is 0.521 atm and the vapor pressure of toluene, C_7H_8, is 0.184 atm. A solution of equimolar amounts of these two liquids is prepared. Find the mole fraction of toluene in the vapor and the total vapor pressure above the solution.

12.44* Two alcohols, isopropyl alcohol and propyl alcohol, have the same molecular formula, C_3H_8O. A solution of the two that is $\frac{2}{3}$ by mass isopropyl alcohol has a vapor pressure of 0.110 atm at 313 K. A solution that is $\frac{1}{3}$ by mass isopropyl alcohol has a vapor pressure of 0.089 atm at 313 K. Calculate the vapor pressure of each pure alcohol at this temperature.

12.45 The vapor pressure of CCl_4 is 0.354 atm and the vapor pressure of chloroform, $CHCl_3$, is 0.526 atm at 316 K. A solution is prepared from equal masses of these two compounds. Calculate the mole fraction of chloroform in the vapor above the solution obtained by condensation of the vapor above the original solution.

12.46 A 1.00-L volume of a solution of 10.0 g of hemoglobin, the oxygen-carrying substance in red blood corpuscles, in water is found to have an osmotic pressure of 2.80 mmHg at 27 °C. Calculate the molecular weight of hemoglobin.

Solution: The osmotic pressure depends on the number of moles of solute particles in a given volume of solution. If we assume hemoglobin does not dissociate, the number of moles in 1.00 L of solution is 10.0 g/MW. Substituting into the osmotic pressure relationship, we have

$$n = \frac{\pi V}{RT} = \frac{10.0 \text{ g}}{\text{MW}} = 2.80 \text{ mmHg} \times \frac{1 \text{ atm}}{760 \text{ mmHg}} \times 1.00 \text{ L}$$

$$\times \frac{1}{0.082} \frac{\text{K-mol}}{\text{L-atm}} \times \frac{1}{273 + 27} \text{K}$$

MW = 66,800

12.47 A water solution of 1 g of catalase, an enzyme found in the liver, has a volume of 100 cm^3 at 27 °C. Its osmotic pressure is found to be 0.745 mmHg. Calculate the molecular weight of catalase.

12.48 Calculate the osmotic pressure of a solution that contains 10.0 g of NaCl in 150 cm^3 of solution at 20 °C.

12.49 What mass of KCl must be dissolved in water to make 1.00 L of a solution with an osmotic pressure of 1.00 atm at 25 °C?

12.50* The ocean is approximately a 3% solution of NaCl in water and has a density of 1.04 g/cm^3. Calculate its osmotic pressure due to the NaCl, at 20 °C, assuming no other salts are dissolved.

12.51 Calculate the osmotic pressure at 298 K of 1.00% by mass solutions of nonionic solutes with the following molecular weights: (a) 1.00×10^2 (b) 1.00×10^4 (c) 1.00×10^6.

12.52 The density of a 9.50% by mass solution of fructose, $C_6H_{12}O_6$, is 1.037 g/mL at 293 K. Calculate the osmotic pressure of the solution.

12.53 Calculate the mass of calcium phosphate that must be dissolved in water to form 125 mL of a solution with an osmotic pressure of 0.0034 atm at 300 K.

Chapter 13

Kinetics

Kinetics deals with the rate at which a process occurs and the way in which the rate is affected by various factors. Two of the most important of these factors are concentration and temperature.

Since the rate of a reaction, in general, depends on the concentration of the constituents of the reaction and since these constituents are changing as the reaction proceeds, the rate of the reaction is also changing as the reaction proceeds. In order to write an expression for the rate of the reaction we must take into account its dependence on time. One approach is to define the average rate over a period of time. For example, suppose we start with a solution in which [A] = 1.00 M and allow the reaction A $\rightarrow$ B to proceed for 100 s, at the end of which time we find that [A] = 0.75 M and [B] = 0.25 M. The average rate for this time can be expressed as "0.25 mol/L of B is formed in 100 s."

We use the symbol Δ to indicate a change in any quantity. By definition, Δ is the final value minus the initial value of the quantity. So ΔB = final [B] $-$ initial [B] = 0.25 M $-$ 0 = 0.25 M and Δt = final t $-$ initial t = 100 s $-$ 0 s = 100 s. The average rate is

$$\frac{\Delta[\mathrm{B}]}{\Delta t} = \frac{0.25\ M}{100\ \mathrm{s}} = 2.5 \times 10^{-3}\,\frac{\mathrm{mol}}{\mathrm{L\text{-}s}}$$

We could also express the rate in terms of Δ[A], which is final [A] $-$ initial [A] = 0.75 $-$ 1.00 = -0.25 M. The average rate is

$$-\frac{\Delta[\mathrm{A}]}{\Delta t} = -\frac{(-0.25)}{100} = 2.5 \times 10^{-3}\,\frac{\mathrm{mol}}{\mathrm{L\text{-}s}}$$

We use a minus sign when the concentration decreases to give a value of the average rate that is always positive.

The stoichiometry of the reaction A $\rightarrow$ B indicates that for every mole of B that is formed a mole of A has reacted and it therefore follows that $\Delta[B]/\Delta t = -\Delta[A]/\Delta t$. For reactions that are not simply mole-for-mole conversions, the coefficients of the chemical equation must be taken into account in writing expressions for the rate and in relating them. For the process

$$H_2(g) + I_2(g) \rightarrow 2HI(g),$$

every time 1 mol of H_2 reacts, 1 mol of I_2 also reacts and 2 mol of HI is produced. Therefore the average rate can be expressed by any one of the following

$$-\frac{\Delta[H_2]}{\Delta t} = -\frac{\Delta[I_2]}{\Delta t} = \frac{1}{2}\frac{\Delta[HI]}{\Delta t}$$

For the general reaction

$$aA + bB \rightarrow cC + dD$$

where the lowercase letters represent coefficients, the expressions for the average rate are

$$-\frac{1}{a}\frac{\Delta[A]}{\Delta t} = -\frac{1}{b}\frac{\Delta[B]}{\Delta t} = \frac{1}{c}\frac{\Delta[C]}{\Delta t} = \frac{1}{d}\frac{\Delta[D]}{\Delta t}$$

As we have seen, expressions involving the use of Δ just give us the average rate over a time interval. Suppose however that we would like to know what the rate is after exactly 50 s for the above hypothetical reaction A $\rightarrow$ B. In general, the value of the average rate that we found, 2.5×10^{-3} mol/L-s, will be only a fair approximation of the rate at any specified time, which is called the *instantaneous rate*. You can see that the approximation will get better as the time interval gets smaller. Thus if the average rate is measured for the time interval from 40 s to 60 s, it will be closer to the instantaneous rate at 50 s. If the time interval is from 49 s to 51 s, it will be better still and the average rate will be exactly the instantaneous rate if the time interval Δt around 50 s becomes infinitely small. This can be expressed as the instantaneous rate and is equal to

$$\lim_{\Delta t \to 0} \frac{\Delta[B]}{\Delta t} \quad \text{or} \quad -\lim_{\Delta t \to 0} \frac{\Delta[A]}{\Delta t}$$

or using the usual notation of the calculus, the instantaneous rate is $d[B]/dt$ or $-d[A]/dt$. The relationship of instantaneous rates to stoichiometry is the same as that of average rates to stoichiometry, so for the general reaction written above,

$$-\frac{1}{a}\frac{d[A]}{dt} = -\frac{1}{b}\frac{d[B]}{dt} = \frac{1}{c}\frac{d[C]}{dt} = \frac{1}{d}\frac{d[D]}{dt}$$

The experimental approach to the study of the rate of a reaction is to measure the concentration of one or more species at different times as the reaction proceeds.

Sometimes instead of measuring concentrations directly we measure some other quantity directly related to concentration. There are a number of such quantities; pressure in the case of gases, intensity of light absorption in the case of colored materials, or radioactivity levels in the case of radioactive substances are examples. We can calculate values for the average rate by calculating from the data the changes in concentration (or the measured related quantity) during measured time intervals. If the time intervals are fairly small, the values of the average rates will be good approximations of the instantaneous rates at times within the intervals. However, in order to calculate actual instantaneous rates from experimental data it is necessary to know exactly how the rate changes with changes in concentration.

The relationship between the rate of a reaction and the concentrations of the species in the system is called the *differential rate law*. In general it is *not* possible just to look at a chemical equation and know what the differential rate law will be. It must be experimentally determined. Sometimes we can guess the details of how the reactants become the products, that is, the mechanism of the reaction, and derive the rate law from our guess.

For the reaction A $\longrightarrow$ B we might find that the rate is proportional to [A]. The differential rate law is $-d[\mathrm{A}]/dt$ or $d[\mathrm{B}]/dt = k[\mathrm{A}]$, where k is a proportionality constant called the specific rate constant. But we might also find that the differential rate law is

$$\frac{-d[\mathrm{A}]}{dt} = k[\mathrm{A}]^2 \text{ or } \frac{-d[\mathrm{A}]}{dt} = k[\mathrm{A}]^{1/2}$$

or a number of other possibilities.

For the general reaction aA + bB $\longrightarrow$ cC + dD many different rate laws are possible. For example

$$-\frac{1}{a}\frac{d[\mathrm{A}]}{dt} = k[\mathrm{A}][\mathrm{B}] \text{ or } k\,[\mathrm{A}]^2[\mathrm{B}] \text{ or } k[\mathrm{A}][\mathrm{B}]^2$$

or, in general, $k[\mathrm{A}]^m[\mathrm{B}]^n$, where m and n are most commonly zero, integers, or half-integers.

The *order* of a reaction is given by the exponents of the concentrations in the rate laws. The overall order is $m + n$, the order with respect to A is m, and the order with respect to B is n.

The experimental determination of the rate law typically requires us to measure the change in rate with change in concentration. For example, if a reaction is first order in [A], doubling the concentration will double the rate; if it is second order in [A], doubling the [A] will quadruple the rate, and so on.

We can also derive a rate law from a postulated mechanism for the reaction. In particular, if a process is presumed to be a single-step or elementary process, the rate law can be written from the stoichiometry by using the coefficients of the chemical equation as exponents in the rate. If A $\rightarrow$ B in a single step, the reaction is first order in [A]. If 2A $\rightarrow$ C in a single step, the reaction is second order in [A]. If A + B $\rightarrow$ D in a single step, the reaction is first order in [A], first order in [B], and second order overall. Single-step processes are never higher than third order overall. This approach works only if the process is an elementary one. Many processes, however, occur by a series of steps.

Since the differential rate laws relate the instantaneous rate to concentration, the instantaneous rate at any concentration can be calculated provided the value of k, the specific rate constant, is known. The value of k can be calculated from rate data in several ways. A differential rate law can be expressed in terms of average rates by defining an average concentration for the time interval of the average rate. For the example A $\rightarrow$ B given above, the $[A]_{av}$ is $(1.00 + 0.75)/2 = 0.88\ M$. If the reaction is first order, then $-\Delta[A]/\Delta t = k[A]_{av}$; if it is a second-order reaction, then $-\Delta A/\Delta t = k[A]^2_{av}$. Expressions such as these can be used to calculate an approximate value of k.

A better value of k can be calculated if we know how concentrations depend on time. This dependence can be obtained by performing the mathematical operation called *integration* on the differential rate law. In many cases this operation is very complicated but in a few cases the integrated rate laws are of simple form. The first-order rate law $-d[A]/dt = k[A]$ has the integrated form

$$\ln \frac{[A_0]}{[A]} = kt \quad \text{or} \quad \ln \frac{[A]}{[A_0]} = -kt \qquad \textbf{(13-1)}$$

where $[A_0]$ is the initial concentration of A when the time t is 0, [A] is the concentration at any time measured after the initial time, and k is the specific rate constant. The second-order rate law $-d[A]/dt = k[A]^2$ has the integrated form

$$\frac{1}{[A]} - \frac{1}{[A_0]} = kt \qquad \textbf{(13-2)}$$

If the reaction under investigation obeys one of these two rate laws, values of k can be obtained by substitution of the data into the appropriate integrated rate law.

With very few exceptions, the rates of chemical processes are accelerated when the temperature is raised. The effect of temperature on rate depends on the value of a quantity called the *activation energy* E_a. The activation energy is the difference between the energy of the reactants and that of the highest energy state leading to the conversion of the reactants to the products. This state is called the *transition state*. The relationship between a specific rate constant k_1 at a temperature T_1 and another specific rate constant k_2 at another temperature T_2 is called the Arrhenius Equation:

$$\ln \frac{k_2}{k_1} = -\frac{E_a}{R}\left(\frac{1}{T_2} - \frac{1}{T_1}\right) \quad \textbf{(13-3)}$$

$$\ln \frac{k_2}{k_1} = -\frac{E_a}{R}\left(\frac{T_1 - T_2}{T_1 T_2}\right) \quad \textbf{(13-4)}$$

PROBLEMS

13.1 A large sample of $CaCO_3(s)$ is placed in an evacuated container and heated to 540 °C, to bring about the reaction

$$CaCO_3(s) \rightarrow CaO(s) + CO_2(g)$$

and the pressure of CO_2 is monitored. The pressure is found to be 100 mmHg after 5 min and 200 mmHg after 10 min. Calculate the average rate of appearance of CO_2 during each 5-min interval and during the 10-min interval.

Solution: The average rate of appearance of CO_2 is $\Delta P_{CO_2}/\Delta t$. Initially $\Delta P_{CO_2} = 0$; after 5 min $\Delta P_{CO_2} = 100$ mmHg. So

$$\frac{\Delta P_{CO_2}}{\Delta t} = \frac{100 - 0}{5 - 0} = 20 \text{ mmHg/min}$$

During the second 5-min period ΔP_{CO_2} increases from 100 mmHg to 200 mmHg and the time increases from 5 min to 10 min

$$\frac{\Delta P_{CO_2}}{\Delta t} = \frac{200 - 100}{10 - 5} = 20 \text{ mmHg/min}$$

For the 10-min period

$$\frac{\Delta P_{CO_2}}{\Delta t} = \frac{200 - 0}{10 - 0} = 20 \text{ mmHg/min}$$

13.2 It is found that $N_2O_3(g)$ decomposes according to $N_2O_3(g) \rightarrow NO_2(g) + NO(g)$. A bulb is filled with N_2O_3 to a pressure of 100 mmHg and the decomposition is monitored by measuring the total pressure. After 75 s the pressure is measured as 150 mmHg and after 150 s the pressure is measured as 175 mmHg. Calculate the average rate of disappearance of $N_2O_3(g)$ during these time intervals.

Solution: The rate of disappearance of $N_2O_3(g)$ is given by $-\Delta P_{N_2O_3}/\Delta t$. Thus we must calculate $P_{N_2O_3}$ after 75 s and after 150 s from the data given on the total pressure. You should note that the pressure increases as the reaction proceeds because each mole of $N_2O_3(g)$ that decomposes produces 2 mol of gaseous products. If we let X = mmHg of $N_2O_3(g)$ that decomposes after 75 s, then at this time $P_{N_2O_3} = 100 - X$, $P_{NO} = P_{NO_2} = X$. Since $P = P_{N_2O_3} + P_{NO} + P_{NO_2}$, then

$$150 = (100 - X) + X + X$$

$$X = 50 \text{ mmHg and } P_{N_2O_3} = 100 - 50 = 50 \text{ mmHg}$$

$$\frac{-\Delta P_{N_2O_3}}{\Delta t} = -\frac{50 - 100}{75} = 0.67 \text{ mmHg/s}$$

After 150 s let Y = the mmHg of N_2O_3 that has decomposed.

Then $175 = (100 - Y) + Y + Y$, $Y = 75$ mmHg, and $P_{N_2O_3} = 100 - 75 = 25$ mmHg.

$$\frac{-\Delta P_{N_2O_3}}{\Delta t} = -\frac{25 - 50}{150 - 75} = 0.33 \text{ mmHg/s}$$

for the interval from 75 s to 150 s. For the entire time interval

$$\frac{-\Delta P_{N_2O_3}}{\Delta t} = -\frac{25 - 100}{150 - 0} = 0.50 \text{ mmHg/s}$$

13.3 Calculate the average rate of appearance of Br^- in the reaction $CH_3Br + I^- \rightarrow CH_3I + Br^-$ during each 5-min time interval from the following data:

$[I^-]$ mol/L	1.00	0.950	0.902	0.857
t min	0	5	10	15

13.4 Calculate the average rate of disappearance of $NO_2(g)$ during each time interval in the reaction $2NO_2(g) \rightarrow N_2O_4(g)$, from the following measurements of total pressure:

P_T mmHg	250	238	224	210
t s	0	200	500	900

13.5 Calculate the average rate of appearance of $N_2O_4(g)$ during the indicated time intervals from the data in Prob. 13.4.

Hint: The average rate of appearance of $N_2O_4(g)$ is one-half the average rate of disappearance of $NO_2(g)$ since 1 mol of $N_2O_4(g)$ forms every time 2 mol of $NO_2(g)$ reacts.

13.6 For a certain time interval it is found that for the process $4NH_3(g) + 5O_2(g) \rightarrow 4NO(g) + 6H_2O(g)$ the average rate of appearance of water is 21.3 mmHg/min. Calculate the average rate of appearance of $NO(g)$ and the average rates of disappearance of $NH_3(g)$ and $O_2(g)$.

13.7 The rate of appearance of $NO_2(g)$ from $NO(g)$ and $O_2(g)$ is found to be 4.8×10^{-4} atm/s during a measured time interval. Find the rate of disappearance of nitric oxide and of oxygen during the interval.

13.8 The rate of disappearance of ozone in the reaction $2O_3(g) \rightarrow 3O_2(g)$ is found to be 8.9×10^{-3} atm/s during a measured time interval. Find the rate of appearance of $O_2(g)$ during this interval.

13.9 The following data are obtained in a study of the rate of the reaction $2CO(g) \rightarrow CO_2(g) + C(s)$ by total pressure measurements:

P_T atm	0.329	0.313	0.295	0.276
t s	0	401	998	1795

Find the average rate of disappearance of CO for the time interval between each measurement. Find the average rate of appearance of CO_2 for the total time interval.

13.10 The gas phase decomposition of N_2O_5 to NO_2 and O_2 is monitored by measurements of total pressure. The following data are obtained:

P_T atm	0.154	0.215	0.260	0.315	0.346
t s	0	52	103	205	309

Find the average rate of disappearance of N_2O_5 for the time interval between each measurement and for the total time interval.

13.11 The reaction $2HI(g) \rightarrow H_2(g) + I_2(g)$ is followed by monitoring the intensity of the purple color of $I_2(g)$ as it appears. The intensity of the purple color, as measured by a quantity called the *optical density*, changes during a 10-min interval from 5.24 to 5.68. In another experiment it is found that the optical density (O.D.) of the I_2 is related to P_{I_2} by $P = 15.2$ O.D. Calculate the average rate of disappearance of HI during the 10-min interval.

13.12 The reaction $2NO(g) + O_2(g) \rightarrow 2NO_2(g)$ is found to follow the differential rate law $-dP_{O_2}/dt = kP_{O_2}P^2_{NO}$. What is the overall order of the reaction and the order with respect to O_2 and NO?

13.13 The reaction $H_2(g) + Br_2(g) \rightarrow 2HBr(g)$ is found to follow the differential law $d[HBr]/dt = k[H_2][Br_2]^{1/2}$. What is the overall order of the reaction and the order with respect to $H_2(g)$ and $Br_2(g)$?

13.14 The reaction $C_4H_9I + H_2O \rightarrow C_4H_{10}O + H^+ + I^-$ is first order in C_4H_9I. Calculate the value of k, the specific rate constant, from the following data obtained by starting with $[C_4H_9I] = 0.200$ M.

$[H^+]$ mol/L	~0	0.0587	0.100	0.130	0.150
t min	0	30	60	90	120

Solution: Since the reaction is first order in C_4H_9I, we can write

$$\frac{-\Delta[C_4H_9I]}{\Delta t} = k[C_4H_9I]_{av} = \frac{\Delta[H^+]}{\Delta t}$$

For each 30-min time interval we can calculate the $[C_4H_9I]_{av}$ since $[C_4H_9I] + [H^+] = 0.200\ M$. We can also calculate the average rate as in Prob. 13.1. At $t = 0$, $[C_4H_9I] = 0.200\ M$.

At $t = 30$ min, $[C_4H_9I] = 0.2 - [H^+] = 0.141\ M$;
$[C_4H_9I]_{av} = (0.200 + 0.141)/2 = 0.17\ M$.
At $t = 60$ min, $[C_4H_9I] = 0.200 - 0.100 = 0.100\ M$
and $[C_4H_9I]_{av}$ from 30 min to 60 min is $(0.100 + 0.141)/2 = 0.121\ M$.
By the same method $[C_4H_9I]_{av}$ is 0.850 M from 60 min to 90 min
and 0.60 M from 90 min to 120 min.

The average rates for the successive time intervals can be calculated directly from $[H^+]$:

$$\frac{\Delta[H^+]}{\Delta t} = \frac{0.0587 - 0}{30} = 1.96 \times 10^{-3} \frac{\text{mol}}{\text{L-min}}$$

for the first 30 min;

$$= \frac{0.100 - 0.0587}{30} = 1.38 \times 10^{-3} \frac{\text{mol}}{\text{L-min}}$$

for the second 30 min; and so on.

To calculate k we substitute each pair of values of $[C_4H_9I]_{av}$ and $\frac{\Delta[H^+]}{\Delta t}$.

Thus $1.96 \times 10^{-3} = k(0.171)$, $k = 0.0115\ \text{min}^{-1}$. (Note that the units mol/L appear on both sides of the equation and therefore cancel each other.) For the second 30-min interval $1.38 \times 10^{-3} = k(0.121)$, $k = 0.0114\ \text{min}^{-1}$. This process can be repeated to obtain two more values of k and the best value of k can be obtained by averaging all the values so obtained. You should note that in order for the values of k obtained from each interval to be close to the correct one, you must know the order of the reaction. Often you can deduce the order of a reaction by testing the data in the way done here for different orders and seeing which order gives the best agreement between the calculated values of k for each time interval.

13.15 Using the data of Prob. 13.3 and assuming the reaction is run in such a way as to be first order in $[I^-]$ and independent of $[CH_3Br]$, calculate the value of k.

13.16 Use the data from Prob. 13.4 to calculate k, given that the reaction is second order in NO_2.

13.17 The reaction $2A \rightarrow B$ is either first or second order in A. Deduce the order of the reaction and then calculate k from the following data:

[A] mol/L	1.00	0.862	0.758	0.678	0.609
t s	0	500	1000	1500	2000

13.18 The following data are obtained for the composition of $N_2O_3(g)$ to nitric oxide and nitrogen dioxide:

$P(N_2O_3)$ (atm)	1.2×10^{-3}	1.8×10^{-3}	2.7×10^{-3}
Initial rate (atm/s)	7.3×10^{-3}	1.1×10^{-2}	1.7×10^{-2}

Find the rate law of this reaction from these data.

13.19 Calculate the specific rate constant from the data of Prob. 13.14 using the integrated form of the first-order rate equation.

Solution: In the form of the integrated rate equation

$$\ln\frac{[A]}{[A_0]} = -kt,$$

[A] and $[A_0]$ refer to C_4H_9I. Since $[C_4H_9I]$ is given as 0.200 M initially, the $[C_4H_9I]$ at any time is $0.200 - [H^+]$. The calculation of k simply involves substitution of the appropriate pairs of values into the equation. For example, at 30 min $[C_4H_9I] = 0.200 - 0.0587 = 0.1413$. Substituting, we have

$$\ln \frac{(0.1413)}{0.200} = -k(30)$$

$$k = 1.16 \times 10^{-2}\ \text{min}^{-1}$$

At 60 min $[C_4H_9I] = 0.200 - 0.100 = 0.100$. Substituting, we have

$$\ln \frac{(0.100)}{(0.200)} = -k(60)$$

$$k = 1.16 \times 10^{-2}\ \text{min}^{-1}$$

This process can be repeated for each measurement and the values of k so obtained averaged to give the best value.

13.20 Use the integrated first-order rate equation to calculate k for the reaction $A \rightarrow B$ from the following data:

[A] mol/L	1.0	0.80	0.63	0.40
t h	0	100	200	400

13.21 The reaction $N_2O_5(g) \rightarrow 2NO_2(g) + \frac{1}{2}O_2(g)$ is first order in $N_2O_5(g)$. The reaction is monitored by measuring the total pressure and the following data are obtained:

P_T mmHg	350	490	592	718	788
t min	0	10	20	40	60

Use the integrated first-order rate equation to calculate the value of k.

Hint: Use a method similar to that of Prob. 13.2 to calculate $P_{N_2O_5}$ from P_T.

13.22 The following data are obtained in a study of the rate of the reaction $SO_2Cl_2(g) \rightarrow SO_2(g) + Cl_2(g)$:

$P(SO_2Cl_2)$(atm)	1.00	0.947	0.895	0.848	0.803
t (s)	0	2500	5000	7500	10,000

The reaction follows a first-order rate law. Calculate the value of the specific rate constant.

13.23* The rate of decomposition of $N_2O_3(g)$ to $NO_2(g)$ and $NO(g)$ is followed by measuring the $[NO_2]$ at different times. The following data are obtained:

$[NO_2]$ (mol/L)	0	0.193	0.316	0.427	0.784
t (s)	0	884	1610	2460	50,000

The reaction follows a first-order rate law. Calculate the specific rate constant.

13.24 The specific rate constant for the first-order decomposition of hydrogen peroxide to water and $O_2(g)$ is found to be $4.9 \times 10^{-6}\ s^{-1}$ at a given temperature. A solution is prepared in which the $[H_2O_2] = 0.986\ M$.

a. Find the length of time required for the $[H_2O_2]$ to fall to $0.329\ M$.
b. Find the length of time for the $[H_2O_2]$ to fall to $0.0986\ M$.
c. Find the $[H_2O_2]$ after 40 hours.

13.25 Use the data of Prob. 13.14 to find how much time is required for the $[C_4H_9I]$ to be reduced from $0.200\ M$ to $0.125\ M$.

Solution: Since this reaction is known to be first order in $[C_4H_9I]$, the integrated first-order rate equation may be used.

$$t = \left(\ln \frac{A}{A_0}\right)\left(-\frac{1}{k}\right)$$
$$= \left(\ln \frac{0.125}{0.200}\right)\left(-\frac{1}{1.16 \times 10^{-2}}\right)$$
$$= 40.5 \text{ min}$$

13.26* The specific rate constant for the first-order decomposition of $N_2O_5(g)$ to NO_2 and O_2 is $7.48 \times 10^{-3}\ s^{-1}$ at a given temperature.

a. Find the length of time required for the total pressure in a system containing N_2O_5 at an initial pressure of 0.100 atm to rise to 0.145 atm.
b. To 0.200 atm.
c. Find the total pressure after 100 s of reaction.

13.27 The reaction $SO_2Cl_2(g) \rightarrow SO_2(g) + Cl_2(g)$ is first order, $k = 2.20 \times 10^{-5}\ s^{-1}$. If 500 mmHg of SO_2Cl_2 is introduced into a bulb, how long will it take for the $P_{SO_2Cl_2}$ to drop to 100 mmHg?

13.28 The specific rate constant for the first-order reaction $N_2O_3(g) \rightarrow NO_2(g) + NO(g)$ at a certain temperature is $8.93 \times 10^{-3}\ min^{-1}$. Calculate how long it will take the pressure to reach 135 mmHg if 100 mmHg of $N_2O_3(g)$ is initially introduced into an evacuated bulb.

13.29 Referring to the reaction of Prob. 13.27 calculate the $P_{SO_2Cl_2}$ after 5 h; calculate the total pressure at this time.

13.30 The following rate data are obtained for a reaction $2A(g) \rightarrow B(l)$ that is second order.

P_A mmHg	300	200	151	99
t s	0	250	500	1000

Calculate the specific rate constant.

Solution: Since the initial concentration and the concentration at different times are given, k can be calculated by substitution into the integrated second-order rate equation.

$$\frac{1}{A} - \frac{1}{A_0} = kt \qquad \text{for } t = 250 \text{ s}$$

$$\frac{1}{200} - \frac{1}{300} = k(250) \quad k = 6.67 \times 10^{-6} \frac{\text{L}}{\text{mol-s}}$$

Repeating for another time, $\frac{1}{151} - \frac{1}{300} = k(500)$

$k = 6.58 \times 10^{-6}$ L/mol-s. The best value of k is then obtained by averaging the values of k obtained from each measurement.

13.31 At a certain temperature the following data are obtained for the second-order reaction $2C_4H_6 \rightarrow C_8H_{12}$

$[C_4H_6]$ mol/L	0.120	0.111	0.104	0.092
t min	0	200	400	800

Calculate the specific rate constant.

13.32 Use the data of Prob. 13.4 and the integrated second-order rate equation to calculate k for the reaction.

13.33 For the reaction of Prob. 13.30, how long does it take for P_A to drop to 50 mmHg?

13.34 For the reaction of Prob. 13.30, what will be the value of P_A after 5000 s?

13.35 The reaction $CH_3CHO(g) \rightarrow CH_4(g) + CO(g)$ is second order in $CH_3CHO(g)$ at 700 K. The following data are obtained from a kinetic study of the reaction:

$[CH_3CHO]$ (mol/L)	0.022	0.020	0.017	0.013
t (s)	0	1000	3000	7000

a. Find the specific rate constant.
b. Find the $[CH_3CHO]$ after 15,000 s.
c. Find the time required for the $[CH_4]$ to reach 0.012 mol/L.

13.36 The reaction $I^-(aq) + ClO^-(aq) \rightarrow IO^-(aq) + Cl^-(aq)$ is second order overall at high $[OH^-]$ and first order in each reactant. To simplify the treatment of the data, a kinetic study is carried out on a solution in which the $[I^-] = [ClO^-] = 0.096\ M$.
a. Indicate why the treatment of the data will be simplified in this case.
b. Use the following data to find the specific rate constant:

$[I^-]$ (M)	0.096	0.049	0.032	0.020
t (s)	0	29	63	146

c. Find the $[IO^-]$ after 200 s given that $[IO^-] = 0$ at $t = 0$.

13.37 Use the integrated form of the first-order rate equation to show that the time necessary for the initial concentration of a substance that reacts in a first-order process to be reduced by half is independent of the value of the initial concentration. Find the relationship between this time, which is called the *half-life*, and the specific rate constant.

Solution: If we start initially with $[A] = A_0$, when half of A_0 reacts $[A] = A_0/2$. Substitution gives

$$\ln \frac{A_0/2}{A_0} = -kt \quad \text{or} \quad \ln \frac{1}{2} = -kt; \quad t = \frac{0.693}{k}$$

13.38 Show that for a first-order reaction the time required for 99.9% of the reaction to take place is 10 times that required for 50% to take place.

13.39 A compound decomposes by a first-order reaction and it is found that after 6 h 20% of the original material remains. Calculate the specific rate constant k, the half-life, and the time it took for the first 20% of the material to react.

13.40* A certain substance A decomposes. It is found that 50% of A remains after 100 s. How much A remains after 200 s if the reaction with respect to A is (a) zero order, (b) first order, and (c) second order?

Hint: For a second-order reaction the half-life depends on A_0. Therefore use the integrated equation to calculate k in terms of A_0 and then calculate [A] at 200 s in terms of A_0.

13.41 A reaction has an activation energy $E_a = 83.7$ kJ and a specific rate constant $k = 5.1 \times 10^{-2}\ s^{-1}$ at 300 K. Calculate the specific rate constant at 400 K.

Hint: Substitute in Eq. (13.4).

13.42 A certain reaction doubles its velocity when the temperature is raised from 298 K to 310 K. Calculate the activation energy.

13.43 The reaction $2N_2O_5 \rightarrow 2N_2O_4 + O_2(g)$ takes place around room temperature in solvents such as CCl_4. The specific rate constant at 293 K is found to be $2.35 \times 10^{-4}\ s^{-1}$. At 303 K it is found to be $9.15 \times 10^{-4}\ s^{-1}$. Find the activation energy of the reaction.

13.44 The following data are obtained for the reaction $H_2C_2O_4(g) \rightarrow CO_2(g) + HCOOH(g)$:

k (s^{-1})	8.09×10^{-5}	1.87×10^{-4}	3.39×10^{-4}
T (K)	407.3	419.6	428.8

Calculate the activation energy of the reaction.

13.45 The activation energy of the reaction $NO_2(g) + CO(g) \rightarrow NO(g) + CO_2(g)$ is 132.0 kJ/mol and the specific rate constant at 500 K is 2.02×10^{-6} L/mol-s. Find the specific rate constant at 550 K.

13.46 Consider the elementary reaction $A + B \rightleftarrows C + D$, which is first order in A and first order in B for the forward reaction and first order in C and first order in D for the back reaction. Let k_1 and k_{-1} be the specific rate constants for the forward and back reactions, respectively. Derive a relationship between these specific rate constants and the equilibrium constant K for the reaction.

Hint: At equilibrium the rate of the forward reaction is equal to the rate of the back reaction.

Chapter 14

Chemical Equilibrium and Equilibrium Constants

Chemical Equilibrium is one of the main focuses of General Chemistry. Understanding this topic and being able to solve problems in chemical equilibrium will be critical for success.

THE EQUILIBRIUM STATE

The equilibrium state is the ultimate goal of every chemical system. A system is in an *equilibrium state* if its *macroscopic properties* do not change *spontaneously*, even over a prolonged period of time. Macroscopic properties are properties of the system as a whole and not of its individual atoms or molecules. Examples are pressure, concentration, mass, volume, and even color. We can usually observe macroscopic properties with our senses or with the aid of simple measuring devices, such as a balance or a measuring stick.

A spontaneous change is a change that takes place without the intervention of an external agent. It is a change that the system undergoes when we leave it alone. If a system is not at equilibrium, it undergoes spontaneous change to get there. If it is at equilibrium, only an external influence can cause it to change.

In addition to what these definitions say, it is also important for you to recognize what is not said. Nothing is said about the time that it takes a system to undergo the spontaneous change that gets it to the equilibrium state. The reason nothing is said about time is that time is not a factor and has absolutely nothing to do with the term "spontaneous." Some systems proceed to equilibrium very rapidly; others proceed immeasurably slowly. For example, at room temperature and atmospheric pressure, graphite is the equilibrium state of carbon. Therefore diamonds are converting to graphite spontaneously, but fortunately very slowly. Left to their own devices, all systems get to the equilibrium state. We shall not be concerned with the time needed for a system to reach equilibrium, but only with the state of the system once it gets there.

A system at equilibrium appears static because its macroscopic properties are not changing. This appearance is deceptive. On the atomic and molecular level, an equilibrium system is *dynamic*; an enormous number of changes occur continuously. But the net result of all these changes on the molecular level is no change on the macroscopic level.

How can this be? The answer comes from kinetics. *At equilibrium any change and the exact reverse of that change take place at the same rate.* We have already discussed this point in connection with phenomena such as the vapor pressure of a liquid. When a liquid is in equilibrium with its vapor (when the vapor pressure equals the atmospheric pressure), the rate of evaporation and the rate of condensation are equal.

So if $A \rightarrow B$ at the same rate that $B \rightarrow A$, nothing changes macroscopically. What happens is that A and B trade places. But since they are molecules (or atoms or ions) the change is not macroscopic and not detectable. It is only when we disturb the system in some way, for example by changing the temperature, that macroscopic changes occur.

CHEMICAL CHANGE AND EQUILIBRIUM

What we want to consider are equilibrium states of systems containing substances that are related by simple chemical changes.

Let's look at the reaction

$$H_2(g) + I_2(g) \rightleftarrows 2HI(g)$$

which we write with a double arrow to indicate that it is a reversible reaction and that it does not go completely to the right.

We find that if we mix H_2 and I_2 at 700 K, HI forms rapidly and the pressures of H_2 and I_2 decrease. It doesn't take very long before the composition of the system stops changing. When it stops we have reached the equilibrium state. We find that there is still some remaining pressure of H_2 and I_2 in the equilibrium state. Not all of the reactants are consumed.

We also find that if we heat HI to 700 K, H_2 and I_2 form rapidly and the pressure of HI decreases. Once again it doesn't take long before the composition of the system stops changing. Once again we have reached the equilibrium state. These two results indicate a crucial aspect of the equilibrium state of substances related by chemical change. It can be reached *from either direction*.

We could also mix all three components at 700 K and observe that the system changes until it reaches the equilibrium state and stops changing. The way in which it changes to reach equilibrium depends on the quantities of the three components with which we start. The composition of equilibrium states as a function of the quantities of the substances in the system has been very thoroughly studied. The results of these studies show that the actual composition of the equilibrium state depends on how much material is in the system. For example if we start with more HI, then the pressure of HI in the equilibrium state will be greater. If we had to describe equilibrium states by reference to how much of the components we start with, this topic would be hopelessly complicated.

But, the results of these studies also show that the equilibrium state of every reacting chemical system can be described by an expression that is independent of starting quantities. This expression is called the *equilibrium constant*, and it is always represented by the letter K. The form of K is based on the equation that we write to describe the chemical change. For a given system at a given temperature, the value of K is fixed. As you can see the first and most important step in the approach to K and the description of an equilibrium state of a reacting chemical system is to write a balanced chemical reaction, including all state symbols, that relates the components of the system.

Once we have done that, we can obtain the expression for K by following a set of rules:

1. Usually K is a fraction whose numerator includes information about the products of the reaction (the stuff on the right of the equation) and whose denominator includes information about the reactants (the stuff on the left of the equation). Clearly the form of K depends on how we choose to write the reaction. Thus K for the reaction $A \rightleftarrows B$ is $1/K$ for the reaction $B \rightleftarrows A$.
2. The expression for K includes terms for gases and substances in solutions. The terms for gases usually are expressed as the concentrations of the gases in moles per liter (mol/L) (C with the formula of the gas as a subscript or the formula of the gas in square brackets [] as if it were a molarity), although the terms for gases can also be expressed as the partial pressures of the gases in atmospheres (atm). The terms for substances in solution, generally aqueous solution, usually express concentration in molarity. By convention, we symbolize molarity by putting the solute in square brackets. So $[OH^-]$ designates the molarity of hydroxide ion in water.
3. Terms for pure liquids and solids are not included in the expression for K. The concentration of a pure liquid or pure solid is essentially 1.
4. The numerator of the expression for K is the product of all the terms (pressure and concentration) written for substances on the right side of the equation. The denominator of the expression for K is the product of all the terms (pressure and concentration) written for substances on the left side of the equation.

5. The concentration of a solute or gas or the pressure of a gas in the expression for K is raised to the power of its coefficient in the equation. If a gas has a coefficient 2 in the reaction, its concentration or pressure in K is squared. If a substance in solution has the coefficient 3 in the equation, its molarity in K is cubed.

The numerical value of the equilibrium constant for a given reaction is obtained by inserting the experimentally determined values of the concentrations in the equilibrium constant expression for the reaction. Thus for the reaction

$$\underset{0.20}{SO_2(g)} + \underset{0.80}{NO_2(g)} \rightleftarrows \underset{0.90}{SO_3(g)} + \underset{1.1}{NO(g)}$$

the calculation, using the equilibrium concentrations given in mol/L, becomes

$$K = \frac{[SO_3] \times [NO]}{[SO_2] \times [NO_2]} = \frac{0.90 \times 1.1}{0.60 \times 0.80} = 2.1$$

For the reaction $\underset{0.20}{2SO_2(g)} + \underset{0.30}{O_2(g)} \rightleftarrows \underset{0.60}{2SO_3(g)}$

$$K = \frac{[SO_3]^2}{[SO_2]^2\,[O_2]} = \frac{(0.60)^2}{(0.20)^2\,(0.30)} = 30$$

Remember that a *solid* or a *liquid* reactant or product is not included in the equilibrium formula. *Only gases or substances in solution are included.* Thus for the equilibrium reaction

$$SiF_4(g) + 2H_2O(g) \rightleftarrows SiO_2(s) + 4HF(g)$$

$$K = \frac{[HF]^4}{[SiF_4] \times [H_2O]^2}$$

and for the reaction

$$LaCl_3(s) + H_2O(g) \rightleftarrows LaClO(s) + 2HCl(g)$$

$$K = \frac{[HCl]^2}{[H_2O]}$$

The *amount* of excess solid present has no effect whatsoever on the *equilibrium* state. The same state of equilibrium is attained whether we have a small excess or a large excess of solid. The *rate* at which equilibrium is attained will be affected by the total surface of the solid. However, once equilibrium has been attained, the removal of some of the excess solid (or liquid) will have no effect on the equilibrium.

You should note that the equilibrium formula as presented in the preceding pages assumes that the behavior of every reacting molecule is completely unaffected by the other molecules that are present in the system; in other words, it assumes that the system is *ideal*. In such an ideal system the *effective concentration* of each species is, in fact, its molar concentration. Hence, we use molar concentrations in the equilibrium constant.

Most gaseous systems and most solutions are not ideal; the behavior of each molecule is affected, to a small degree, by its neighbors. As a consequence, the *effective concentration* of a given species, called its *activity*, is slightly less than its *molar concentration*; it is equal to the product of its molar concentration and its *activity coefficient* in the particular system. This difference is especially serious in aqueous solutions. *Activity* or *effective concentration*, rather than molar concentration, should appear in the equilibrium formulas for nonideal systems.

We will assume, in the problems given in this book, that all systems are ideal. In such systems the activity coefficient of each species has a value of 1 and the molar concentration is, in fact, the effective concentration. Making this assumption will in no way detract from the value of a problem as a vehicle for developing logical thinking and reasoning.

PROBLEMS

14.1 Write the expressions for K for the following chemical equations.

a. $C(s) + CO_2(g) \rightleftarrows 2CO(g)$

b. $NH_3(g) + HCN(l) \rightleftarrows NH_4^+(aq) + CN^-(aq)$

c. $5Fe^{2+}(aq) + MnO_4^-(aq) + 8H^+(aq) \rightleftarrows 5Fe^{3+}(aq) + Mn^{2+}(aq) + 4H_2O(l)$

Solution:

a. Solids do not appear in the expression for K, and the coefficient of the CO appears as an exponent:

$$K = C_{CO}^2/C_{CO_2} \text{ or } P_{CO}^2/P_{CO_2}$$

b. Liquids do not appear in the expression for K:

$$K = [NH_4^+][CN^-]/C_{NH_3}$$

c. The concentration of water the solvent is not included in K:

$$K = [Mn^{2+}][Fe^{3+}]^5/[MnO_4^-][Fe^{2+}]^5[H^+]^8$$

14.2 Write net ionic equations and the corresponding expressions for K for the following processes:

a. the decomposition of calcium carbonate to calcium oxide and carbon dioxide

b. the dissolution of silver sulfate in water

c. the reaction of a solution of sodium carbonate with a solution of hydrochloric acid, liberating carbon dioxide

d. the conversion of dinitrogen trioxide to nitric oxide and nitrogen dioxide

Solution:

a. $CaCO_3(s) \rightleftarrows CaO(s) + CO_2(g)$

$K = C_{CO_2} = P_{CO_2}$

b. $Ag_2SO_4(s) \rightleftarrows 2Ag^+(aq) + SO_4^{2-}(aq)$

$K = [Ag^+]^2[SO_4^{2-}]$

Note that the equation for the precipitation of silver sulfate from aqueous solution would be the reverse reaction whose $K = 1/[Ag^+]^2[SO_4^{2-}]$.

c. $CO_3^{2-}(aq) + 2H^+(aq) \rightleftarrows H_2O(l) + CO_2(g)$

$K = C_{CO_2}/[CO_3^{2-}][H^+]^2$ or with the gas as a P

d. $N_2O_3(g) \leftrightarrows NO(g) + NO_2(g)$

$$K = \frac{C_{NO}/C_{NO_2}}{C_{N_2O_3}}$$ or with the gas as P.

14.3 The following processes differ only in the states of the reactants and products. Write the expressions for K for each of them:

a. pure ammonia and pure hydrogen chloride form ammonium chloride

b. pure ammonia and hydrochloric acid form ammonium chloride

c. a solution of ammonia and pure hydrogen chloride form ammonium chloride

d. a solution of ammonia and hydrochloric acid form ammonium chloride

Solution:

The substances of interest are

pure ammonia—$NH_3(g)$

hydrogen chloride—$HCl(g)$

a solution of ammonia—$NH_3(aq)$

hydrochloric acid—$H^+(aq) + Cl^-(aq)$

ammonium chloride—$NH_4Cl(s)$

ammonium chloride in solution—$NH_4^+(aq) + Cl^-(aq)$.

Now we can write the chemical equations and the expressions for K

a. $NH_3(g) + HCl(g) \rightleftarrows NH_4Cl(s)$; $K = 1/C_{NH_3}C_{HCl}$

b. $NH_3(g) + H^+(aq) \rightleftarrows NH_4^+(aq)$; $K = [NH_4^+]/C_{NH_3}[H^+]$

c. $NH_3(aq) + HCl(g) \rightleftarrows NH_4^+(aq) + Cl^-(aq)$;

$K = [NH_4^+][Cl^-]/[NH_3]C_{HCl}$

d. $NH_3(aq) + H^+(aq) \rightleftarrows NH_4^+(aq)$; $K = [NH_4^+]/[NH_3][H^+]$

14.4 At equilibrium at a given temperature and in a 1-L reaction vessel, HI is 20 mol % dissociated into H_2 and I_2 according to the equation $2HI(g) \rightleftarrows H_2(g) + I_2(g)$. If 1 mol of pure HI is introduced into a 1-L reaction vessel at the given temperature, what amount of each component will be present when equilibrium is established?

Solution: The amount of HI that dissociates = 20% of 1 mol = 0.2 mol. The amount of HI that is not dissociated is $1 - 0.2 = 0.8$ mol. The equation $2HI(g) \rightleftarrows H_2(g) + I_2(g)$ tells us that 2 mol of HI yields 1 mol of H_2 and 1 mol of I_2. Therefore, 0.2 mol of HI yields 0.1 mol of H_2 and 0.1 mol of I_2.

14.5 PCl_5 is 20 mol % dissociated into PCl_3 and Cl_2 at equilibrium at a given temperature and in a 1-L vessel in accordance with the equation $PCl_5(g) \rightleftarrows PCl_3(g) + Cl_2(g)$. One mole of pure PCl_5 was introduced into a 1-L reaction vessel at the given temperature. What amount of each component was present at equilibrium?

14.6 A reaction vessel in which the reaction $CO + Cl_2 \rightleftarrows COCl_2$ had reached a state of equilibrium was found, on analysis, to contain 0.30 mol of CO, 0.20 mol of Cl_2, and 0.80 mol of $COCl_2$ in 1 L of mixture. Calculate the equilibrium constant for the reaction.

Solution: First, write the equilibrium formula.

$$K = \frac{[COCl_2]}{[CO] \times [Cl_2]}$$

The notation $[COCl_2]$, by definition, means concentration of $COCl_2$ in moles of $COCl_2$ per liter. Likewise, [CO] and $[Cl_2]$ mean, respectively, moles of CO per liter and moles of Cl_2 per liter. Substitution in the above equation gives

$$K = \frac{0.80}{(0.30)(0.20)} = 13$$

14.7 A reaction vessel with a capacity of 1 L in which the reaction $2SO_2(g) + O_2(g) \rightleftarrows 2SO_3(g)$ had reached a state of equilibrium was found to contain 0.6 mol of SO_3, 0.2 mol of SO_2, and 0.3 mol of O_2. Calculate the equilibrium constant.

Solution: First, write the equilibrium formula.

$$K = \frac{[SO_3]^2}{[SO_2]^2 \times [O_2]}$$

Substitute the concentrations of O_2, SO_3, and SO_2 in this formula to give

$$K = \frac{(0.6)^2}{(0.2)^2(0.3)} = 3 \times 10^1$$

14.8 A quantity of PCl_5 was heated in a 1-L vessel at 250 °C. At equilibrium the concentrations of the gases in the vessel were as follows: PCl_5 7.05 mol/L; PCl_3 0.54 mol/L; Cl_2 0.54 mol/L. Calculate the equilibrium constant K for the dissociation of PCl_5 at 250 °C.

14.9 An equilibrium mixture of N_2, H_2, and NH_3, in a 1-L vessel, which reacts according to the equation $N_2(g) + 3H_2(g) \rightleftarrows 2NH_3(g)$, was found to consist of 0.800 mol of NH_3, 0.300 mol of N_2, and 0.200 mol of H_2. Calculate the equilibrium constant.

14.10 An equilibrium mixture, $CO(g) + Cl_2(g) \rightleftarrows COCl_2(g)$, contained 1.50 mol of CO, 1.00 mol of Cl_2, and 4.00 mol of $COCl_2$ in a 5.00-L reaction vessel at a specific temperature. Calculate the equilibrium constant for the reaction at this temperature.

Hint: In calculating the equilibrium constant, concentration must be expressed in *moles per liter*.

14.11 An equilibrium mixture, $2SO_2(g) + O_2(g) \rightleftarrows 2SO_3(g)$, contained in a 2.00-L reaction vessel at a specific temperature was found to contain 96.0 g of SO_3, 25.6 g of SO_2, and 19.2 g of O_2. Calculate the equilibrium constant for the reaction at this temperature.

Hint: Concentrations must be expressed in *moles* per *liter*.

14.12 When the system $H_2(g) + CO_2(g) \rightleftarrows H_2O(g) + CO(g)$ is at equilibrium at 1260 K, it is found to contain 0.19 mol of CO_2, 22.6 mol of H_2, and 2.6 mol of CO and of H_2O. Find the value of K.

14.13 The equilibrium constant for the reaction $2SO_2(g) + O_2(g) \rightleftarrows 2SO_3(g)$ is 4.5 at 600 °C. A quantity of SO_3 gas was placed in a 1-L reaction vessel at 600 °C. When the system reached a state of equilibrium, the vessel was found to contain 2.0 mol of O_2 gas. What amount of SO_3 gas was originally placed in the reaction vessel?

Hint: Note that when SO_3 decomposes to yield SO_2 and O_2, the products are formed in the ratio of 2 mol of SO_2 and 1 mol of O_2 for every 2 mol of SO_3 decomposed. Since in this problem there are 2 mol of O_2 in the vessel, there must also be 4 mol of SO_2 present, and 4 mol of SO_3 must have decomposed to yield 4 mol of SO_2 and 2 mol of O_2. That means that at equilibrium the reaction vessel must contain 4 mol less of SO_3 than was originally introduced. With these facts and the knowledge that the equilibrium constant is 4.5, the amount of SO_3 originally added can be calculated.

14.14 The equilibrium constant for the reaction $N_2(g) + 3H_2(g) \rightleftarrows 2NH_3(g)$ is 2.00 at 300 °C. A quantity of NH_3 gas was introduced into a 1-L reaction vessel at 300 °C. When equilibrium was established, the vessel was found to contain 2.00 mol of N_2. What amount of $NH_3$was originally introduced into the vessel?

14.15 A 1-L reaction vessel in which the reaction

$$C(s) + H_2O(g) \rightleftarrows CO(g) + H_2(g)$$

has reached a state of equilibrium contains 0.16 mol of C, 0.58 mol of H_2O, 0.15 mol of CO, and 0.15 mol of H_2. Calculate the equilibrium constant for the reaction.

Solution: Since C is a solid it is not included in the equilibrium formula

$$K = \frac{[CO] \times [H_2]}{[H_2O]} = \frac{[0.15] \times [0.15]}{[0.58]} = 3.9 \times 10^{-2}$$

14.16 The equilibrium mixture $SO_2(g) + NO_2(g) \rightleftarrows SO_3(g) + NO(g)$ in a 1-L vessel was found to contain 0.600 mol of SO_3, 0.400 mol of NO, 0.100 mol of NO_2, and 0.800 mol of SO_2. What amount of NO would have to be forced into the reaction vessel, volume and temperature being kept constant, in order to increase the amount of NO_2 to 0.300 mol?

Solution: First calculate the equilibrium constant.

$$K = \frac{[SO_3] \times [NO]}{[SO_2] \times [NO_2]} = \frac{0.6 \times 0.4}{0.8 \times 0.1} = 3.00$$

Let X = moles of NO that must be added. We can see from the equation $SO_2(g) + NO_2(g) \rightleftarrows SO_3(g) + NO(g)$ that in order to produce 0.2 mol more of NO_2 we must also produce 0.2 mol more of SO_2 and we must use up 0.2 mol of SO_3 and 0.2 mol of NO. Therefore, when we have 0.3 mol of NO_2, we will have 1 mol

(0.8 + 0.2) of SO_2, 0.4 mol (0.6 − 0.2) of SO_3, and X + 0.2 mol (0.4 + X − 0.2) of NO. Substitution of these values in the equilibrium formula gives

$$\frac{0.4 \times (X + 0.2)}{1 \times 0.3} = 3.00$$

$$X = 2.05 \text{ mol}$$

14.17 An equilibrium mixture, $CO(g) + H_2O(g) \rightleftarrows CO_2(g) + H_2(g)$, contains 0.2 mol of H_2, 0.80 mol of CO_2, 0.10 mol of CO, and 0.40 mol of H_2O. What amount of CO_2 would have to be added at constant temperature and volume to increase the amount of CO to 0.20 mol?

14.18 A reaction system in equilibrium according to the equation $2SO_2(g) + O_2(g) \rightleftarrows 2SO_3(g)$ in a 1-L reaction vessel at a given temperature was found to contain 0.11 mol of SO_2, 0.12 mol of SO_3, and 0.050 mol of O_2. Another 1-L reaction vessel contains 64 g of SO_2 at the same temperature. What mass of O_2 must be added to this vessel in order that, at equilibrium, half of the SO_2 will be oxidized to SO_3?

Hint: Calculate K from the data in the first sentence. In the second situation 1 mol (64 g) of SO_2 is initially present. Let X = g of O_2 added. Then $X/32$ = moles of O_2 added. At equilibrium $[SO_2] = 0.5$, $[SO_3] = 0.5$, $[O_2] = X/32 - 0.25$. The value of X can then be calculated.

14.19 A 0.252-mol amount of $NO_2(g)$ is placed in a sealed 2.00-L container and heated to 700 K. At equilibrium it is found that there remains 0.179 mol of $NO_2(g)$. Calculate K for the reaction $2NO_2(g) \leftrightarrows 2NO(g) + O_2(g)$.

14.20 The equilibrium constant for the reaction $SO_2(g) + NO_2(g) \leftrightarrows SO_3(g) + NO(g)$ is 3.0. Find the amount of NO_2 that must be added to 2.4 mol of SO_2 in order to form 1.2 mol of SO_3 at equilibrium.

14.21* The equilibrium constant for the reaction $PCl_5(g) \rightleftarrows PCl_3(g) + Cl_2(g)$ at 250 °C is 0.041. Set up, but do not solve, an algebraic equation in one unknown X that, if solved for X, will give the mass of Cl_2 in grams that will be present at equilibrium when 0.3 mol of PCl_5 is heated in a 1-L vessel at 250 °C.

14.22* A mixture of 2 mol of CH_4 gas and 1 mol of H_2S gas was placed in an evacuated container, which was then heated to and maintained at a temperature of 727 °C. When equilibrium was established in the gaseous reaction $CH_4(g) + 2H_2S(g) \rightleftarrows CS_2(g) + 4H_2(g)$, the total pressure in the container was 0.92 atm and the partial pressure of the hydrogen gas was 0.20 atm. What was the volume of the container?

Solution: To calculate V when T is constant we must know P and n. Before any reaction has occurred, we know the total amount, 3 moles, but we do not know the pressure. At equilibrium, we know $P = 0.92$ atm, but we do not know the amount. Therefore, to calculate V we must calculate either the amount at equilibrium or the pressure at the start.

To calculate amount at equilibrium, let X = moles of CS_2. Then $4X$ = moles of H_2, $2 - X$ = moles of CH_4, and $1 - 2X$ = moles of H_2S. Total moles at equilibrium = the sum of the above quantities = $3 + 2X$. In a system at constant V and T the amount is directly proportional to the pressure. Therefore,

$$\frac{\text{total moles}}{\text{moles of } H_2} = \frac{\text{total pressure}}{\text{partial pressure of } H_2}$$

$$\frac{3 + 2X}{4X} = \frac{0.92}{0.20}$$

Solving $X = 0.183$ mol, we have total moles = 3.366. We now have P, T, and n and we can use $PV = nRT$ to find $V = 300$ L.

14.23* To determine the equilibrium constant at a given temperature for the gas-phase reaction $N_2(g) + 3H_2(g) \rightleftarrows 2NH_3(g)$, 0.326 mol of H_2 and 0.439 mol of N_2 were mixed in a 1.00-L vessel. At equilibrium the system was found to contain a total of 0.657 mol of gas.

a. Calculate the equilibrium constant for the reaction as written above and state the "units" in which this value is expressed.

b. Call the constant calculated in part (a) K_1. Call the constant for the following reaction K_2: $NH_3(g) \rightleftarrows \frac{1}{2}N_2(g) + \frac{3}{2}H_2(g)$. State the algebraic relation between K_1 and K_2.

14.24* When N_2O_5 gas is heated, it dissociates into N_2O_3 gas and O_2 gas according to the reaction $N_2O_5(g) \rightleftarrows N_2O_3(g) + O_2(g)$. K_1 for this reaction at a specific temperature is 7.75. The N_2O_3 dissociates to give N_2O gas and O_2 gas according to the reaction $N_2O_3(g) \rightleftarrows N_2O(g) + O_2(g)$. K_2 for this reaction at the same specific temperature is 4.00. When 4.00 mol of N_2O_5 is heated in a 1.00-L reaction vessel at the same temperature, the concentration of O_2 at equilibrium is 4.50. Calculate the concentrations in moles per liter of all other species in the equilibrium system.

Hint: Let X equal the concentration of O_2 derived from the N_2O_5 and Y equal the concentration of O_2 derived from the N_2O_3. Then

$$[O_2] = X + Y \qquad [N_2O_5] = 4.00 - X$$
$$[N_2O_3] = X - Y \qquad [N_2O] = Y$$

Since three equations involving X and Y are available, the value of X and of Y, and hence the concentrations of all species, can be calculated.

14.25* At 1227 °C the equilibrium constant for the reaction $CaCO_3(s) \rightleftarrows CaO(s) + CO_2(g)$ is 0.50. Also, at 1227 °C CO_2 decomposes according to the reaction $CO_2(g) \rightleftarrows CO(g) + \frac{1}{2}O_2(g)$. One mole of solid $CaCO_3$ is placed in an evacuated 1-L container and heated to 1227 °C. When equilibrium is established, the mole fraction of O_2 in the gaseous mixture in the container is 0.15. What amount of CaO is in the container at equilibrium?

Hint: Moles of CaO = moles of CO_2 + moles of CO.

14.26* At 1227 °C the partial pressure of the CO_2 gas that is in equilibrium with solid CaO and solid $CaCO_3$ in a 1-L reaction vessel according to the equation $CaCO_3(s) \rightleftarrows CaO(s) + CO_2(g)$ is 61.5 atm. Also, at 1227 °C CO_2 gas decomposes according to the equilibrium reaction $CO_2(g) \rightleftarrows CO(g) + \frac{1}{2}O_2(g)$. One mole of solid $CaCO_3$ is placed in an evacuated 1-L container and heated to 1227 °C. When equilibrium is established, the mole fraction of the CO_2 in the gaseous mixture of CO_2, CO, and O_2 is 0.55. What amount of solid $CaCO_3$ is present at equilibrium?

EQUILIBRIUM CONSTANTS FROM PRESSURES

When the gas law equation $PV = nRT$ is written in the form $n/V = P/RT$, the term n/V represents the concentration of the gas in moles per liter. The equation $n/V = P/RT$ tells us that at constant temperature the concentration of a gas in moles per liter is directly proportional to its partial pressure. It follows, therefore, that for a gaseous equilibrium reaction an equilibrium constant K_p can be written in terms of the partial pressures of the gases in the chemical reaction. The forms of the K_p expressions for some typical equilibria are given in Table 14.1.

TABLE 14.1

Reaction	Equilibrium Expression
(1) $CO + H_2O \rightleftarrows CO_2 + H_2$	$K_p = \dfrac{P_{CO_2} \times P_{H_2}}{P_{CO} \times P_{H_2O}}$
(2) $COCl_2 \rightleftarrows CO + Cl_2$	$K_p = \dfrac{P_{CO} \times P_{Cl_2}}{P_{COCl_2}}$
(3) $2NH_3 \rightleftarrows N_2 + 3H_2$	$K_p = \dfrac{P_{N_2} \times (P_{H_2})^3}{(P_{NH_3})^2}$
(4) $4H_2 + CS_2 \rightleftarrows CH_4 + 2H_2S$	$K_p = \dfrac{P_{CH_4} \times (P_{H_2S})^2}{(P_{H_2})^4 \times P_{CS_2}}$

Note that the form of the K_p expression is the same as that of the K_c expression (concentration expressed in mole units) except that P_a is substituted for $[a]$. Just as concentration is measured in moles/L when we find the value of K, so pressure is usually measured in atmospheres when we find K_p.

Since $n/V = P/RT$, and since n/V represents concentration in moles per liter, it follows that at constant temperature P/RT can be substituted for the concentration terms in the K_c expression. Using this approach we can show that for any gas-phase equilibrium

$$K = K_p \times \left(\frac{1}{RT}\right)^{\Delta n}$$

where Δn is the change in the number of moles of gas when the reaction goes from left to right.

PROBLEMS

14.27 In an equilibrium mixture of N_2, H_2, and NH_3, contained in a 5.00-L reaction vessel at 450 °C and a total pressure of 332 atm, the partial pressures of the gases were: N_2, 47.55 atm; H_2, 142.25 atm; and NH_3, 142.25 atm. Calculate the equilibrium constant for the reaction $2NH_3(g) \leftrightarrows N_2(g) + 3H_2(g)$.

Hint: To calculate the equilibrium constant as a function of concentrations we must know the concentration of each reactant in *moles* per *liter*. To this end we first calculate the moles of each substance per liter by use of the formula $PV = nRT$.

14.28 The equilibrium constant for the reaction $CO(g) + H_2O(g) \rightleftarrows CO_2(g) + H_2(g)$ is 4.0 at a given temperature. An equilibrium mixture of the above substances at the given temperature was found to contain 0.60 atm of CO, 0.20 atm of steam, and 0.50 atm of CO_2 in a liter. What pressure of H_2 was in the mixture?

14.29 Ammonia at exactly 1 atm was introduced into a 1-L reaction vessel at a certain high temperature. When the reaction $2NH_3(g) \rightleftarrows N_2(g) + 3H_2(g)$ had reached a state of equilibrium, 0.6 atm of H_2 was found to be present. Calculate the equilibrium constant for the reaction.

Hint: When NH_3 dissociates to give N_2 and H_2, the products are formed in the ratio of 1 atm of N_2 and 3 atm of H_2 for every 2 atm of NH_3 that dissociates. To yield 0.6 atm of H_2, 0.4 atm of NH_3 must have dissociated. This leaves 0.6 atm of undissociated NH_3. The 0.4 atm of NH_3 that dissociates will yield 0.2 atm of N_2 and 0.6 atm of H_2. The equilibrium mixture will contain 0.6 atm of NH_3, 0.2 atm of N_2, and 0.6 atm of H_2.

14.30 Sulfur trioxide at 1 atm was placed in a reaction vessel at a certain temperature. When equilibrium was established in the reaction $2SO_3(g) \rightleftarrows 2SO_2(g) + O_2(g)$, the vessel was found to contain 0.60 atm of SO_2. Calculate the equilibrium constant for the reaction.

14.31* A 7.24-g sample of IBr is placed in a container of volume 0.225 L and is heated to 500 K. At this temperature the following reaction takes place $2IBr(g) \leftrightarrows I_2(g) + Br_2(g)$. The measured pressure of Br_2 at equilibrium is 3.01 atm. Find the value of K_p.

14.32 In an equilibrium mixture of $CO_2(g) + H_2(g) \rightleftarrows CO(g) + H_2O(g)$ contained in a 6.0-L reaction vessel at 1007 °C, the partial pressures of the reactants are: $CO_2 = 63.1$ atm, $H_2 = 21.1$ atm, $CO = 84.2$ atm, $H_2O = 31.6$ atm. Enough CO_2 is then removed from the vessel to reduce the partial pressure of the CO to 63.0 atm, the temperature being kept constant.

a. Calculate the partial pressure of the CO_2 in the new equilibrium system.

b. For the above reaction, how does the numerical value of K_c, in which concentration is expressed in moles per liter, compare with the numerical value of K_p, in which concentrations are expressed in atmospheres?

c. If the volume of the new equilibrium system were reduced to 3 L by depressing a piston, what would the partial pressure of the CO_2 be?

Solution:

a.

$$\overset{63.1\text{ atm}}{CO_2(g)} + \overset{21.1\text{ atm}}{H_2(g)} \rightleftarrows \overset{84.2\text{ atm}}{CO(g)} + \overset{31.6\text{ atm}}{H_2O(g)}$$

$$K_p = \frac{P_{CO} \times P_{H_2O}}{P_{CO_2} \times P_{H_2}} = \frac{84.2\text{ atm} \times 31.6\text{ atm}}{63.1\text{ atm} \times 21.1\text{ atm}} = 2.0$$

Let X = the partial pressure of the CO_2 in the system after removal of CO_2. Since the partial pressure of the CO is reduced to 63.0 atm, a quantity of CO with a partial pressure of 21.2 atm must have reacted with H_2O to form CO_2 and H_2. Therefore, 21.2 atm worth of H_2O must have reacted and 21.2 atm worth of both CO_2 and H_2 must have been produced. The partial pressures of each reactant in the new equilibrium system will then be

$$\overset{X\text{ atm}}{CO_2(g)} + \overset{42.3\text{ atm}}{H_2(g)} \rightleftarrows \overset{63.0\text{ atm}}{CO(g)} + \overset{10.4\text{ atm}}{H_2O(g)}$$

$$K_p = \frac{63.0\text{ atm} \times 10.4\text{ atm}}{X\text{ atm} \times 42.3\text{ atm}} = 2.0$$

Solving, we have $X = 7.7$ atm.

b. Since there is no change in the number of moles, $K = K_p$.

c. Since in the gaseous equilibrium represented by the reaction $CO_2 + H_2 \leftrightarrows CO + H_2O$ there is no change in the number of moles, increasing the pressure by reducing the volume to one-half of its original value will not shift the equilibrium. All that will happen will be that the partial pressure of each reactant will be doubled. Therefore, the partial pressure of the CO_2 will be 2×7.7 atm, or 15.4 atm.

14.33 In an equilibrium mixture of $CO_2(g) + H_2(g) \rightleftarrows CO(g) + H_2O(g)$, the partial pressures were 0.60 atm of CO_2, 0.20 atm of H_2, 0.80 atm of CO, and 0.30 atm of H_2O. What pressure of CO_2 would have to be removed from the system at constant volume and temperature in order to reduce the pressure of CO to 0.70 atm?

A GENERAL METHOD FOR SOLVING EQUILIBRIUM PROBLEMS

The best way for you to solve problems in chemical equilibrium, indeed all general chemistry problems, is to understand the problem and all its ramifications and then to use chemical reasoning, one step at a time, to reach the solution. We have tried to use that approach, as much as possible, in the example problems on chemical equilibrium so far. But sometimes the chemical reasoning can be complex and require a number of steps. There are methods to help you progress through these steps in an organized way.

One common method is to prepare what is called an "ICE" table for the equilibrium problem. "ICE" is an acronym that stands for Initial, Change, Equilibrium. Each cell of the table contains a concentration in moles/liter or a pressure in atm for a gas. Each of these terms is either an actual number or an expression using an algebraic unknown.

a. The first line of the table has the relevant chemical equation. Each substance in this equation is the head of a column.

b. The second line contains information about the initial conditions, usually explicitly stated in the problem. The initial conditions are generally not equilibrium conditions.

c. The third line contains information about the changes that take place as the system proceeds from the initial conditions to the equilibrium state. These changes are generally found by considering the stoichiometry of the chemical reaction.

d. The fourth line contains information about the equilibrium state.

Here is an example of an ICE table:

Reaction	A +	B ⇆	C +	D
Initial				
Change				
Equilibrium				

To prepare an ICE table start by filling in all the information that is given. Then, depending on what has already been given and what you are asked to find, you will have to assign one or more unknowns. In order to solve for these unknowns you will need to have at least one additional piece of information, given in the problem, for each unknown.

The solutions to the next two problems illustrate the use of an ICE table. If you wish you can go back through the previous problems in this chapter and construct an ICE table for any of them you found especially difficult.

PROBLEMS

14.34 The value of K_p for the reaction in which two moles of NO_2 form 1 mol of N_2O_4 at 451 K is 5.33×10^{-3}. A 0.460-g sample of NO_2 is heated to 451 K in a reaction vessel whose volume is 0.500 L. Calculate the pressure of N_2O_4 that forms at equilibrium.

Solution: Since we are given K_p we must first convert the initial quantity of NO_2 to a pressure using $PV = nRT$. At 451 K, $P_{NO_2} = 0.741$ atm. Now we prepare an ICE table, using the relevant chemical equation:

Reaction	$2NO_2(g) \leftrightarrows$	$N_2O_4(g)$
Initial	0.741 atm	0
Change	$-2x$	$+x$
Equilibrium	0.741 atm – $2x$	X

The initial line shows the pressures of the components before any reaction has taken place. The initial pressure of NO_2 is not given explicitly in the problem, so we did a $PV = nRT$ calculation to convert the actual data given. Before any reaction takes place there is no N_2O_4. Let $x = P_{N_2O_4}$ that forms at equilibrium. Thus the change in the N_2O_4 pressure is $+x$ and the change in the P_{NO_2} is $-2x$ according to the coefficient of the equation. Now we add the I and C lines to get the E line. In order for us to solve for x, the problem must give us an additional piece of information. In this case the information is the value of K_p.

We now substitute the equilibrium pressures in the expression for K_p to solve for x

$$K_p = \frac{P_{N_2O_4}}{P^2_{NO_2}} = \frac{x}{(0.741 - 2x)^2} = 5.33 \times 10^{-3}$$

Using the quadratic formula, we find $x = 2.89 \times 10^{-3}$ atm, which as the E line in the table shows is the pressure of N_2O_4.

14.35 When phosgene, $COCl_2$, is heated to 600 K, it partially decomposes to form CO and Cl_2. The value of K_p for this process is 4.10×10^{-3}. Calculate the pressures at equilibrium of the components of the system produced by a starting pressure of phosgene of 0.124 atm.

Solution: The ICE table for this problem is

	$COCl_2(g) \leftrightarrows$	$CO(g) +$	$Cl_2(g)$
Initial	0.124 atm	0	0
Change	$-x$	$+x$	$+x$
Equilibrium	0.124 atm $- x$	X	X

The additional piece of information is that $K_p = 4.10 \times 10^{-3}$ for the reaction. We use this value and the quadratic formula to find that $x = 2.06 \times 10^{-2}$ atm. We can then use this value of x and the expressions for the pressure expression for each component on the E line to find the required pressures.

14.36 At 300 K the K_p for the reaction $2NH_3(g) \leftrightarrows N_2(g) + 3H_2(g)$ is 1.7×10^{-6}.

a. Calculate the composition of the equilibrium state reached from an initial pressure of NH_3 of 1.0 atm

b. Calculate the fraction of NH_3 that decomposes

14.37 A mixture of $NH_3(g)$ at $P = 0.50$ atm and $H_2(g)$ at $P = 0.40$ atm is placed in a reaction vessel at 300 K. Using the data given in Prob. 14.36, find the pressure of $N_2(g)$ at equilibrium.

14.38 A system at equilibrium at 700 K contains 0.18 atm of $H_2(g)$, 0.29 atm of $CH_4(g)$, and some graphite. An additional 0.11 atm of H_2 is introduced. Calculate the composition of the system when it returns to equilibrium.

14.39* An equilibrium mixture of $H_2(g) + I_2(g) \rightleftarrows 2HI(g)$ contains 3 mol of H_2, 2 mol of I_2, and 2 mol of HI in a 1-L vessel. What amount of I_2 must be added at constant temperature to have half the added I_2 react to form HI? Let X equal moles of I_2 added. Set up the equilibrium expression but do not solve.

14.40* To the system $LaCl_3(s) + H_2O(g) \rightleftarrows LaClO(s) + 2HCl(g)$, already at equilibrium, we add more water vapor without changing either the temperature or the volume of the system. When equilibrium is reestablished, the pressure of water vapor is found to have been doubled. Hence, the pressure of HCl present in the system has been multiplied by what factor?

Solution:

$$K = \frac{[HCl]^2}{[H_2O]}$$

Assume that in the first equilibrium $[HCl] = a$ and $[H_2O] = b$. Then $K = a^2/b$. Since T is constant, K will have the value a^2/b in the second equilibrium.

Let X equal the factor by which the concentration of HCl is multiplied in the second equilibrium; since at constant volume and temperature the concentration of a gas is directly proportional to its partial pressure, this will be the factor by which its partial pressure is multiplied. Then at the second equilibrium, $[HCl] = Xa$ and $[H_2O] = 2b$

$$K = \frac{[Xa]^2}{2b} = \frac{X^2a^2}{2b}, \text{ but } K = a^2/b$$

Therefore, $\frac{X^2a^2}{2b} = \frac{a^2}{b}$. Solving, we have $X = \sqrt{2} = 1.41$.

14.41* The equilibrium mixture $SO_2(g) + NO_2(g) \leftrightarrows SO_3(g) + NO(g)$ was found to contain 0.60 mol of SO_3, 0.40 mol of NO, 0.80 mol of SO_2, and 0.10 mol of NO_2. One mole of NO was then forced into the reaction vessel, temperature and volume being kept constant. Calculate the amount of each gas in the new equilibrium mixture.

Solution: First calculate the equilibrium constant. Its value is 3.0. We can see from the equation $SO_2 + NO_2 \leftrightarrows SO_3 + NO$ that if the concentration of NO is increased, the equilibrium will be shifted to the left. Let X be the additional amount of SO_2 formed as a result of this shift. There will then be $0.8 + X$ moles of SO_2. But when X new moles of SO_2 is formed, X new moles of NO_2 will also be formed; X moles of SO_3 and X moles of NO will be used up. There will, therefore, be present in the new mixture $0.8 + X$ moles of SO_2, $0.1 + X$ moles of NO_2, $0.6 - X$ moles of SO_3, and $1 + 0.4 - X$, or $1.4 - X$, moles of NO. If we insert these values in the equilibrium formula, for which the constant 3 has been calculated, we have

$$\frac{[SO_3] \times [NO]}{[SO_2] \times [NO_2]} = \frac{(0.6 - X) \times (1.4 - X)}{(0.8 + X) \times (0.1 + X)} = 3.0$$

Solving, we have $X = 0.12$ mol. Substituting this value of X, we find the amounts of the four reactants to be: SO_3, 0.48 mol; NO, 1.3 mol; SO_2, 0.92 mol; and NO_2, 0.22 mol.

14.42* At a given temperature at equilibrium, a 1-L vessel contained $COCl_2$ at 0.60 atm, CO at 0.30 atm, and Cl_2 at 0.10 atm. $CO(g) + Cl_2(g) \leftrightarrows COCl_2(g)$. An amount of Cl_2 at a partial pressure of 0.40 atm was added to the vessel at constant temperature and volume. Calculate the pressure of CO in the new equilibrium system.

14.43* The equilibrium constant for the reaction

$$CO(g) + H_2O(g) \rightleftarrows CO_2(g) + H_2(g)$$

is 4.00 at a given temperature. Carbon monoxide at 0.400 atm and steam at 0.600 atm were brought together at this temperature. What is the pressure of CO_2 when the system reaches a state of equilibrium?

14.44* A reaction vessel at 27 °C contains a mixture of SO_2 and O_2 in which the partial pressures of SO_2 and O_2 are 3.00 atm and 1.00 atm, respectively. When a catalyst is added, the reaction $2SO_2(g) + O_2(g) \rightleftarrows 2SO_3(g)$ occurs. At equilibrium at 27 °C, the total pressure is 3.75 atm. Calculate K_p and K.

14.45* Pure water vapor is present at a pressure of 1.5 atm in a reaction vessel. To the vessel we add, without a change of volume or temperature, excess solid $LaCl_3$. When equilibrium is established, the total pressure in the vessel is found to be 2.0 atm. What is K_p for the reaction?

$$LaCl_3(s) + H_2O(g) \rightleftarrows LaClO(s) + 2HCl(g)$$

14.46* A sample of gas that was initially pure NO_2 is heated to a temperature of 337 °C. The NO_2 partially dissociates according to the equation $2NO_2(g) \rightleftarrows 2NO(g) + O_2(g)$. At equilibrium, the observed density of the gas mixture at 0.750 atm pressure is 0.520 g/L. Calculate K and K_p for this reaction.

14.47* A 1-L reaction vessel in which the reaction $A(g) + B(g) \rightleftarrows AB(g)$ has reached a state of equilibrium at 727 °C contains 0.0200 mol of solid B. The partial pressures of the gaseous reactants in the equilibrium system are: A, 8.20 atm; B, 4.92 atm; and AB, 11.48 atm. Calculate the minimum amount of A that must be added to the above equilibrium system at 727 °C in order that no solid B will be present at equilibrium.

14.48* A reaction vessel at 850 °C contains $SrCO_3(s)$, $SrO(s)$, and $C(s)$ in equilibrium with $CO_2(g)$ and $CO(g)$. The total pressure of the CO_2 and CO is 169 mmHg. K_p for the reaction $SrCO_3(s) \rightleftarrows SrO(s) + CO_2(g)$ is 2.45 mmHg at 850 °C. Calculate K_p for the reaction $C(s) + CO_2(g) \rightleftarrows 2CO(g)$ at 850 °C based on expressing the pressures in mmHg.

14.49* Pure PCl_5 gas is introduced into an evacuated reaction vessel and comes to equilibrium at 250 °C, the reaction being $PCl_5(g) \rightleftarrows PCl_3(g) + Cl_2(g)$ and all substances being gases. The total pressure is 2.00 atm and the mole fraction of the Cl_2 is 0.407.

a. What are the partial pressures of PCl_3 and PCl_5?

b. Calculate K_p for the reaction at 250 °C.

Chapter 15

Acid and Base Equilibria I

All weak electrolytes are incompletely ionized in water solution, the ionization reaching a state of equilibrium as represented by an equation such as

$$HF(aq) \rightleftarrows H^+(aq) + F^-(aq) \tag{15-1}$$

Most of the weak electrolytes we will deal with are either weak acids or weak bases. An *acid* can be defined as a proton [H^+] donor and a *base* as a proton acceptor. This is called the Brønsted definition. With this definition in mind we can recognize that Eq. (15-1) is a convenient simplification of a reaction between an acid and a base in which a proton is transferred from the acid to the base. The acid is HF, hydrofluoric acid, and the base is H_2O. The reaction can also be written as

$$HF(aq) + H_2O \rightleftarrows H_3O^+(aq) + F^-(aq) \tag{15-2}$$

to emphasize this point. In practice it is more convenient to write the reaction of a weak acid with water in the form of Eq. (15-1). But you should always remember that any acid–base reaction is a reaction in which a proton is transferred from the acid, in this case HF, to the base, in this case water. When water is the base we will often omit it from the chemical equation for convenience. So in water solution the symbol H^+ represents protonated water. Sometimes protonated water is written as H_3O^+ to remind us that a proton transfer has occurred.

Since the ionization reaches a state of equilibrium, it can, like all reactions that reach a state of equilibrium, be represented by an equilibrium constant K, called in this instance an *ionization constant*.

$$K_i = \frac{[H^+] \times [F^-]}{[HF]} \tag{15-3}$$

This equilibrium constant is also referred to as K_a, where *a* stands for acid. It is also called the *acid dissociation constant*. The bracketed formulas $[H^+]$, $[F^-]$, and $[HF]$ represent molarities, concentrations in moles per liter. The numerical value of the ionization constant for HF at 25 °C has been determined to be 6.9×10^{-4}.

Weak bases also undergo partial ionization in aqueous solution as the result of an acid–base reaction in which water acts as an acid or proton donor and the weak base acts as a proton acceptor. The most common weak base is ammonia, NH_3. Its ionization is best represented by the equation

$$NH_3(aq) + H_2O \rightleftarrows NH_4^+(aq) + OH^-(aq) \qquad \textbf{(15-4)}$$

and the equilibrium constant for this reaction is

$$K_i = \frac{[NH_4^+] \times [OH^-]}{[NH_3]} \qquad \textbf{(15-5)}$$

This equilibrium constant is also referred to as K_b, where *b* stands for base. It is also called the *base dissociation constant*. You should note that $[H_2O]$ does not appear in the equilibrium-constant expression since its concentration does not undergo any appreciable changes as the result of any of these equilibria.

The anion of an acid is a Brønsted base. Since the strength of a base is determined by its affinity for protons, it follows that *the weaker the acid, the greater the basic strength of its anion*. Therefore, the OH^- ion is the strongest base among the anions of the acids listed in Table A.2 of the appendix. The order of strength of the five strongest bases in this table is

$$OH^- > S^{2-} > AsO_4^{3-} > PO_4^{3-} > CO_3^{2-}$$

The numerical values of the ionization constants for a number of weak electrolytes are given in Table A.2 of the appendix. The smaller the ionization constant, the weaker the electrolyte.

Pure water itself is also slightly ionized:

$$H_2O \rightleftarrows H^+(aq) + OH^-(aq)$$

The concentration of H^+ ions has been shown experimentally to be 1.0×10^{-7} mol/L at 25 °C. The concentration of OH^- ions is the same as the concentration of H^+ ions, 1.0×10^{-7} mol/L; for this reason we say pure water is neutral.

The ionization constant for water is expressed by the familiar equation

$$K_w = [H^+] \times [OH^-] \qquad \textbf{(15-6)}$$

At 25 °C $K_w = 1.0 \times 10^{-14}$, which means that at this temperature for any aqueous solution at equilibrium the product of $[H^+]$ and $[OH^-]$ *must* be 1.0×10^{-14}.

If the H^+ concentration is greater than 10^{-7} mol/L (for example, 10^{-6}), the solution is acidic, while if the hydrogen-ion concentration is smaller than 10^{-7} mol/L (for example, 10^{-8}), the solution is alkaline.

We shall find it convenient in many of our treatments of acid–base systems to make use of the symbol *p*. The symbol *p* precedes a quantity and means "take the negative logarithm of that quantity." For example, pH is the negative logarithm of $[H^+]$, pOH is the negative logarithm of $[OH^-]$, and pK is the negative logarithm of the equilibrium constant. The negative logarithm of a very small number is usually a much simpler expression than an exponential.

To convert an exponential to its negative logarithm we use a calculator or the relationship $-\log(a \times 10^{-b}) = -(\log a + \log 10^{-b}) = b - \log a$.

For example, if

$$[H^+] = 2.0 \times 10^{-4}, \text{pH} = -\log[H^+]$$
$$= -\log(2.0 \times 10^{-4}) = 4 - \log 2 = 3.7.$$

A solution whose pH is 7 is neutral. A solution whose pH is greater than 7 is alkaline, while one whose pH is less than 7 is acidic. It is important to remember that an increase in pH represents a decrease in hydrogen-ion concentration, that is, a decrease in acidity.

PROBLEMS

15.1 Calculate the OH^- concentration in mol of OH^-/L of a solution that contains 1.0×10^{-2} mol of H^+/L. Will the solution be neutral, acidic, or alkaline?

Solution: In any water solution the product of the concentration of H^+ and the concentration of OH^-, when these concentrations are expressed in moles per liter, is always equal to 1×10^{-14}. That is,

$$[H^+] \times [OH^-] = 1.0 \times 10^{-14}$$
$$[OH^-] = \frac{1.0 \times 10^{-14}}{[H^+]}$$
$$= \frac{1.0 \times 10^{-14}}{1.0 \times 10^{-2} \text{ mol H}^+/\text{L}}$$
$$= 1.0 \times 10^{-12} \text{ mol OH}^-/\text{L}$$

If the concentration of H^+ is 1.0×10^{-7} mol of H^+/L, the concentration of OH^- will also be 1.0×10^{-7} mol of OH^-/L, and the solution will be neutral. If the concentration of H^+ is greater than 1.0×10^{-7} mol/L, the solution will be acidic; if less, it will be alkaline.

Since 1.0×10^{-2} mol of H^+/L is a higher concentration than 1.0×10^{-7} mol of H^+/L, the solution will be acidic.

15.2 Calculate the OH^- concentration, in grams of OH^-/L, of a solution containing 1.0×10^{-10} mol of H^+/L.

Solution:

$$[H^+] \times [OH^-] = 1.0 \times 10^{-14}$$

$$[OH^-] = \frac{1.0 \times 10^{-14}}{[H^+]} = \frac{1.0 \times 10^{-14}}{1.0 \times 10^{-10} \text{ mol } H^+/L}$$

$$= 1.0 \times 10^{-4} \text{ mol } OH^-/L$$

$$1.0 \times 10^{-4} \text{ mol } OH^-/L \times \frac{17 \text{ g } OH^-}{1 \text{ mol } OH^-} = 1.7 \times 10^{-3} \text{ g } OH^-/L$$

15.3 Calculate the OH^- concentration, in grams of OH^-/L, of a solution whose H^+ concentration is

a. 1.0×10^{-6} mol of H^+/L

b. 3.0×10^{-4} g of H^+/L

15.4 Calculate the H^+ concentration, in grams of H^+/L, of a solution whose OH^- concentration is

a. 2.0×10^{-5} mol of OH^-/L

b. 3.4×10^{-2} g of OH^-/L

15.5 Calculate the pH of a solution that contains 1.0×10^{-5} mol of H^+/L.

Solution: By definition, pH is the negative logarithm of the hydrogen-ion concentration when this concentration is expressed in moles of H^+/L.

$$pH = -\log[H^+] = -\log 10^{-5} = 5.0$$

15.6 Calculate the pH of a solution that contains 3.0×10^{-4} mol of H^+/L.

Solution:

$$pH = \log \frac{1}{[H^+]} = \log \frac{1}{3.0 \times 10^{-4}} = \log \frac{10^4}{3}$$

$$pH = \log \frac{10^4}{3.0} = \log 10^4 - \log 3.0$$

$$\log 10^4 = 4$$

$$\log 3.0 = 0.477$$

$$pH = 4 - 0.477 = 3.5$$

15.7 Calculate the pH of a solution that contains
a. 1.0×10^{-8} mol of H^+/L
b. 0.0020 g of H^+/L
c. 0.0030 mol of H^+/L
d. 0.00017 g of OH^-/100 cm^3
e. 2.0×10^{-3} mol of OH^-/L

15.8 Calculate the H^+ concentration in moles of H^+/L, of a solution whose pH is 5.0.

Solution: pH is, by definition, the negative of the logarithm of the H^+ concentration. Since the pH is 5.0, $[H^+]$ must be $10^{-5.0}$ mol/L.

15.9 Calculate the H^+ concentration in mol/L of a solution whose pH is 4.80.

Solution: Since pH is 4.80, $[H^+]$ must be $10^{-4.80}$ mol/L. But $10^{-4.80} = 10^{-5} \times 10^{0.20}$

$$10^{0.20} = 1.59 \text{ (log } 10^{0.20} = 0.20 \text{ and antilog of } 0.20 = 1.59)$$

Therefore, $10^{-4.80} = 1.59 \times 10^{-5}$

$$[H^+] = 1.6 \times 10^{-5} \text{ mol/L}$$

15.10 Calculate the H^+ concentration in moles of H^+/L of a solution whose
a. pH is 1.5
b. pH is 13.6

15.11 Calculate the OH^- concentration in moles of OH^-/L of a solution whose
a. pH is 3.6
b. pH is 6.2

15.12 Which is more strongly acidic?
a. a solution with a pH of 2 or
b. a solution containing 0.020 g of H^+ per liter

THE CONCEPT OF FORMALITY AND A SECOND DEFINITION OF MOLARITY

We learned in Chapter 11 that 1.6 M K_2CO_3 is prepared by dissolving 1.6 mol of K_2CO_3 in enough water to give 1 L of solution. But we know that when K_2CO_3 is dissolved in water, it immediately dissociates completely into K^+ and CO_3^{2-} ions; furthermore, the CO_3^{2-} ions react to a limited extent with water to form HCO_3^-, H_2CO_3, H^+, and OH^-. Accordingly, a solution formed by dissolving 1.6 mol of K_2CO_3 in enough water to form 1 L of solution in fact contains several species, none of which has a concentration of precisely 1.6 mol/L.

To provide for this situation the concepts of *formality* and *formal solutions* have been introduced. According to these concepts the term *formality* (abbreviated F) is used to designate *the amount of solute that is used in preparing 1 liter of solution*; the term *molarity* is reserved for designating *the actual concentration, in moles per liter of solution, of a particular species that is present in the solution*. Thus the above solution would be 1.6 F in K_2CO_3, but it would be 3.2 M in K^+ and slightly less than 1.6 M in CO_3^{2-}.

Formality, as defined above, is synonymous with *molarity* as that term was defined and used in Chapter 11. From this point on in this book we will use the concept of formality and with it the new definition of molarity. *Formality*, abbreviated F, will always designate *the amount of solute used in preparing 1 liter of solution*; there should never be any ambiguity about the meaning of the notations 0.25 F Na_2CO_3 and 0.10 F $HC_2H_3O_2$. The concentration of a particular species in a solution will always be expressed in terms of *molarity*, denoted by the abbreviation M; *mola rity* in this usage *represents the amount of the particular species present in 1 liter of solution*.

PROBLEMS

(See Table A.2 of the appendix for ionization constants.)

15.13 A 0.010 F solution of $HC_2H_3O_2$ (acetic acid) is 4.17% ionized. Calculate the ionization constant of $HC_2H_3O_2$.

Solution: We will use the abbreviation OAc (acetate) for $C_2H_3O_2$.

$$HOAc(aq) \rightleftarrows H^+(aq) + OAc^-(aq)$$

$$K = \frac{[H^+] \times [OAc^-]}{[HOAc]}$$

We express 4.17% in decimal form as 0.0417.

$$0.0417 \times 0.010 \text{ mol} = 0.00042 \text{ mol HOAc ionized}$$

Since 1 mol of HOAc yields 1 mol of H^+ and 1 mol of OAc^-, the 4.2×10^{-4} mol of HOAc will yield 4.2×10^{-4} mol each of H^+ and OAc^-; $[H^+]$ and $[OAc^-]$ will each be 4.2×10^{-4} mol/L. The concentration of un-ionized HOAc molecules will then be $0.010 - 0.00042$ or 0.0096 mol/L. Substituting these values in the equilibrium formula, we have

$$K = \frac{4.2 \times 10^{-4} \times 4.2 \times 10^{-4}}{9.6 \times 10^{-3}}$$

$$= 1.8 \times 10^{-5}$$

In calculations of the ionization constants of acids and bases in water solution, the small concentrations of H^+ or OH^- ions due to the ionization of water are generally ignored.

While it should be easy to solve this problem by this kind of reasoning, it can also be approached using an ICE table. The 4.17% ionization is the piece of information that allows us to solve for x on the C line.

15.14 From the facts given calculate the ionization constant of each substance:
a. A 0.10 F solution of NH_3 is 1.3% ionized.
b. A 0.0010 F solution of HOAc is 12.6% ionized.
c. A 0.01 F solution of HCN is 0.02% ionized.

15.15 The ionization constant of hypoiodous acid, HIO, is 2.3×10^{-11}. Find the percent ionization of a 0.10 F solution. Find the percent ionization of a 0.0010 F solution.

15.16 The ionization constant for HCN is 4×10^{-10} at 25 °C. Calculate the formality and the H^+ ion concentration of a solution of HCN that is 0.010% ionized.

Solution:

$$HCN(aq) \rightleftarrows H^+(aq) + CN^-(aq)$$

$$K = \frac{[H^+] \times [CN^-]}{[HCN]} = 4 \times 10^{-10}$$

Let X = formality. Since the solution is 0.010% ionized, and since 0.010% expressed as a decimal is 0.00010, the concentration of H^+ will be $0.00010X$. The concentration of CN^- will also be $0.00010X$. The concentration of un-ionized HCN will be $X - 0.00010X$. Substituting these values in the above equilibrium formula, we have

$$\frac{0.00010X \times 0.00010X}{X - 0.00010X} = 4 \times 10^{-10}$$

Solving, we obtain $X = 4 \times 10^{-2}$ = the formality.

$$[H^+] = 0.00010X = 1 \times 10^{-4} \times 4 \times 10^{-2}$$
$$= 4 \times 10^{-6}\ M$$

Although the data given in this problem are somewhat different than in earlier problems, we can still use an ICE table to help set up the solution. Notice that we can ignore the very small (10^{-7}) concentration of H^+.

	$HCN \leftrightarrows$	H^+ +	CN^-
Initial	X	0	0
Change	$-0.00010X$	$+0.00010X$	$+0.00010X$
Equilibrium	$X - 0.00010X$	$0.00010X$	$0.00010X$

15.17 The ionization constant for NH_3 is 1.8×10^{-5}. Calculate the formality and OH^- concentration of a solution in which the NH_3 is 1.3% ionized.

15.18 The ionization constant for HOAc is 1.8×10^{-5}. Calculate the hydrogen-ion concentration of 0.01 *F* HOAc.

Solution:

$$HOAc(aq) \leftrightarrows H^+(aq) + OAc^-(aq)$$

Let $X = [H^+] = [OAc^-]$.

$$0.01 - X = [HOAc]$$

$$K = \frac{[H^+] \times [OAC^-]}{[HOAc]} = 1.8 \times 10^{-5}$$

Substituting the values of $[H^+]$, $[OAc^-]$, and [HOAc] in the equilibrium formula, we have

$$\frac{X^2}{0.01 - X} = 1.8 \times 10^{-5}$$

The term X can be dropped from the expression $0.01 - X$ in the denominator if the value of X is so small that, within the limits imposed by the number of allowable significant figures, $0.01 - X = 0.01$. We can estimate the value of X as follows: Drop X from the expression, $0.01 - X$. As a result, $X^2 = 1.8 \times 10^{-7}$, X is about 4×10^{-4} or 0.0004, and $0.01 - 0.0004 = 0.0096$. When rounded off to one significant figure, 0.0096 becomes 0.01. Therefore, X can be dropped safely from the denominator.

$$\frac{X^2}{0.01} = 1.8 \times 10^{-5}$$

$$X = 4 \times 10^{-4}\,M = [H^+]$$

This example illustrates the rule that a term n *that is added to or subtracted from* a term m in an expression $m + n$ or $m - n$ can be dropped if it is so small that $m - n$ or $m + n$, when rounded off to the permissible number of significant figures, is equal to m.

It should be emphasized that a small term can be dropped only in expressions involving its *addition to* or *subtraction from* a large term, never in an expression involving its multiplication or division by a large term.

15.19 The ionization constant for NH_3 is 1.8×10^{-5}. Calculate the hydroxide-ion concentration of 0.10 F NH_3.

15.20 A 0.0010 F solution of HF has a pH of 4.00. Calculate the percent ionization of the HF.

15.21 Calculate the pH of a solution prepared from 0.14 mol of formic acid and enough water to make 1.0 L of solution.

15.22 Calculate the $[H^+]$ in a solution prepared from 0.86 mol of ammonia and enough water to make 0.25 L of solution.

15.23 A solution prepared from 0.13 mol of an organic acid and enough water to form 1 L of solution is found to have a pH of 5.45. Find K_a of the acid.

15.24 Calculate the fraction of HNO_2 that dissociates in a solution that has an initial $[HNO_2] = 0.115$ F at the following pHs: (a) 7, (b) 4, (c) 1.

CONJUGATE ACIDS AND BASES

Since a base can be defined as a proton acceptor, it follows that the anion derived from an acid is a base. This point is underscored by the common practice of referring to the species remaining after an acid donates a proton as the *conjugate base* of the acid. For example, the conjugate base of acetic acid HOAc is the acetate ion OAc^-. The conjugate base of an acid can accept a proton from an acid, for example water. In aqueous solution, an equilibrium is established in which water acts as an acid and donates a proton to the conjugate base of the acid.

$$OAc^-(aq) + H_2O \rightleftarrows HOAc(aq) + OH^-(aq) \qquad \textbf{(15-7)}$$

Although the base is an anion, this reaction is no different from the other acid–base reactions we have discussed so far. It is a proton-transfer reaction from an acid (water) to a base (the anion). Salts in which the anion is acetate (or the conjugate base of any weak acid) and the cation is not the conjugate acid of a weak base will form basic solutions when they dissolve in water.

An analogous situation arises with the species formed when a base accepts a proton. For example, when NH_3 is dissolved in water, some NH_4^+ is formed. This species is called the *conjugate acid* of the base, since it is a proton donor. The NH_4^+ cation can donate a proton to a base. For example in aqueous solution, the reaction is

$$NH_4^+(aq) + H_2O \rightleftarrows NH_3(aq) + H_3O^+(aq) \qquad \textbf{(15-8)}$$

or written more simply,

$$NH_4^+(aq) \rightleftarrows NH_3(aq) + (H)^+(aq) \qquad \textbf{(15-9)}$$

Thus salts in which the cation is ammonium (or the conjugate acid of any weak base) and the anion is not the conjugate base of a weak acid will form acidic solutions when they dissolve in water.

The values of the equilibrium constants for these types of processes are found from the corresponding K_a or K_b of the neutral member of the conjugate acid–base pair. For Eq. (15-7)

$$K = \frac{[\text{HOAc}][\text{OH}^-]}{[\text{OAc}^-]} \tag{15-10}$$

and for Eq. (15-8)

$$K = \frac{[\text{NH}_3][\text{H}^+]}{[\text{NH}_4^+]} \tag{15-11}$$

The numerical value of the equilibrium constant for a particular ion can be calculated from the ion product constant for water and the ionization constant for the conjugate neutral weak acid or weak base in the following way:

In Eq. (15-10) we multiply both numerator and denominator by $[H^+]$, obtaining thereby

$$K = \frac{[\text{HOAc}]}{[\text{H}^+][\text{OAc}^-]} \times [\text{OH}^-][\text{H}^+] \tag{15-12}$$

The first term to the right of the equal sign is the reciprocal of the acid dissociation constant K_a for HOAc and the rest of the expression is the ion-product constant K_w for water.

If we substitute the numerical values of K_w and K_a we obtain the numerical value of this K.

$$K = \frac{K_w}{K_a} = \frac{1.0 \times 10^{-14}}{1.8 \times 10^{-5}} = 5.6 \times 10^{-10} \tag{15-13}$$

You can see from these calculations that *the weaker the acid* (or base), *the stronger its conjugate base* (or acid).

If the salt is derived from a weak acid and a weak base, both the cation and the anion will donate or accept a proton in aqueous solution. Thus for the salt NH_4F the reactions are

$$\text{NH}_4^+(aq) \rightleftarrows \text{NH}_3(aq) + \text{H}^+(aq) \tag{15-14}$$

$$\text{F}^-(aq) + \text{H}_2\text{O} \rightleftarrows \text{HF}(aq) + \text{OH}^-(aq) \tag{15-15}$$

The H^+ and OH^- shown as products in Eqs. (15-14) and (15-15) will react:

$$\text{H}^+(aq) + \text{OH}^-(aq) \rightleftarrows \text{H}_2\text{O} \tag{15-16}$$

Equations (15-14), (15-15), and (15-16), when added, give the net equation for the proton transfer,

$$NH_4^+(aq) + F^-(aq) \rightleftarrows NH_3(aq) + HF(aq) \qquad \textbf{(15-17)}$$

The equilibrium constant for this reaction is

$$K = \frac{[NH_3][HF]}{[NH_4^+][F^-]} \qquad \textbf{(15-18)}$$

By multiplying both numerator and denominator by $[H^+] \times [OH^-]$, we resolve Eq. (15-18) into three separate equilibria,

$$K = \frac{[NH_3]}{[NH_4^+][OH^-]} \times \frac{[HF]}{[H^+][F^-]} \times [H^+][OH^-] \qquad \textbf{(15-19)}$$

$$K = \frac{1}{K_{NH_3} K_{HF}} \times K_w = \frac{K_w}{K_{NH_3} K_{HF}} \qquad \textbf{(15-20)}$$

By substituting in Eq. (15-20) the numerical values of the three constants, we can calculate the numerical value of K to be 8.0×10^{-7}.

OTHER EQUILIBRIA OF WEAK ELECTROLYTES

The technique of resolving a given equilibrium constant into its component constants that was used in calculating reactions of conjugate acids and bases can be applied to other systems. To illustrate, when sodium formate ($NaCHO_2$) is added to a solution of hydrocyanic acid, the following equilibrium is set up.

$$CHO_2^-(aq) + HCN(aq) \rightleftarrows HCHO_2(aq) + CN^-(aq)$$

The equilibrium constant for this reaction is

$$K = \frac{[HCHO_2] \times [CN^-]}{[CHO_2^-] \times [HCN]}$$

By multiplying both numerator and denominator by $[H^+]$ we obtain

$$K = \frac{[HCHO_2]}{[H^+] \times [CHO_2^-]} \times \frac{[H^+] \times [CN^-]}{[HCN]}$$

$$K = \frac{1}{K_{HCHO_2}} \times K_{HCN}$$

$$= \frac{4.0 \times 10^{-10}}{2.1 \times 10^{-4}} = 1.9 \times 10^{-6}$$

PROBLEMS

15.25 Calculate the OH^- concentration of 0.2 *F* KCN.

Solution: The net equation for the proton transfer from water is

$$CN^-(aq) + H_2O \rightleftarrows HCN(aq) + OH^-(aq)$$

$$K = \frac{[HCN] \times [OH^-]}{[CN^-]} = \frac{[HCN]}{[H^+] \times [CN^-]} \times [H^+] \times [OH^-]$$

$$= \frac{K_w}{K_{HCN}} = \frac{1.0 \times 10^{-14}}{4.0 \times 10^{-10}} = 2.5 \times 10^{-5}$$

Let $X = [OH^-] = [HCN]$

$$0.2 - X = [CN^-]$$

$$\frac{X^2}{0.2 - X} = 2.5 \times 10^{-5}$$

$$X^2 = 5 \times 10^{-6}$$

$$X = 2 \times 10^{-3}\, M = [OH^-]$$

15.26 Calculate the pH of 0.10 *F* $KCHO_2$ (potassium formate).

15.27 Calculate the H^+ ion concentration of 0.10 *F* NH_4Cl.

15.28 Calculate the concentration of HCN and of OH^- in 0.20 *F* NH_4CN.

15.29 In a 0.5 *F* solution of KClO, the OH^- ion concentration is 3×10^{-4} *M*. What is the ionization constant for HClO?

15.30 Find the pH and the pOH of a solution prepared from 0.25 mol of NH_4Cl and enough water to make 1 L of solution.

15.31 A solution of the sodium salt of ascorbic acid (HAscorbate), commonly known as Vitamin C, is prepared from 0.025 mol of the salt in enough water to form 0.50 L of solution. The $[OH^-]$ is found to be 2.5×10^{-6} *M*. Calculate K_a of ascorbic acid.

15.32 Calculate the amount of sodium acetate that must be dissolved in 0.25 L of water to produce a solution of pH = 8.90.

15.33 Calculate the amount of ammonium chloride that must be dissolved in 0.50 L of water to produce a solution of pH = 4.79.

15.34 Calculate the amount of ammonia that must be dissolved in 1.0 L of a solution prepared from 0.087 mol of ammonium sulfate to produce a solution with pH = 8.88.

15.35 The ionization constant for HOAc is 1.8×10^{-5}. What amount of hydrogen ions will there be in 1.0 L of 0.10 *F* HOAc containing 0.20 mol of NaOAc?

Solution:

$$HOAc(aq) \rightleftarrows H^+(aq) + OAc^-(aq)$$

$$\text{Let } X = \text{concentration of } H^+ \text{ at equilibrium}$$

$$0.20 + X = \text{concentration of } OAc^- \text{ at equilibrium}$$

$$0.10 - X = \text{concentration of HOAc at equilibrium}$$

$$K = \frac{[H^+][OAc^-]}{[HOAc]} = 1.8 \times 10^{-5}$$

Substituting the values of $[H^+]$, $[OAc^-]$, and [HOAc] in the above equation, we have

$$\frac{X(0.20 + X)}{0.10 - X} = 1.8 \times 10^{-5}$$

In solving for X in the above equation we assume that, since acetic acid is weak, the value of X (the concentration of H^+ ions) will be very much less than 0.10. If this assumption is true, X can be dropped from the terms $0.20 + X$ and $0.10 - X$. That leaves the expression

$$\frac{0.20X}{0.10} = 1.8 \times 10^{-5}$$

$$X = 9.0 \times 10^{-6} \text{ mol } H^+/L$$

The fact that the value of X turns out to be very much less than 0.10 means that the assumption made above is justified.

15.36 The ionization constant for NH_3 is 1.8×10^{-5}. What amount of OH^- is there in 1.0 L of 0.10 *F* NH_3 that contains 0.10 mol of NH_4Cl?

15.37 What is the OH^- concentration of a solution prepared by dissolving 0.25 mol of NH_3 and 0.75 mol of NH_4Cl in enough water to make 1.0 L of solution?

15.38 What amount of NaCN must be dissolved in 1.0 L of 0.2 *F* HCN to yield a solution with a hydrogen-ion concentration of 1.0×10^{-6} *M*?

POLYPROTIC ACIDS

For polyprotic acids, such as H_3PO_4, H_2S, and H_2CO_3, the ionization takes place in steps and each step has its own ionization constant. For H_3PO_4 the three steps, with their ionization constants, are

$$H_3PO_4(aq) \rightleftarrows H^+(aq) + H_2PO_4^-(aq) \qquad K_1 = \frac{[H^+] \times [H_2PO_4^-]}{[H_3PO_4]} = 7.5 \times 10^{-3} \qquad \textbf{(15-21)}$$

$$H_2PO_4^-(aq) \rightleftarrows H^+(aq) + HPO_4^{2-}(aq) \qquad K_2 = \frac{[H^+] \times [HPO_4^{2-}]}{[H_2PO_4^-]} = 6.2 \times 10^{-8} \qquad \textbf{(15-22)}$$

$$HPO_4^{2-}(aq) \rightleftarrows H^+(aq) + PO_4^{3-}(aq) \qquad K_3 = \frac{[H^+] \times [PO_4^{3-}]}{[HPO_4^{2-}]} = 1.0 \times 10^{-12} \qquad \textbf{(15-23)}$$

The overall ionization constant for a polyprotic acid is the product of the constants for the separate steps. Thus for H_2S the total ionization is

$$H_2S(aq) \rightleftarrows 2H^+(aq) + S^{2-}(aq)$$

and the two steps are

$$H_2S(aq) \rightleftarrows H^+(aq) + HS^-(aq) \qquad K_1 = 1.0 \times 10^{-7}$$
$$HS^-(aq) \rightleftarrows H^+(aq) + S^{2-}(aq) \qquad K_2 = 1.3 \times 10^{-13}$$

and

$$K_i = K_1 \times K_2 = \frac{[H^+] \times [HS^-]}{[H_2S]} \times \frac{[H^+] \times [S^{2-}]}{[HS^-]} = \frac{[H^+]^2 \times [S^{2-}]}{[H_2S]}$$
$$= 1.0 \times 10^{-7} \times 1.3 \times 10^{-13} = 1.3 \times 10^{-20}$$

PROBLEMS

15.39 Referring to Table A.2 in the appendix for the ionization constants, calculate the concentrations of HCO_3^- and CO_3^{2-} ions in a 0.034 *F* solution of CO_2 in water.

Solution: The reactions that occur when CO_2 is dissolved in water are

$$CO_2(aq) + H_2O \rightleftarrows H_2CO_3(aq) \quad \textbf{(a)}$$
$$H_2CO_3(aq) \rightleftarrows H^+(aq) + HCO_3^-(aq) \quad \textbf{(b)}$$
$$HCO_3^-(aq) \rightleftarrows H^+(aq) + CO_3^{2-}(aq) \quad \textbf{(c)}$$

The ionization constant for reaction (b) is
$\dfrac{[H^+] \times [HCO_3^-]}{[H_2CO_3]} = 4.2 \times 10^{-7}$, usually designated by K_1 (*In this formula $[H_2CO_3]$ represents the total of CO_2 and H_2CO_3 since reaction (a) does not go to completion.)

and that for reaction (c) is

$$\frac{[H^+] \times [CO_3^{2-}]}{[HCO_3^-]} = 4.8 \times 10^{-11}\text{, usually designated by } K_2$$

Since the two equilibrium reactions (b) and (c) occur in the same solution, the $[H^+]$ that appears in the formula for K_1 must be the same $[H^+]$ that appears in the formula for K_2 and this $[H^+]$ must be the sum of the hydrogen ions provided by reactions (b) and (c). But since the equilibrium constant for reaction (c) is so much smaller than that for reaction (b), the amount of H^+ provided by reaction (c) is so very small compared with that provided by reaction (b) that, within the limitations imposed by the number of allowable significant figures, it can be neglected. Accordingly, we will proceed as follows: Referring to reaction (b) and its ionization constant K_1, let $X = [HCO_3^-]$; then $X = [H^+]$ and $0.034 - X = [H_2CO_3]$.

Substituting these values in the equation for K_1, we have

$$\frac{X^2}{0.034 - X} = 4.2 \times 10^{-7}$$

X is so small by comparison with 0.034 that it can be dropped from the expression $0.034 - X$.

$$X^2 = 1.43 \times 10^{-8}$$
$$X = 1.2 \times 10^{-4}\,M = [HCO_3^-] = [H^+]$$

Since K_2 is so much smaller than K_1, the value of $[H^+]$ calculated from K_1 will be the $[H^+]$ of the solution and can be inserted in the formula for K_2. Letting $Y = [CO_3^{2-}]$ and $1.2 \times 10^{-4} = [H^+]$

$$\frac{1.2 \times 10^{-4}\, Y}{1.2 \times 10^{-4} - Y} = 4.8 \times 10^{-11}$$

Since K_2 is very small, Y is so small by comparison with 1.2×10^{-4} that it can be dropped from the term $1.2 \times 10^{-4} - Y$. Therefore

$$\frac{1.2 \times 10^{-4}\, Y}{1.2 \times 10^{-4}} = 4.8 \times 10^{-11}$$

$$Y = 4.8 \times 10^{-11}\, M = [CO_3^{2-}]$$

Note: The equation for the overall ionization is

$$H_2CO_3(aq) \rightleftarrows 2H^+(aq) + CO_3^{2-}(aq)$$

The overall ionization constant, K_i, is

$$K_i = \frac{[H^+]^2 \times [CO_3^{2-}]}{[H_2CO_3]} = K_1 \times K_2 = 2.0 \times 10^{-17}$$

It should be emphasized that this overall ionization equation and constant cannot be used to solve for $[CO_3^{2-}]$, $[HCO_3^-]$, and $[H^+]$ in a solution of CO_2 in pure water. Use of the overall equation assumes that $[H^+] = 2 \times [CO_3^{2-}]$. This is not a valid assumption.

It is worth noting, however, that although the overall ionization constant *cannot* be used to calculate correctly $[H^+]$ and $[CO_3^{2-}]$, the value of this overall ionization constant *must be satisfied*.

Substituting the values $[H^+] = 1.2 \times 10^{-4}\, M$, $[CO_3^{2-}] = 4.8 \times 10^{-11}\, M$, and $[H_2CO_3] = 3.4 \times 10^{-2}\, M$ into the equation for the overall ionization, we see that this is indeed the case.

If the H^+ *concentration is fixed* by the addition of a strong acid to a solution of CO_2 in water, K_i can then be used in solving for $[CO_3^{2-}]$. This case is presented in Prob. 15.43.

15.40 Calculate the concentrations of $H_2PO_4^-$, HPO_4^{2-}, and PO_4^{3-} ions in 0.10 *F* H_3PO_4.

Solution: The three equilibria and their constants are

1. $H_3PO_4(aq) \rightleftarrows H^+(aq) + H_2PO_4^-(aq)$ $\quad K_1 = \dfrac{[H^+] \times [H_2PO_4^-]}{[H_3PO_4]} = 7.5 \times 10^{-3}$

2. $H_2PO_4^-(aq) \rightleftarrows H^+(aq) + HPO_4^{2-}(aq)$ $\quad K_2 = \dfrac{[H^+] \times [HPO_4^{2-}]}{[H_2PO_4^-]} = 6.2 \times 10^{-8}$

3. $HPO_4^{2-}(aq) \rightleftarrows H^+(aq) + PO_4^{3-}(aq)$ $\quad K_3 = \dfrac{[H^+] \times [PO_4^{3-}]}{[HPO_4^{2-}]} = 1.0 \times 10^{-12}$

Since all three equilibria occur in the same solution, the value of $[H^+]$ must be the same in each. Since K_1 is so much larger than K_2 and K_3, it will determine the value of $[H^+]$. Likewise, K_1 will determine the value of $[H_2PO_4^-]$. Accordingly, we first calculate $[H^+]$ and $[H_2PO_4^-]$ from K_1. Let $X = [H^+]$. Then $X = [H_2PO_4^-]$ and $0.10 - X = [H_3PO_4]$.

$$\frac{X^2}{0.10 - X} = 7.5 \times 10^{-3}$$

It is obvious from the size of K_1 that X is too large in comparison with 0.10 to allow it to be dropped. Solving this quadratic, we obtain $X = 2.4 \times 10^{-2}\ M = [H^+] = [H_2PO_4^-]$.

We next substitute these values in K_2, letting $Y = [HPO_4^{2-}]$

$$K_2 = \frac{2.4 \times 10^{-2} \times Y}{2.4 \times 10^{-2} - Y} = 6.2 \times 10^{-8}$$

Since K_2 is very small, Y is so small by comparison with 2.4×10^{-2} that it can be dropped from the term $2.4 \times 10^{-2} - Y$. This leaves

$$\frac{2.4 \times 10^{-2}\ Y}{2.4 \times 10^{-2}} = 6.2 \times 10^{-8}$$

$$Y = 6.2 \times 10^{-8}\ M = [HPO_4^{2-}]$$

We then substitute the calculated values of $[H^+]$ and $[HPO_4^{2-}]$ in K_3, letting $Z = [PO_4^{3-}]$.

$$K_3 = \frac{(2.4 \times 10^{-2})(Z)}{(6.2 \times 10^{-8} - Z)} = 1.0 \times 10^{-12}$$

Since K_3 is extremely small, Z is so small by comparison with 6.2×10^{-8} that it can be dropped from the term $6.2 \times 10^{-8} - Z$.

That leaves

$$\frac{2.4 \times 10^{-2}\,Z}{6.2 \times 10^{-8}} = 1.0 \times 10^{-12}$$

$$Z = 2.5 \times 10^{-18}\,M = [PO_4^{3-}]$$

15.41 Calculate the concentration of S^{2-} in 0.10 $F\,H_2S$.

15.42 Calculate the concentration of CrO_4^{2-} in 0.10 $F\,H_2CrO_4$.

15.43 The hydrogen-ion concentration of a 0.034 F solution of CO_2 in dilute HCl is 0.10 M. Calculate the molar concentration of CO_3^{2-}.

Solution: This problem differs from 15.39 in that H^+ has been added (in the form of HCl) to give a total $[H^+]$ of 0.10 M. Using the first ionization constant for H_2CO_3, we first solve for $[HCO_3^-]$.

$$H_2CO_3(aq) \rightleftarrows H^+(aq) + HCO_3^-(aq)$$

Let $X = [HCO_3^-]$

$0.10 = [H^+]$

$0.034 - X = [H_2CO_3]$

$$K_1 = \frac{[H^+] \times [HCO_3^-]}{[H_2CO_3]} = 4.2 \times 10^{-7} = \frac{0.10X}{0.034 - X}$$

Since K_1 is very small and $[H^+]$ is high (0.10 M), $[HCO_3^-]$ will be very small in comparison with $[H^+]$ and $[H_2CO_3]$. Therefore, we can drop the X in the term $0.034 - X$. That leaves

$$\frac{0.10X}{0.034} = 4.2 \times 10^{-7}$$

$$0.10X = 1.4 \times 10^{-8}$$

$$X = 1.4 \times 10^{-7}\,M = [HCO_3^-]$$

We can now substitute this value of $[HCO_3^-]$ and the value of 0.10 M for $[H^+]$ in the second ionization constant to give

$$K_2 = \frac{[H^+] \times [CO_3^{2-}]}{[HCO_3^-]} = 4.8 \times 10^{-11} = \frac{0.10 \times Y}{1.4 \times 10^{-7} - Y}$$

In this formula $Y = [CO_3^{2-}]$.

Since Y will be very small in comparison with 1.4×10^{-7}, we can drop it from the term $1.4 \times 10^{-7} - Y$. This leaves

$$\frac{0.10Y}{1.4 \times 10^{-7}} = 4.8 \times 10^{-11}$$

$$0.10Y = 6.7 \times 10^{-18}$$

$$Y = 6.7 \times 10^{-17}\,M = [CO_3^{2-}]$$

Note: You should realize that when the hydrogen-ion concentration of a solution of a weak polyprotic acid (such as H_2CO_3, H_2S, or H_3PO_4) is fixed by the addition of a strong acid, the overall ionization constant K_i can be used in solving for $[CO_3^{2-}]$. Thus

$$K_i = K_1 \times K_2 = \frac{[H^+]^2 \times [CO_3^{2-}]}{[H_2CO_3]} = 2.0 \times 10^{-17}$$

If we substitute the values of $[H^+]$ and $[H_2CO_3]$ in this equation, we have

$$\frac{(0.10)^2\,[CO_3^{2-}]}{0.034} = 2.0 \times 10^{-17}$$

$$[CO_3^{2-}] = 6.8 \times 10^{-17}\,M$$

This, you can see, is practically the same value for $[CO_3^{2-}]$ that was obtained when the calculation was made via K_1 and K_2.

15.44 Calculate the sulfide-ion concentration of a 0.10 F solution of H_2S in 0.10 F HCl.

15.45 Calculate the molar concentrations of $H_2AsO_4^-$, $HAsO_4^{2-}$, and AsO_4^{3-} in 0.20 F H_3AsO_4 that is 0.10 F in HCl.

15.46 One liter of solution prepared by dissolving H_2SO_4 in pure water has a pH of 3.00.

a. What amount of H_2SO_4 was dissolved? The first ionization of H_2SO_4 is complete. The ionization constant for HSO_4^- is 1.20×10^{-2}.

b. Calculate the molarity of each species in solution.

Solution: Let X equal the moles of H_2SO_4 dissolved. Let Y equal the moles of SO_4^{2-}.

Then, since the first ionization of H_2SO_4 is complete,

$$\begin{array}{ccccc} X & & X & & X - Y \\ H_2SO_4 & \rightarrow & H^+(aq) & + & HSO_4^-(aq) \\ X - Y & & Y & & Y \end{array} \qquad \textbf{(a)}$$

$$HSO_4^-(aq) \rightleftarrows H^+(aq) + SO_4^{2-}(aq) \qquad \textbf{(b)}$$

$$X + Y = 1.00 \times 10^{-3} = [H^+] \qquad \textbf{(c)}$$

$$\frac{Y}{(X - Y)} = \frac{[SO_4^{2-}]}{[HSO_4^-]} = \frac{K_2}{[H^+]} = 12.0 \qquad \textbf{(d)}$$

From (c) and (d) we find that

$$X = 5.20 \times 10^{-4} \text{ mol} = \text{moles of } H_2SO_4 \text{ dissolved}$$

$$Y = 4.80 \times 10^{-4} \text{ mol/L} = [SO_4^{2-}]$$

$$X - Y = 4.0 \times 10^{-5} \text{ mol/L} = [HSO_4^-]$$

$$\frac{1.00 \times 10^{-14}}{1.00 \times 10^{-3}} = 1.00 \times 10^{-11} = [OH^-]$$

15.47 Find the amount of HCl that must be added to a solution prepared from 0.20 mol of H_2S and 1.0 L of water in order to obtain $[S^{2-}] = 2.4 \times 10^{-19}$ *M*.

15.48* Calculate the *K* of the proton transfer reaction of HCO_3^- with itself and use this reaction to calculate the pH of a 0.10 *M* solution of sodium bicarbonate.

Chapter 16

Acids and Bases II

NEUTRALIZATION AND EQUIVALENTS

In the most general sense a *neutralization* is a reaction between an acid and a base. In a more limited sense the term *neutralization* is applied by chemists to the reaction between an acid stronger than water and a base stronger than water, carried out with stoichiometric quantities of each. The neutralization procedure is used very often in the chemical laboratory and can provide important information about the composition of samples of acid or base.

We can divide neutralization reactions into four types based on the strengths of the acid and the base.

1. Strong acid and strong base:

$$H^+(aq) + OH^-(aq) \rightleftarrows H_2O \quad K = \frac{1}{K_w} = 10^{14}$$

2. Strong acid and weak base:

$$H^+(aq) + NH_3(aq) \rightleftarrows NH_4^+(aq) \quad K = \frac{K_b}{K_w}$$

3. Weak acid and strong base:

$$HCN(aq) + OH^-(aq) \rightleftarrows H_2O + CN^-(aq) \quad K = \frac{K_a}{K_w}$$

4. Weak acid and weak base:

$$HCN(aq) + NH_3(aq) \rightleftarrows NH_4^+(aq) + CN^-(aq) \quad K = K_a\frac{K_b}{K_w}$$

All these reactions have the same fundamental stoichiometry. The amount of H^+ donated by the acid is equal to the amount of H^+ accepted by the base. Therefore, if we know the amount of acid we have, we can find the amount of base. If instead we know the amount of base we have, we can find the amount of acid. But there is one additional complication. Many acids can donate more than 1 mol of H^+ for each mole of acid and many bases can accept more than 1 mol of H^+ for each mole of base. This information must also be included when we find stoichiometric relationships in neutralization reactions.

For example, suppose we find that a sample of HCl requires 1 mol of NaOH for neutralization. We can then say that we have 1 mol of HCl. But if a sample of H_2SO_4 requires 1 mol NaOH for neutralization, then we have only $\frac{1}{2}$ mol of H_2SO_4. If a sample of H_3PO_4 requires 1 mol NaOH for neutralization, then we have only $\frac{1}{3}$ mol of H_3PO_4. In a similar way, we can say that 1 mol of NaOH or $\frac{1}{2}$ mol of $Ba(OH)_2$ or $\frac{1}{3}$ mol of $Al(OH)_3$ will be neutralized by 1 mol HCl.

We can conveniently approach the calculations of neutralization by the use of a quantity called an *equivalent*, which is closely related to the mole. One equivalent of an acid is 1 mol of acid divided by the number of moles of H^+ available per mole of acid. One equivalent of a base is 1 mol of base divided by the amount of H^+ that can be accepted by 1 mol of base, or in the case of hydroxides, the number of OH^- groups per mole of base. For example, 1 mol of H_2SO_4 has a mass of 98 g, so 1 equivalent has a mass of $98/2 = 49$ g; 1 mol of $Al(OH)_3$ has a mass of 78 g, so 1 equivalent has a mass of $78/3 = 26$ g. Just as the mass of 1 mol is called the molecular weight, the mass of one equivalent is called the *equivalent weight.*

By analogy with molarity and formality, a unit of concentration called *normality* (*N*) is defined as the number of equivalents of solute per liter of solution. From the definition of an equivalent it follows that *one equivalent of an acid reacts with one equivalent of a base*.

$$\text{equivalents}_a = \text{equivalents}_b \qquad \textbf{(16-1)}$$

We can write a number of relationships that we can use to handle the stoichiometry of neutralization. Since number of equivalents equals mass of substance divided by equivalent weight,

$$\frac{m_a}{\text{eq.wt.}_a} = \frac{m_b}{\text{eq.wt.}_b} \qquad \textbf{(16-2)}$$

Also, equivalents $=$ normality $\times$ volume of solution, so that for a solution of acid reacting with a solution of base

$$N_a V_a = N_b V_b \qquad \textbf{(16-3)}$$

Equations (16-1), (16-2), (16-3), or appropriate combinations of them can be used to simplify neutralization calculations.

In the laboratory we generally carry out neutralizations by a procedure called *titration*, in which a measured volume of one solution is carefully added to a known amount of the other solution. In a typical procedure we put a measured volume of a solution of base of unknown concentration into a flask. We then add a solution of acid of known concentration slowly from a *buret*, a graduated tube with a stop-cock at the bottom. When the required number of equivalents of acid has been added, we stop the addition and read the volume of added solution from the buret. Since this procedure gives us two volumes and one concentration, we can calculate the unknown concentration.

We still need a method of detecting the *equivalence point*, the point at which the required number of equivalents has been added and the addition should be stopped. The most common method is one in which we monitor the pH of the solution, stopping the addition when the pH of the solution reaches the pH at the equivalence point. The pH at the equivalence point is determined by the nature of the acid and the base and the concentration of the solution. If both the acid and base are strong, the pH will be 7 at the equivalence point. But if the acid is strong and the base is weak, the pH will be below 7, while if the acid is weak and the base is strong, the pH will be above 7 at the equivalence point. We can calculate the value of the pH from the value of K_a or K_b (see Probs. 16.17–16.26).

One way to monitor the pH is to use a substance called an *indicator*, which is a dye whose color depends on pH. Many indicators are weak acids whose dissociation can be represented as $HIn \rightleftarrows H^+ + In^-$. HIn and In^- have different colors and therefore the color of the solution depends on the ratio $[In^-]/[HIn]$, which depends on $[H^+]$, since

$$K_{In} = \frac{[In^-][H^+]}{[HIn]} \quad \text{or} \quad \frac{[In^-]}{[HIn]} = \frac{K_{In}}{[H^+]}$$

At some pH of the solution the indicator will change from the color of HIn to the color of In^-. At this point $[In^-] = [HIn]$, $[In^-]/[HIn] = 1$, and $[H^+] = K_{In}$ or pH = pK_{In}. When monitoring a neutralization with an indicator, it is desirable to choose an indicator whose pK_{In} is close to the pH at the equivalence point.

At first it may seem surprising to us that something as imprecise as a color change can be used to monitor the pH of a solution during titration. Not only are our eyes not very accurate instruments for measuring color changes, but we generally cannot find an indicator that changes color precisely at the equivalence point of the titration. If we calculate the pH at various points during the course of a titration (Probs. 16.25 and 16.26), we find that the pH changes very slowly with the addition of solution from the buret until we get close to the equivalence point. Then the pH changes very rapidly. A reasonably small error in estimating the pH around the equivalence point causes a negligible error in the volume of added solution.

The changes in pH that take place as a titration is carried out can conveniently be represented graphically. A titration curve is a plot of pH against volume of added

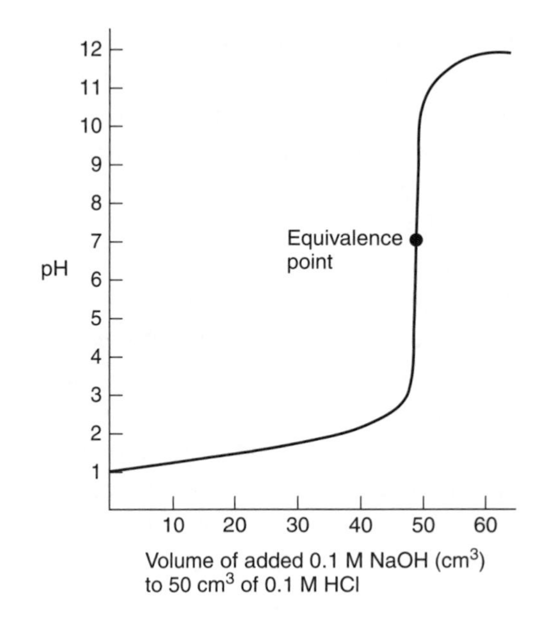

titrant. In the curve shown in the figure for the titration of a strong acid with a strong base, the rapid change of pH near the equivalence point is reflected in the steepness of the curve. Titration curves for weak acids or weak bases do not rise quite as steeply. Generally, for these titrations we will wish to choose an indicator whose color change occurs fairly close to the equivalence point.

PROBLEMS

16.1 If 12 g of NaOH is required to neutralize 400 cm^3 of a solution of HCl in water, calculate the concentration of the HCl solution in equivalents of HCl per liter.

Solution: Equivalents NaOH = equivalents HCl. The number of equivalents of NaOH in 12 g of NaOH equals the number of equivalents of HCl in 400 cm^3 of solution. The equivalent weight of NaOH is equal to its formula weight, 40. Therefore, 12 g of NaOH is 12/40 or 0.30 equivalents of NaOH, and 0.30 equivalents of NaOH will neutralize 0.30 equivalents of HCl. That means that there is 0.30 equivalents of HCl in 400 cm^3 of solution. The number of equivalents per liter will then be

$$1000\ \text{cm}^3 \times \frac{0.30\ \text{equivalent}}{400\ \text{cm}^3} = 0.75\ \text{equivalent/L}$$

16.2 What mass of KOH will be required for the preparation of 500 cm^3 of 0.400 N KOH for use in neutralization reactions?

Solution: A 0.400 *N* KOH solution contains 0.400 equivalent of KOH per liter (1000 cm^3). Therefore, 500 cm^3 will contain 0.200 equivalent of KOH. The equivalent weight of KOH is its formula weight, 56.1.

$$0.200 \text{ equivalent KOH} \times \frac{56.1 \text{ g}}{1 \text{ equivalent}} = 11.2 \text{ g KOH}$$

The problem can be solved in one operation:

$$500 \text{ cm}^3 \times \frac{0.400 \text{ equivalent}}{1000 \text{ cm}^3} \times \frac{56.1 \text{ g KOH}}{1 \text{ equivalent}} = 11.2 \text{ g KOH}$$

16.3 What is the normality of a solution of NaOH that contains 8 g of NaOH per 400 cm^3 of solution?

Solution: The solution, 8 g of NaOH in 400 cm^3, has the same concentration as 20 g of NaOH in 1000 cm^3 of solution. One equivalent weight of NaOH is 40 g; 20 g of NaOH is 0.5 of an equivalent weight. Since 0.5 equivalent of NaOH is present in a liter of solution, the normality is 0.5 *N*. The problem can be solved in one operation:

$$\frac{1000 \text{ cm}^3}{\text{L}} \times \frac{8 \text{ g NaOH}}{400 \text{ cm}^3} \times \frac{1 \text{ equivalent NaOH}}{40 \text{ g NaOH}} = \frac{0.5 \text{ equivalent NaOH}}{\text{L}}$$

16.4 A sample of 50 cm^3 of hydrochloric acid was required to react with 0.40 g of NaOH. Calculate the normality of the hydrochloric acid.

Solution: To find the normality we must find the number of equivalents of HCl in a liter of acid

$$\text{equivalents HCl} = \text{equivalents NaOH}$$

The equivalent weight of NaOH is 40. Therefore 0.40 g of NaOH is 0.010 equivalent of NaOH. Since there is 0.010 equivalent of NaOH there must be 0.010 equivalent of HCl in the 50 cm^3 of hydrochloric acid that was required. To find the number of equivalents per 1000 cm^3, which is the normality,

$$1000 \text{ cm}^3 \times \frac{0.010 \text{ equivalent}}{50 \text{ cm}^3} = 0.20 \text{ equivalent}$$

Therefore, the solution is 0.20 *N*.

16.5 What mass of KOH will be required to react with 100 cm^3 of 0.80 *N* HCl?

Solution:

$$\text{equivalents KOH} = \text{equivalents HCl.}$$

$$\text{equivalents KOH} = \frac{\text{mass KOH}}{56.1 \text{ g KOH per equivalent}}$$

$$\text{equivalents HCl} = 100 \text{ cm}^3 \text{ HCl} \times \frac{0.80 \text{ equivalent HCl}}{1000 \text{ cm}^3 \text{ HCl}}$$

$$\frac{\text{mass KOH}}{56.1 \text{ g KOH per equivalent}} = 100 \text{ cm}^3 \text{ HCl} \times \frac{0.80 \text{ equivalent HCl}}{1000 \text{ cm}^3 \text{ HCl}}$$

$$\text{mass KOH} = \frac{56.1 \times 100 \times 0.80}{1000} = 4.5 \text{ g KOH}$$

Since normality is defined as equivalents of solute per liter of solution, since equivalents of KOH equals equivalents of HCl, and since 100 cm^3 is 0.10 L, the entire calculation can take the simple form,

$$0.10 \text{ L} \times \frac{0.80 \text{ equivalent}}{1 \text{ L}} \times \frac{56.1 \text{ g}}{1 \text{ equivalent}} = 4.5 \text{ g}$$

16.6 What volume of 0.30 N HNO_3 will be required to react with 24 cm^3 of 0.25 N KOH?

Solution:

equivalents HNO_3 = equivalents KOH.

$$\frac{\text{cm}^3 \text{ HNO}_3 \times 0.30 \text{ equivalent}}{1000 \text{ cm}^3} = \frac{24 \text{ cm}^3 \text{ KOH} \times 0.25 \text{ equivalent}}{1000 \text{ cm}^3}$$

$$\text{Volume HNO}_3 = \frac{24 \text{ cm}^3 \times 0.25 \text{ equivalent}}{0.30 \text{ equivalent}} = 20 \text{ cm}^3$$

Since, in the above calculation, equivalents/1000 cm^3 represents normality, cm^3 of HNO_3 × normality HNO_3 = cm^3 of KOH × normality of KOH.

16.7 If 18 cm^3 of 0.1 *F* H_2SO_4 was required to liberate the CO_2 from 82 cm^3 of sodium carbonate solution, calculate the normality of the sodium carbonate solution.

Hint: cm^3 of H_2SO_4 × normality of H_2SO_4 = cm^3 of Na_2CO_3 × normality of Na_2CO_3.

16.8 What volume of 0.250 *F* HCl will be required to neutralize 500 cm^3 of solution containing 8.00 g of NaOH?

16.9 Calculate the normality of a H_3PO_4 solution 40 cm^3 of which neutralized 120 cm^3 of 0.53 *N* NaOH.

16.10 What volume of 0.25 *N* NaOH will be required to neutralize 116 cm^3 of 0.0625 *N* H_2SO_4?

16.11 It took 40 cm^3 of 0.10 *F* H_2SO_4 to precipitate completely the Ba^{2+} ion (as $BaSO_4$) from a $BaCl_2$ solution. Calculate the mass of $BaCl_2$ that was originally present in the $BaCl_2$ solution.

16.12 If 0.664 g of phthalic acid ($C_8H_6O_4$) was required to neutralize 20.0 cm^3 of 0.400 *N* NaOH, calculate the equivalent weight of the acid.

16.13 It is found that 30.0 cm^3 of 0.200 *F* H_2SO_4 is required to neutralize 2.61 g of an unknown base. Calculate the equivalent weight of the base.

16.14 What volume of 0.100 *F* H_3PO_4 solution is required to neutralize 1.35 g of a base whose equivalent weight is 120 g?

16.15 A sample of 200 cm^3 of 1.000 *N* H_2SO_4 was treated with an excess of Na_2CO_3. What volume of dry CO_2 gas, measured at STP, was given off?

16.16 A volume of 600 cm^3 of HCl of a certain normality was mixed with 400 cm^3 of NaOH of the same normality. The resulting solution had a pH of 1. Calculate the normality of the HCl and NaOH.

16.17 Calculate the pH at the equivalence point when 50 cm^3 of 0.1 *F* $HC_2H_3O_2$ is neutralized by 50 cm^3 of 0.1 *F* NaOH.

Hint: When the two solutions are mixed, the volume of the solution is 100 cm^3. Assuming the neutralization equilibrium of the weak acid–strong base lies completely to the right,

$$\text{HOAc}(aq) + \text{OH}^-(aq) \rightleftarrows \text{H}_2\text{O} + \text{OAc}^-(aq)$$

The amount of OAc^- produced equals amount of HOAc = molarity × volume = (0.1)(0.05) and $[OAc^-] = (0.1)(0.05)/(0.05 + 0.05)$. Now the problem is treated as if it read: Calculate the pH of 0.05 *M* OAc^- solution.

16.18 Calculate the pH at the equivalence point when 40 cm^3 of 0.2 *F* NH_3 is neutralized by 80 cm^3 of 0.1 *F* HCl.

16.19 Calculate the pH at the equivalence point when 100 cm^3 of 0.220 *F* HCN is neutralized by 0.125 *F* NaOH solution.

16.20 A sample of fruit juice requires 12.3 mL of 0.125 *M* sodium hydroxide solution for neutralization. Find the total amount of available protons in the sample of fruit juice.

16.21* Morphine is a base that accepts one proton per molecule. Its molecular formula is $C_{17}H_{19}NO_3$. The major source of morphine is opium. A 0.682-g sample of opium is found to require 8.92 mL of a 0.0116 *M* solution of sulfuric acid for neutralization. Assuming that morphine is the only acid or base in opium, calculate the percentage of morphine in the sample of opium.

16.22* A 25.0-mL volume of a sodium hydroxide solution requires 19.6 mL of a 0.189 *M* hydrochloric acid solution for neutralization. A 10.0-mL volume of a phosphoric acid solution requires 34.9 mL of the sodium hydroxide solution for complete neutralization. Calculate the concentration of the phosphoric acid solution.

16.23* When a solution of a polyprotic acid is neutralized by slow addition of a solution of base, the protons are neutralized one at a time. For H_2CO_3 we can write

$$H_2CO_3(aq) + OH^-(aq) \rightleftarrows HCO_3^-(aq) + H_2O$$
$$HCO_3^-(aq) + OH^-(aq) \rightleftarrows CO_3^{2-}(aq) + H_2O$$

Calculate the *difference* in pH between the equivalence point for the neutralization of the first proton and the second when a 0.2 *F* solution of NaOH is added to a 0.2 *F* solution of H_2CO_3.

16.24 What should be the pK_{In} of an indicator used to monitor the neutralization of (a) 0.1 *M* HCl by 0.1 *M* NaOH; (b) 0.1 *M* HCl by 0.1 *M* NH_3; (c) 0.1 *M* HCN by 0.1 *M* NaOH?

16.25* A common experimental procedure for carrying out a neutralization is titration. Suppose that we place 100 cm^3 of 0.10 *M* NaOH in a flask and slowly add 0.10 *M* HCl solution from a buret. Calculate the pH of the original NaOH solution and the pH after the addition of 50 cm^3 of HCl, 90 cm^3 of HCl, 99 cm^3 of HCl, 99.9 cm^3 of HCl, and 100 cm^3 of HCl. Show from these calculations that a suitable indicator can have a p*K* in a range around the equivalence point.

Hint: Calculate the amount of OH^- and the total volume of the solution after each of the specified volumes has been added.

16.26* Perform the same calculations as Prob. 16.25 for the addition of 0.1 *M* HCl solution to 100 cm^3 of 0.1 *M* NaCN.

Hint: The reaction of interest for the titration is $H^+ + CN^- \rightleftarrows HCN$, which allows calculation of $[CN^-]$ and [HCN] after each addition.

BUFFERS

Buffers are solutions of a weak acid and the salt of its conjugate base or of a weak base and the salt of its conjugate acid whose pH will stay relatively constant when relatively small amounts of acid or base are added. The amount of added acid or base that the pH of a buffer can resist is called the *capacity of the buffer*. Buffers are also solutions whose pH does not change when they are diluted or concentrated. The pH of the buffer solution depends only on the ratio of the two species composing the buffer. For example, a buffer made from 1 mol of hydrocyanic acid and 1 mol of sodium cyanide has $[H^+] = 4 \times 10^{-10}$ no matter what the volume of the solution, since

$$[H^+] = K_a \frac{[HCN]}{[CN^-]} \quad \text{and} \quad K_a = 4 \times 10^{-10}$$

Similarly a buffer of 1 mol of ammonia and 1 mol of ammonium chloride has $[OH^-] = 1.8 \times 10^{-5}$, since

$$[OH^-] = K_b \frac{[NH_3]}{[NH_4^+]} \quad \text{and} \quad K_b = 1.8 \times 10^{-5}$$

Buffers are best prepared from roughly comparable molar quantities of the two required species. Thus in preparing a buffer solution of a desired pH, one uses a weak acid–salt combination whose pK_a is close to the desired pH.

PROBLEMS

(See Table A.2 of the appendix for ionization constants.)

16.27 Calculate the pH of 1 L of solution that is 0.1 F in NaOAc and 0.001 F in HCl. K_a for HOAc $= 1.8 \times 10^{-5}\ M$.

Solution: 0.1 F NaOAc will yield 0.1 M Na^+ and 0.1 M OAc^-. 0.001 F HCl will yield 0.001 M H^+ and 0.001 M Cl^-. H^+ is a very strong acid; OAc^- is a moderately strong base.

The 0.001 mol of H^+ will combine with 0.001 mol of OAc^- to form 0.001 mol of HOAc.

That will leave 0.1 − 0.001 or 0.1 mol of OAc^-.

Therefore, $[OAc^-] = 0.1\ M$ and $[HOAc] = 0.001\ M$.

$$K = \frac{[H^+] \times [OAc^-]}{[HOAc]} = 1.8 \times 10^{-5}\ M$$

Substituting in the above formula, we have

$$\frac{[H^+] \times 0.1\ M}{0.001\ M} = 1.8 \times 10^{-5}\ M$$

$$[H^+] = 2 \times 10^{-7}\ M$$

$$pH = 6.7$$

16.28 Calculate the pH of a solution that is 0.2 *F* in NaF and 0.002 *F* in HCl. K_a for HF = 6.9×10^{-4}.

16.29 What concentrations of NaOAc and HOAc must be used in preparing a solution with a pH of 6.0?

16.30 Which acid and its salt listed in Table A.2 of the appendix would be suitable for preparing a buffer of pH = 3.2?

16.31 What is the pH of a buffer prepared from 40.0 g of NaH_2PO_4 and 47.3 g of Na_2HPO_4 in 3 L of water?

Hint: Use K_2 for H_3PO_4.

16.32 Calculate the amount of Na_3PO_4 that should be added to 1 L of 0.2 *M* Na_2HPO_4 solution to produce a buffer whose pH is 11.7.

16.33* What amount of HCl gas must be added to 1 L of a buffer solution that is 2 *M* in acetic acid and 2 *M* in sodium acetate to produce a solution whose pH = 4.0? What would be the pH if this quantity of HCl were dissolved in 1 L of pure water?

16.34* In a solution prepared by dissolving NaOAc and HOAc in pure water, the sum of the formalities of the two solutes is 1.0. The pH of the solution is 5.0.

a. Calculate the formality of each of the two solutes.

b. Calculate the molarity of each of the species in solution. K_i for HOAc is 1.8×10^{-5} *M*.

Solution: Since $[H^+]$ is 1.0×10^{-5} *M* and $K = 1.8 \times 10^{-5}$, the ratio $[OAc^-]/[HOAc]$ will be 1.8. Since we know that $[OAc^-] + [HOAc] = 1.0$, the values of $[OAc^-]$ and [HOAc] can be calculated to be 0.64 *M* and 0.36 *M*, respectively. The fact that $[H^+]$ is 1×10^{-5} means that a negligible amount of HOAc ionizes and a negligible amount of OAc^- is produced by this ionization. Therefore, the molarity of the OAc^- equals the formality of the NaOAc and the molarity of the HOAc equals its formality.

Note: The solution of this problem illustrates the following fact that can be very useful in solving problems: When the pH or $[H^+]$ or $[OH^-]$ of a solution is known, the *ratio* of an anion to its acid can be calculated provided the ionization constant of the acid is known. This *ratio* can often be used in solving the problem. If the ratio of two quantities and the sum of the two quantities are known, the value of each quantity can be calculated.

16.35* What volume of 0.200 *F* NaOH must be added to 100 cm^3 of 0.150 *F* HOAc to give a solution with a pH of 4.046?

16.36* What volume of each of 1.00 *F* solutions of NaOH and HOAc must be mixed to give 1 L of solution having a pH of 4.00?

16.37 Calculate the relative masses of phenol and sodium phenoxide (C_6H_5ONa) necessary to prepare a buffer solution of pH = 9.89.

16.38 Find the pH of 1.0 L of a buffer solution prepared from 1.0 mol of lactic acid and 1.0 mol of sodium lactate before and after the addition of 0.10 mol of HCl.

16.39 Repeat the calculation of prob. 16.38 for a buffer solution prepared from 0.10 mol of each component. Discuss your result in terms of buffer capacity.

16.40* What amount of solid NaOH must be added to 1 L of 0.10 *F* H_2CO_3 to produce a solution whose hydrogen-ion concentration is 3.2×10^{-11} *M*? There is no measurable change in volume when the NaOH is dissolved.

Solution: The situation in this problem is complicated by the fact that two reactions are possible, namely,

$$H_2CO_3(aq) + OH^-(aq) \rightleftarrows HCO_3^-(aq) + H_2O$$
$$H_2CO_3(aq) + 2OH^-(aq) \rightleftarrows CO_3^{2-}(aq) + 2H_2O$$

Since $[H^+]$ is 3.2×10^{-11} *M*, we can calculate from K_1 and K_2 for H_2CO_3 that

$$\frac{[CO_3^{2-}]}{[HCO_3^-]} = \frac{4.8 \times 10^{-11}}{3.2 \times 10^{-11}} = 1.5 \text{ and } \frac{[HCO_3^-]}{[H_2CO_3]} = \frac{4.2 \times 10^{-7}}{3.2 \times 10^{-11}} = 1.3 \times 10^4$$

This means that all the H_2CO_3 is consumed.

Let X = moles of HCO_3^- formed; $1.5X$ = moles of CO_3^{2-} formed.
$2.5X$ = total moles of carbonate = 0.10 mol.
Solving, we obtain $X = 0.040$ mol of HCO_3^- and $1.5X = 0.060$ mol of CO_3^{2-}.
To form 0.040 mol of HCO_3^- requires 0.040 mol of OH^-.
To form 0.060 mol of CO_3^{2-} requires 0.120 mol of OH^-.
Total moles of OH^- consumed = 0.16 mol.
Since $[H^+]$ is 3.2×10^{-11} *M*, the amount of excess OH^- added was $1.0 \times 10^{-14}/3.2 \times 10^{-11} = 3.1 \times 10^{-4}$ or 0.00031 mol.
Therefore, total moles of NaOH added = 0.16 + 0.00031 = 0.16.

THE MAIN REACTION APPROXIMATION

Many equilibrium states involve a number of simultaneous equilibria that can make the quantitative treatment of such systems complex. But, very often we find that one of these simultaneous equilibria will have a *K* much larger than any of the others. This equilibrium is called the *main reaction* and when its *K* is at least 100 times greater than any other, problem solving can be greatly simplified by the use of the *main reaction approximation*. This approximation states that only the main reaction need be considered in calculation of the equilibrium concentrations of all the species that are involved in it and that the concentrations of other species in the system can be calculated from those in the main reaction by using the principle that at equilibrium the values of K for all equilibria must be satisfied.

A problem will specify that certain components are present in solution or are being mixed. All these components should be listed in their correct states before any equilibria are established or at the instant of mixing according to the following rules:

1. Insoluble or slightly soluble substances are written out of solution as pure, undissolved components with the appropriate symbols to indicate their state. Thus we write $AgCl(s)$ or $H_2S(g)$ or $Br_2(l)$ or $Ca(OH)_2(s)$.
2. Soluble substances fall into various categories:
   a. Salts are written as dissociated ions. Thus calcium nitrate $Ca(NO_3)_2$ is written as $Ca^{2+}(aq) + NO_3^-(aq)$ since these are the species present. Sodium cyanide is $Na^+(aq) + CN^-(aq)$.
   b. Strong acids are written as dissociated. Thus hydrochloric acid is $H^+(aq) + Cl^-(aq)$, sulfuric acid is $H^+(aq) + HSO_4^-(aq)$.
   c. Strong bases (that are soluble) are written as dissociated. Sodium hydroxide is $Na^+(aq) + OH^-(aq)$.
   d. Weak acids and weak bases are written as undissociated. Thus acetic acid is $HOAc(aq)$, ammonia is $NH_3(aq)$, phosphoric acid is $H_3PO_4(aq)$.
   e. Water is written as undissociated.

EXAMPLE 1

Calculate the pH of a 0.1 M $NaHCO_3$ solution, given $K_1 = 4.2 \times 10^{-7}$, $K_2 = 4.8 \times 10^{-11}$ for H_2CO_3.

The species present under the conditions specified are Na^+, HCO_3^-, and H_2O. No others may be considered.

Step 1 List all the possible reactions that may occur between the species present. These reactions will fall primarily into one of several categories:

1. Reactions between an acid and a base. This will often be the most important category.
2. Dissolution of insoluble materials.
3. Formation of precipitates from species in solution.
4. Formation or dissociation of complex ions.
5. Evolution of gases.
6. Oxidation-reductions.
7. Combinations of the above.

In the problem that we are considering there are only three species to consider: $Na^+(aq)$, $HCO_3^-(aq)$, and H_2O. In general, Na^+ (or K^+ or other group I cations) never enter into the kinds of reactions we are discussing and can be ignored. Thus we need only consider HCO_3^- and H_2O. The only reactions possible in this system are acid–base reactions. In order to list them we must identify what is an acid and what is a base. As we have already seen, water is both an acid and a base. Similarly, HCO_3^- is both an acid and a base—an acid because it can donate a proton, and a base because it can accept one. The reactions that can occur are

$$H_2O \rightleftarrows H^+(aq) + OH^-(aq) \tag{16-4}$$

(water with itself),

$$H_2O + HCO_3^-(aq) \rightleftarrows H_3O^+(aq) + CO_3^{2-}(aq) \tag{16-5}$$
$$HCO_3^-(aq) \rightleftarrows H^+(aq) + CO_3^{2-}(aq)$$

(water as a base, bicarbonate as an acid),

$$H_2O + HCO_3^-(aq) \rightleftarrows H_2CO_3(aq) + OH^-(aq) \tag{16-6}$$

(water as an acid, bicarbonate as a base), and

$$HCO_3^-(aq) + HCO_3^-(aq) \rightleftarrows H_2CO_3(aq) + CO_3^{2-}(aq) \tag{16-7}$$

(bicarbonate with itself as an acid and a base).

Step 2 Evaluate the value of K for each reaction from the information given. Often this will require combination of various given K's and their associated reactions. Remember chemical equations can be added, in which case their K's are multiplied. Equations can be subtracted, in which case their K's are divided. Equations can be multiplied by a numerical factor, in which case their K's are raised to the power of the numerical factor. Evaluating K's:

$$K \text{ for (16-4) is } K_w = 10^{-14}$$
$$K \text{ for (16-5) is } K_2\text{, which is given as } 4.8 \times 10^{-11}$$

Equation (16-5) represents simply the second dissociation of H_2CO_3. K for Eq. (16-6) can be evaluated by the following procedure: Equation (16-6) has HCO_3^- on the left and H_2CO_3 on the right. Are we given data for an equilibrium that relates HCO_3^- and H_2CO_3? Yes, K_1 refers to

$$H_2CO_3(aq) \rightleftarrows H^+(aq) + HCO_3^-(aq) \tag{a}$$

which has H_2CO_3 on the left and HCO_3^- on the right. Since we want to reverse this equation it will have to be subtracted from some other equation.

Equation (16-6) also has OH^- on the right. Do we have the K of a reaction with OH^- on the right? Yes, K refers to

$$H_2O \rightleftarrows H^+(aq) + OH^-(aq). \quad \textbf{(b)}$$

If we now take (b) − (a) we get Eq. (16-6). Note that subtraction of an equation is like reversing this equation and then adding. If we get an equation from (b) − (a), then its equilibrium constant is K_b/K_a or in this case,

$$\frac{K_w}{K_1} = \frac{1 \times 10^{-14}}{4.2 \times 10^{-7}} = 2.4 \times 10^{-8}$$

Notice that this is simply the reaction of the salt of a weak acid with water. The K is K_w/K_a. K for Eq. (16-7) must also be evaluated in this way. An equation relating HCO_3^- and CO_3^{2-} is

$$HCO_3^-(aq) \rightleftarrows H^+(aq) + CO_3^{2-}(aq) \quad K_2 \quad \textbf{(c)}$$

and one relating HCO_3^- and H_2CO_3 is

$$H_2CO_3(aq) \rightleftarrows H^+(aq) + HCO_3^-(aq) \quad K_1 \quad \textbf{(d)}$$

when subtracted,

$$\text{(c)}-\text{(d)} = 2HCO_3^-(aq) \rightleftarrows H_2CO_3(aq) + CO_3^{2-}(aq)$$

$$K = \frac{K_2}{K_1} = \frac{4.8 \times 10^{-11}}{4.2 \times 10^{-7}} = 1.1 \times 10^{-4}$$

Step 3 Examine all the calculated K's. If one of them is at least 10^2 greater than the others, the main reaction approximation will work. Select the largest K and its associated reaction as the main reaction and temporarily ignore all the others. Here the main reaction is (16-7).

Step 4 Calculate the equilibrium concentrations of everything in the main reaction:

$$2HCO_3^-(aq) \leftrightarrows H_2CO_3(aq) + CO_3^{2-}(aq)$$

start	0.1	0	0
equil.	$0.1 - 2x$	x	x

$$K = \frac{[H_2CO_3][CO_3^{2-}]}{[HCO_3^-]^2} = \frac{(x)(x)}{(0.1 - 2x)^2} = 1.1 \times 10^{-4}$$

$$x = 1.1 \times 10^{-3}\,M = [CO_3^{2-}] = [H_2CO_3] \text{ (assuming } 0.1 - 2x = 0.1)$$

$$[HCO_3^-] = 0.98(0.100 - 2(0.001)) = 9.8 \times 10^{-2}\,M$$

Step 5 To calculate the concentration of species not in the main reaction, choose one of the secondary reactions that includes the species of interest and other species of known concentration. Substitute in the expression for K.

To get $[H^+]$ we can use $HCO_3^-(aq) \leftrightarrows H^+(aq) + CO_3^{2-}(aq)$

$$K_2 = \frac{[H^+][CO_3^{2-}]}{[HCO_3^-]} = \frac{[H^+](1.1 \times 10^{-3})}{(9.8 \times 10^{-2})} = 4.8 \times 10^{-11}$$

$$[H^+] = 4.3 \times 10^{-9}\,M$$

$$pH = 9 - \log 4.3 = 8.37$$

To get OH^- use $[H^+] \times [OH^-] = 1.0 \times 10^{-14} = 4.3 \times 10^{-9} \times [OH^-]$

$$[OH^-] = 2.3 \times 10^{-6}\,M$$

$$pOH = 14 - pH = 5.63$$

EXAMPLE 2

Given for H_3PO_4 that $K_1 = 7.5 \times 10^{-3}$, $K_2 = 6.2 \times 10^{-8}$, $K_3 = 1.0 \times 10^{-12}$, calculate the pH of 0.1 M NaH_2PO_4 and the concentration of all species at equilibrium.

Species present: $Na^+(aq)$, $H_2PO_4^-(aq)$, H_2O

Reactions: (ignore Na^+)

1. $H_2O \rightleftarrows H^+(aq) + OH^-(aq)$ $\quad K_w = 10^{-14}$
2. $H_2PO_4^-(aq) \rightleftarrows H^+(aq) + HPO_4^{2-}(aq)$ $\quad K_2 = 6.2 \times 10^{-8}$
3. $H_2PO_4^-(aq) \rightleftarrows 2H^+(aq) + PO_4^{3-}(aq)$ $\quad K = K_2 K_3 = 6.2 \times 10^{-20}$
4. $H_2PO_4^-(aq) + H_2O \rightleftarrows H_3PO_4(aq) + OH^-(aq)$ $\quad K = \frac{K_w}{K_1} = \frac{1 \times 10^{-14}}{7.5 \times 10^{-3}}$
   $= 1.3 \times 10^{-12}$
5. $H_2PO_4^-(aq) + H_2PO_4^-(aq) \rightleftarrows H_3PO_4(aq) + HPO_4^{2-}(aq)$
   $$K = \frac{K_2}{K_1} = \frac{6.2 \times 10^{-8}}{7.5 \times 10^{-3}}$$
   $= 8.3 \times 10^{-6}$
6. $H_2PO_4^-(aq) + 2H_2PO_4^-(aq) \rightleftarrows 2H_3PO_4(aq) + PO_4^{3-}(aq)$
   This is a combination of

$$H_2PO_4^-(aq) \rightleftarrows H^+(aq) + HPO_4^{2-}(aq) \qquad K_2 \qquad \textbf{(a)}$$
$$HPO_4^{2-}(aq) \rightleftarrows H^+(aq) + PO_4^{3-}(aq) \qquad K_3 \qquad \textbf{(b)}$$
$$2H_3PO_4(aq) \rightleftarrows 2H^+(aq) + 2H_2PO_4^-(aq) \qquad K_1^2 \qquad \textbf{(c)}$$

$$\text{(a)} + \text{(b)} - \text{(c)} = \frac{K_2K_3}{K_1^2}$$
$$= \frac{(6.2 \times 10^{-8})(1.0 \times 10^{-12})}{(7.5 \times 10^{-3})^2}$$
$$= 1.1 \times 10^{-15}$$

So reaction 5 is the main reaction.

$$2H_2PO_4^-(aq) \rightleftarrows H_3PO_4(aq) + HPO_4^{2-}(aq)$$

start	0.1	0	0
equil.	$0.1 - 2x$	x	x

$$\frac{[H_3PO_4] \times [HPO_4^{2-}]}{[H_2PO_4^-]^2} = 8.3 \times 10^{-6} = \frac{(x)(x)}{(0.1 - 2x)^2}$$

Assume that $2x << 0.1$. Then

$$x = 2.9 \times 10^{-4} = [H_3PO_4] = [HPO_4^{2-}]$$
$$[H_2PO_4^-] = 0.1 - 2(0.0003) = 0.099 = 9.9 \times 10^{-2}$$

To get $[H^+]$ we can use

$$\frac{[H^+] \times [HPO_4^{2-}]}{[H_2PO_4^-]} = 6.2 \times 10^{-8} = \frac{[H^+] \times 2.9 \times 10^{-4}}{9.9 \times 10^{-2}}$$
$$[H^+] = 2.1 \times 10^{-5}\,M$$
$$\text{pH} = 5 - \log 2.1 = 4.7$$

Notice that if we check this $[H^+]$ by substituting, we obtain

$$\frac{[H^+] \times [H_2PO_4^-]}{[H_3PO_4]} = \frac{2.1 \times 10^{-5} \times 9.9 \times 10^{-2}}{2.9 \times 10^{-4}} = 7.2 \times 10^{-3}$$

which agrees well with the value of K_1 given.

To get $[PO_4^{3-}]$ we use $K_3 = \dfrac{[PO_4^{3-}] \times [H^+]}{[HPO_4^{2-}]} = \dfrac{[PO_4^{3-}] \times 2.1 \times 10^{-5}}{2.9 \times 10^{-4}}$

To get $[OH^-]$ we use $[H^+] \times [OH^-] = 10^{-14} = 2.1 \times 10^{-5} \times [OH^-]$

$$[OH^-] = 4.8 \times 10^{-10}\,M$$

Thus all the equilibrium concentrations are

$$[Na^+] = 0.1\ M$$
$$[H_2PO_4^-] = 9.9 \times 10^{-2}\ M$$
$$[H_3PO_4] = [HPO_4^{2-}] = 2.9 \times 10^{-4}\ M$$
$$[PO_4^{3-}] = 1.4 \times 10^{-11}\ M$$
$$[H^+] = 2.1 \times 10^{-5}\ M$$
$$[OH^-] = 4.8 \times 10^{-10}\ M$$

PROBLEMS

16.41 Calculate the OH^- ion concentration of 0.10 F K_2SO_3. Calculate the concentrations of H^+, HSO_3^-, SO_3^{2-}, H_2SO_3, and K^+.

16.42 Calculate the pH of 0.20 F Na_2CO_3.

16.43 Calculate the formate-ion concentration of a 0.2 F solution of NaF in 0.1 F HCOOH (formic acid).

Hint: The main reaction is

$$F^-(aq) + HCOOH(aq) \rightleftarrows HF(aq) + HCOO^-(aq)$$

16.44 Calculate the CN^- ion concentration of a 0.1 F solution of NaOAc in 0.1 F HCN.

16.45 Calculate the pH of
a. 0.10 F $NaHSO_3$
b. 0.10 F K_2HPO_4
c. 0.10 F KH_2PO_4

16.46* Calculate the concentration of NH_3 in a solution that is 0.1 F in HOAc and 0.1 F in NH_4Cl.

Hint: The principal reactions with their equilibrium constants are

$$HOAc(aq) \rightleftarrows H^+(aq) + OAc^-(aq) \qquad K_i = 1.8 \times 10^{-5}$$
$$NH_4^+(aq) \rightleftarrows H^+(aq) + NH_3(aq) \qquad K = \frac{K_w}{K_b} = 5.6 \times 10^{-10}$$

Since K is so much smaller than K_i, the amount of H^+ derived from the NH_4^+ is negligible in comparison with the amount derived from the ionization of HOAc. Therefore, calculate the $[H^+]$ derived from the ionization of HOAc as in Prob. 15.18. The value of $[H^+]$ thus calculated is $1.3 \times 10^{-3}\ M$.

To calculate $[NH_3]$ substitute the above value of $[H^+]$, $1.3 \times 10^{-3}\ M$, for $[H^+]$ and 0.1 M for $[NH_4^+]$ in K.

Note: In the reaction of NH_4^+, as in all equilibrium reactions, the magnitude of the equilibrium constant tells us the extent of completeness of the reaction. In many problems that will be encountered it will be very useful to have this information.

16.47* Calculate the molar concentration of each species in a solution that is 0.20 *F* in NH_3, and 0.20 *F* in NaCN.

Hint: The principal reactions with their equilibrium constants are

$$NH_3(aq) + H_2O \rightleftarrows NH_4^+(aq) + OH^-(aq)$$

$$K_i = \frac{[NH_4^+] \times [OH^-]}{[NH_3]} = 1.8 \times 10^{-5} \quad \textbf{(16-8)}$$

$$CN^-(aq) + H_2O \leftrightarrows HCN(aq) + OH^-(aq)$$

$$K = \frac{[HCN] \times [OH^-]}{[CN^-]} = 2.5 \times 10^{-5} \quad \textbf{(16-9)}$$

Since the two equilibria occur in the same solution, the value of $[OH^-]$ is the same for each. Therefore, K_i and K can be solved for $[OH^-]$ and the $[OH^-]$ equated to give

$$\frac{1.8 \times 10^{-5} \times [NH_3]}{[NH_4^+]} = \frac{2.5 \times 10^{-5} \times [CN^-]}{[HCN]} \quad \textbf{(16-10)}$$

But $[NH_3]$ is 0.20 *M* and $[CN^-]$ is 0.20 *M*. Substituting these values in Eq. (16-10), we have

$$\frac{[HCN]}{[NH_4^+]} = \frac{2.5 \times 10^{-5}}{1.8 \times 10^{-5}} \quad \textbf{(16-11)}$$

or

$$[HCN] = 1.4 \times [NH_4^+] \quad \textbf{(16-12)}$$

Let $X = [NH_4^+]$; then $1.4X = [HCN]$.

Since the OH^- derived in the ionization of NH_3 is, according to Eq. (16-8), equal to $[NH_4^+]$ and since the OH^- derived from proton transfer to the CN^- from water is, according to Eq. (16-9), equal to [HCN], the total $[OH^-]$ will be equal to $[NH_4^+]$ + [HCN], or $2.4X$.

Substituting X for $[NH_4^+]$ $2.4X$ for $[OH^-]$, and 0.20 M for $[NH_3]$ in the formula for K_i, we can calculate the value of X and hence the molar concentration of each species in solution.

16.48* Calculate the molar concentration of each species in a solution that is 0.20 *F* in NH_3 and 0.20 *F* in $NaC_2H_3O_2$.

16.49* A mixture of 500 cm^3 of 1.0 *F* HNO_3 and 100 cm^3 of 15 *F* NH_3 was diluted with water to 1.0 L. Calculate the $[H^+]$ of the solution.

16.50* A volume of 100 cm^3 of a certain solution of NH_3 in water was mixed with 400 cm^3 of 1.00 *F* HCl. The resulting solution was diluted with water to a volume of 1 L; this liter of solution was found to have a hydrogen-ion concentration of 2.22×10^{-9} *M*. Calculate the formality of the original solution of NH_3.

16.51* What amount of solid KOH must be added to 1 L of 0.20 *F* H_2SO_4 to yield a solution whose pH is 7.85? There is no change of volume when the KOH is added.

16.52* A solution that is 0.020 *F* in oxalate has a pH of 4.0. Calculate the molarity of each species in the solution. For $H_2C_2O_4$, $K_1 = 6.5 \times 10^{-2}$ *M* and $K_2 = 6.1 \times 10^{-5}$ *M*.

Hint: Since $[H^+]$ is 1.0×10^{-4}, and since the ionization constants are known, the ratios $[HC_2O_4^-]/[H_2C_2O_4]$, $[C_2O_4^{2-}]/[HC_2O_4^-]$, and $[C_2O_4^{2-}]/[H_2C_2O_4]$ can be calculated. From these ratios and the fact that $[H_2C_2O_4] + [HC_2O_4^-] + [C_2O_4^{2-}] = 0.020$, the molarity of each species in solution can be calculated.

16.53* A mixture of solid Na_2CO_3 and $NaHCO_3$ has a mass of 59.2 g. The mixture is dissolved in enough water to give 2.00 L of solution. The pH of this solution is found to be 10.62. What mass of Na_2CO_3 was in the mixture?

16.54* Calculate the pH of a solution that is 0.10 *F* in HCl and 0.35 *F* in NaCN.

Solution: Since both HCl and NaCN are strong electrolytes and since CN^- is, as the ionization constant of the acid of 4.0×10^{-10} attests, a base, the principal reaction is

$$H^+(aq) + CN^-(aq) \rightleftarrows HCN(aq)$$

Since an excess of CN^- is present, all but a very small amount *X* of the H^+ from the HCl is converted into HCN by reaction with CN^-. Therefore, at equilibrium $[H^+] = X$, $[HCN] = 0.10 - X$, and $[CN^-] = 0.25 + X$.

$$K = \frac{[H^+] \times [CN^-]}{[HCN]} = \frac{X(0.25 + X)}{0.10 - X} = 4.0 \times 10^{-10}$$

Since X will obviously be very small, it can be dropped from the terms $0.25 + X$ and $0.10 - X$.

$$\frac{0.25X}{0.10} = 4.0 \times 10^{-10}$$

$$X = 1.6 \times 10^{-10} = [H^+]$$

$$pH = 10 - \log 1.6 = 10 - 0.2 = 9.8$$

It is important to recognize in all problems, as in this problem, that the anion of a weak acid is a base and will therefore have a strong tendency to combine with the strong acid H^+.

16.55* A volume of 4.0 cm^3 of 0.10 F NaCN was mixed with 2.0 cm^3 of hydrochloric acid and 4.0 cm^3 of water to give 10.0 cm^3 of solution with a hydrogen ion concentration of 0.10 M. What was the formality of the hydrochloric acid?

16.56* **a.** A water solution is 0.10 F in the soluble strong electrolyte NaCN. Calculate the pH of this solution.

b. The above 0.10 F solution of NaCN is made 0.10 F in HOAc by the addition of pure acetic acid; assume there is no change in the volume of the solution when the pure acetic acid is added. Calculate the pH of the resulting solution.

16.57* A mixture of solid KCN and solid $KHSO_4$ totals 0.40 mol. When this mixture is dissolved in enough water to form 1 L of solution, the pH of this solution is 10. Calculate the amount of KCN in the mixture of solids.

Hint: What is the net equation for the main reaction? What is the numerical value of the equilibrium constant for this reaction; what does its value tell us about the completeness of the reaction? Which solute is present in excess? How does the amount of $KHSO_4$ in the mixture relate to the amount of one of the species present at equilibrium?

16.58* **a.** Calculate the pH of 0.10 F H_2S.

b. To 1.00 L of 0.10 F H_2S is added solid KOH until the pH is 7.00. Compute the amount of KOH added.

c. What is the pH of the solution when 0.090 mol of KOH has been added all told?

d. Calculate how much KOH must be added (in total) to bring the pH to 13.00.

16.59* A saturated CO_2 solution in pure water is 3.4×10^{-2} M in CO_2. What amount of CO_2 will dissolve in 1.00 L of 0.100 F NaOH?

Hint: Write the net equation for the reaction that takes place when excess CO_2 is added to a solution of a strong base. How many moles of OH^- is present in a liter of 0.100 *F* NaOH? With how many moles of CO_2 will this OH^- react? What, then, is the total solubility of CO_2 in 0.100 *F* NaOH?

16.60* A solution is prepared by dissolving 1.07 mol of NaH_2PO_4 and 3.32 mol of Na_3PO_4 in enough water to make 1 L of solution. What is the pH of the solution? What are the molar concentrations of $H_2PO_4^-$, HPO_4^{2-}, PO_4^{3-}, and H_3PO_4?

Hint: The main equilibrium is

$$PO_4^{3-}(aq) + H_2PO_4^-(aq) \leftrightharpoons 2HPO_4^{2-}(aq)$$

16.61* To what volume must a liter of a solution of the weak acid HZ be diluted with water in order to give a hydrogen-ion concentration one-half that of the original solution?

Solution:

$$HZ \rightleftarrows H^+ + Z^-$$

Let $X = [H^+]$ at original equilibrium; then $X = [Z^-]$ at original equilibrium.

Let $Y = [HZ]$ at original equilibrium.

$$K = \frac{X^2}{Y}$$

Since diluting the solution does not affect the value of K,

$$\frac{X^2}{Y} = \frac{(X/2)^2}{Y/V} \qquad (V = \text{volume in liters of diluted solution})$$

Solving, we have $V = 4$ L.

16.62* HA is a weak acid with an ionization constant of 1.0×10^{-8} *M*. HA forms the ion HA_2^-. The equilibrium constant for the reaction $HA_2^- \leftrightharpoons HA + A^-$ is 0.25 *M*. Calculate $[H^+]$, $[A^-]$, and $[HA_2^-]$ in 1.0 *F* HA.

Solution:

$$HA(aq) \leftrightharpoons H^+(aq) + A^-(aq)$$

$$A^-(aq) + HA(aq) \leftrightharpoons HA_2^-(aq)$$

$$K_i = \frac{[H^+] \times [A^-]}{[HA]} = 1.0 \times 10^{-8}\ M$$

$$K = \frac{[HA] \times [A^-]}{[HA_2^-]} = 0.25\ M$$

Since K_i is very small, the molar quantity of HA that ionizes will be so small that [HA] will be 1.0 M.

Therefore,

$$[H^+] \times [A^-] = 1 \times 10^{-8}$$

Since $K = 0.25\ M$ and $[HA] = 1.0$, $[HA_2^-] = 4 \times [A^-]$.

Also, we see that $[H^+] = [A^-] + [HA_2^-]$.

Therefore, if we let $X = [A^-]$, then $[HA_2^-] = 4X$, and $[H^+] = 5X$.

Substituting in K_i, we have

$$5X^2 = 1 \times 10^{-8}\ M^2$$

$$X = 4.5 \times 10^{-5} = [A^-]$$

$$5X = 2.2 \times 10^{-4} = [H^+]$$

$$4X = 1.8 \times 10^{-4} = [HA_2^-]$$

Chapter 17

Solubility Products and Complex Ions

When a saturated solution of sugar is prepared by shaking an excess of sugar with water, the following equilibrium is set up:

$$\text{solid sugar} \rightleftarrows \text{sugar molecules in solution}$$

When a saturated solution of the sparingly soluble salt AgCl is prepared by shaking excess AgCl with water, a somewhat different type of equilibrium is set up. AgCl is a salt; hence, we assume it is 100% ionized. Therefore, we also assume that the saturated solution which is in equilibrium with solid AgCl contains silver ions and chloride ions but no un-ionized AgCl molecules. These silver ions and chloride ions are in equilibrium with the excess solid AgCl. We may, therefore, represent the equilibrium that exists in such a saturated solution of AgCl as follows:

$$\text{AgCl}(s) \rightleftarrows \text{Ag}^+(aq) + \text{Cl}^-(aq)$$

Since this reaction is a true equilibrium, it will have an equilibrium constant. As has already been noted in Chapter 15, solid reactants are not involved in the equilibrium equation. Therefore

$$K = [\text{Ag}^+] \times [\text{Cl}^-]$$

This equation tells us that the product of the concentrations of the solute ions in a saturated solution of a sparingly soluble electrolyte is constant at a given temperature. This constant, K, is called the *solubility product constant*, or simply the *solubility product*, and is usually designated by the notation K_{sp} or SP.

If we prepare a solution in which the concentration of silver ions and the concentration of chloride ions, each expressed in moles per liter, is such that their product is less than K_{sp} for AgCl or is just barely equal to K_{sp}, no precipitate of AgCl will form. If, on the other hand, the product of the concentrations of Ag^+ and Cl^- is greater than K_{sp}, silver chloride will precipitate; furthermore, AgCl will keep on precipitating until enough Ag^+ ions and Cl^- ions have been removed to lower the product of their concentrations to the value of K_{sp}.

The solubility product represents a typical ionic equilibrium. As such it behaves exactly like the other ionic equilibria discussed in Chapter 15. In all problems involving solubility products we will assume, as we did in Chapter 15, that all systems are ideal, and the solubility product can be expressed, correctly, as a function of the molar concentrations of the ions involved.

The solubility product constants used are normally those determined at 20 °C. In certain problems the calculations are based on data obtained at temperatures other than 20 °C. Because the recorded solubility product is a function of the temperature as well as the accuracy of the determination, the value of K_{sp} for a certain substance may not be the same in all problems. Since such lack of agreement will affect the values calculated but not the process of the calculation, it need not be the source of any concern here.

It should be emphasized that the solubility product concept applies only to *sparingly soluble strong electrolytes*. It does not apply to *highly soluble* strong electrolytes such as NaCl, $MgSO_4$, and KOH, or to *weak* electrolytes, regardless of their solubility. The solubility product should not be confused with the solubility. By *solubility* we mean the quantity of solute that will go into solution when the pure solid is shaken with pure water at a given temperature until a saturated solution is obtained.

PROBLEMS

(See Tables A.2 and A.4 of the appendix for values of equilibrium constants.)

17.1 A solution in equilibrium with a precipitate of AgCl was found, on analysis, to contain 1.0×10^{-4} mol of Ag^+/L and 1.7×10^{-6} mol of Cl^-/L. Calculate the solubility product of AgCl.

Solution: The solubility product is, by definition, the product of the concentrations of the ions in equilibrium with a precipitate of a sparingly soluble ("insoluble") substance.

For AgCl,

$$\begin{aligned} K_{sp} &= [Ag^+] \times [Cl^-] \\ &= (1.0 \times 10^{-4}) \times (1.7 \times 10^{-6}) \\ &= 1.7 \times 10^{-10} \end{aligned}$$

Note: In the equilibrium with which this problem is concerned, the concentration of Ag^+ ions is not the same as the concentration of Cl^- ions. As was pointed out in Chapter 14, the reacting substances in an equilibrium need not be present in the exact ratio called for by the equation. They can be present in an unlimited combination of ratios. However, when they react, they always do so in the mole ratio represented by the equation. Thus when Ag^+ ions react with Cl^- ions to form AgCl, they always do so in the ratio of 1 mol of Ag^+ to 1 mol of Cl^-. However, the solution that is in equilibrium with the solid AgCl can contain Ag^+ and Cl^- ions in any ratio. The only requirement is that the product of $[Ag^+]$ and $[Cl^-]$ at the particular temperature must always equal the K_{sp}.

17.2 A solution in equilibrium with a precipitate of Ag_2S was found on analysis to contain 6.3×10^{-18} mol of S^{2-}/L and 1.26×10^{-17} mol of Ag^+/L. Calculate the solubility product of Ag_2S.

Solution:

$$K_{sp}\,Ag_2S = [Ag^+]^2 \times [S^{2-}]$$
$$= (1.26 \times 10^{-17})^2 \times (6.3 \times 10^{-18})$$
$$= 1.0 \times 10^{-51}$$

17.3 A solution in equilibrium with a precipitate of Ag_3PO_4 was found on analysis to contain 1.6×10^{-15} mol of PO_4^{3-}/L and 4.8×10^{-5} mol of Ag^+/L. Calculate the solubility product of Ag_3PO_4.

17.4 A solution in equilibrium with a precipitate of $Pb_3(PO_4)_2$ was found, on analysis, to contain 3.4×10^{-7} mol of PO_4^{3-}/L and 5.1×10^{-7} mol of Pb^{2+}/L. Calculate the solubility product of $Pb_3(PO_4)_2$.

17.5 A solution in equilibrium with a precipitate of Ag_3PO_4 was found on analysis to contain 1.52×10^{-3} g of PO_4^{3-}/L and 5.18×10^{-3} g of Ag^+/L. Calculate the solubility product of Ag_3PO_4.

Hint: First find the concentration of each ion in moles per liter. Then solve as in Prob. 17.3.

17.6 A 450 cm^3 volume of 1.00×10^{-4} F $BaCl_2$ is placed in a beaker. In order to just start precipitation of $BaSO_4$ it is necessary to add, with constant stirring, 350 cm^3 of 2.00×10^{-4} F K_2SO_4. What is the solubility product of $BaSO_4$?

Hint: Note that the final volume is 800 cm^3. You should note in this problem as well as in later problems that when the anion SO_4^{2-} of the weak acid HSO_4^- is dissolved in water, proton transfer occurs according to the equation

$$SO_4^{2-}(aq) + H_2O \rightleftarrows HSO_4^-(aq) + OH^-(aq)$$

Therefore, 2.00×10^{-4} *F* K_2SO_4 will not be *exactly* 2.00×10^{-4} *F* in SO_4^{2-}. The value of the equilibrium constant for accepting a proton from water, calculated as in Prob. 14.33, is 8.3×10^{-13}. That means that the amount of proton transfer is so small that, within the limits of the precision of our measurements, it can be neglected. Only when the anion is a very strong base, such as S^{2-}, PO_4^{3-}, and CN^-, and the precision is very high must the effects of proton transfer from water be considered.

17.7 In each of the following a saturated solution was prepared by shaking the pure solid compound with pure water. The solubilities obtained are given. From these solubilities, calculate the solubility product of each solute. (In each instance ignore the acid–base reactions of each ion.)

a. AgCl; 1.67×10^{-5} mol AgCl/L

Solution: Since AgCl ionizes completely according to $AgCl(s) \rightleftarrows Ag^+(aq) + Cl^-(aq)$, a saturated solution containing 1.67×10^{-5} mol of AgCl/L will contain 1.67×10^{-5} mol of Ag^+/L and 1.67×10^{-5} mol of Cl^-/L.

$$\begin{aligned} K_{sp} &= [Ag^+] \times [Cl^-] \\ &= 1.67 \times 10^{-5} \text{mol } Ag^+/L \times 1.67 \times 10^{-5} \text{ mol } Cl^-/L \\ &= 2.8 \times 10^{-10} \end{aligned}$$

b. AgI; 2.2×10^{-3} mg AgI/L

c. AgBr; 5.7×10^{-10} equivalent AgBr/cm^3

d. $BaCrO_4$; 1.4×10^{-5} mol $BaCrO_4$/L

e. Ag_2SO_4; 1.4×10^{-2} mol Ag_2SO_4/L

Solution: From the equation for the ionization of Ag_2SO_4,

$$Ag_2SO_4(s) \rightleftarrows 2Ag^+(aq) + SO_4^{2-}(aq)$$

it is evident that 1 mol of Ag_2SO_4 produces 2 mol of Ag^+ and 1 mol of SO_4^{2-}. There fore, 1.4×10^{-2} mol of Ag_2SO_4 will produce 2.8×10^{-2} mol of Ag^+ and 1.4×10^{-2} mol of SO_4^{2-}.

$$\begin{aligned} K_{sp} &= [Ag^+]^2 \times [SO_4^{2-}] \\ &= (2.8 \times 10^{-2})^2 \times 1.4 \times 10^{-2} \\ &= 1.1 \times 10^{-5} \end{aligned}$$

f. PbI_2; 1.28×10^{-3} *M*

17.8 What concentration of Ag^+ must be present to just start precipitation of AgCl from a solution containing 1.0×10^{-4} mol of Cl^-/L? The solubility product of AgCl is 2.8×10^{-10}.

Solution: A substance will start to precipitate when the product of the concentrations of its ions equals (or just barely exceeds) the solubility product. No precipitate will form until the product of the concentrations of its ions equals the solubility product. In this particular problem, precipitation of AgCl will not begin until the product of the molar concentrations of the ions involved equals the solubility product of AgCl.

$$K_{sp} = [Ag^+] \times [Cl^-] = 2.8 \times 10^{-10}$$

$$[Ag^+] = \frac{2.8 \times 10^{-10}}{[Cl^-]} = \frac{2.8 \times 10^{-10}}{1 \times 10^{-4} \text{ mol } Cl^-/L}$$

$$= 2.8 \times 10^{-6} \text{ mol } Ag^+/L$$

17.9 What concentration of OH^- is necessary to start precipitation of $Fe(OH)_3$ from a solution containing 2×10^{-6} mol of Fe^{3+}/L? The solubility product of $Fe(OH)_3$ is 6×10^{-38}.

17.10 What concentration of sulfide ion must be present to just start precipitation of the sulfide of the metal from each of the following solutions? The solubility product of each of the sulfides precipitated is given.

a. 1.0 *F* $CuCl_2$; 4×10^{-36}
b. 0.2 *F* $FeCl_2$; 4×10^{-17}
c. 0.0010 *F* $CdCl_2$; 6×10^{-27}
d. 0.1 *F* $BiCl_3$; 1×10^{-70}

17.11 The solubility of PbI_2 in water is 2×10^{-3} mol/L. What concentration of lead ion would be required to just start precipitation of PbI_2 from 0.002 *F* KI?

17.12 The solubility product of Ag_3PO_4 is 1.8×10^{-18}. Assuming that a precipitate can be seen as soon as it begins to form, what is the minimum concentration of PO_4^{3-} in milligrams per liter that can be detected by the addition of Ag^+ until the solution is 0.010 *M* in silver ions?

17.13 Amounts of 0.00005 mol of soluble, 100% ionized iron(III) sulfate and 0.00001 mol of soluble, 100% ionized barium hydroxide are added to enough water to give 1 L solution. Will a precipitate form?

$$K_{sp} \text{ of } BaSO_4 = 1.5 \times 10^{-9}\,; K_{sp} \text{ of } Fe(OH)_3 = 6 \times 10^{-38}$$

17.14 To a solution containing 0.010 mol of Ag^+, Cl^- was added; the final volume was 1 L and 7.0×10^{-3} mol of AgCl precipitated. How much Cl^- remained in solution? K_{sp} of AgCl is 2.8×10^{-10}.

17.15 A suspension of calcium hydroxide in water was found to have a pH of 12.3. Calculate the solubility product of $Ca(OH)_2$.

Hint: How does $[OH^-]$ compare with $[Ca^{2+}]$?

17.16 The molar concentration of the Cd^{2+} in a solution in equilibrium with a precipitate of CdS was found to be 4 times as great as the molar concentration of the S^{2-}. K_{sp} of CdS is 6×10^{-27}. What was the concentration of the Cd^{2+}?

Solution:

$$K_{sp} = [Cd^{2+}] \times [S^{2-}] = 6 \times 10^{-27}$$

Let X = concentration of Cd^{2+}

$$\frac{X}{4} = \text{concentration of } S^{2-}$$

Substituting these values in the equation for K_{sp}, we have

$$X \times \frac{X}{4} = 6 \times 10^{-27}$$

$$X^2 = 2.4 \times 10^{-26}$$

$$X = 1.6 \times 10^{-13}\ M$$

17.17 One liter of solution that is in equilibrium with a precipitate of $Cd(OH)_2$ contains 4 times as many moles of OH^- as Cd^{2+}. What amount of OH^- is present? K_{sp} of $Cd(OH)_2$ is 1.6×10^{-14}.

17.18* You are given equal volumes of two lead salt solutions in which the concentration of Pb^{2+} is exactly the same. To one is added 4.00×10^{-1} mol of KCl and to the other is added 1.00×10^{-3} mol of Na_2SO_4. The final volume is in each case 1 L. A total of 0.103 g of $PbCl_2$ precipitates from one solution and 0.136 g of $PbSO_4$ precipitates from the other. The solubility product of $PbSO_4$ is 1.10×10^{-8}. Formula weights: $PbCl_2$ = 278; $PbSO_4$ = 303. Calculate the solubility product of $PbCl_2$.

17.19 From the respective solubility products at 20 °C, calculate the solubility of each of the following in moles per liter. The solubility product of each solute is given directly after its formula. (Ignore acid–base reactions.)

a. AgSCN; 1×10^{-12}

Solution: When AgSCN dissolves, it is 100% dissociated into Ag^+ and SCN^-. Therefore,

$$\text{moles of AgSCN dissolved} = \text{moles of } Ag^+ = \text{moles of } SCN^-$$

Let X = moles of AgSCN dissolved. Then

$$X = \text{moles of } Ag^+ \quad X = \text{moles of } SCN^-$$

$$K_{sp} = [Ag^+] \times [SCN^-] = X^2 = 1 \times 10^{-12}$$

$$X = 1 \times 10^{-6}\ M$$

b. AgCl; 2.8×10^{-10}
c. $Mg(OH)_2$; 8.9×10^{-12}

Solution:

$$Mg(OH)_2 \rightleftarrows Mg^{2+}(aq) + 2OH^-(aq)$$

Let X = moles of $Mg(OH)_2$ that dissolve. Then

$$X = \text{moles of } Mg^{2+} \quad 2X = \text{moles of } OH^-$$

$$K_{sp} = [Mg^{2+}] \times [OH^-]^2 = X \times (2X)^2 = 8.9 \times 10^{-12}$$

$$4X^3 = 8.9 \times 10^{-12}$$

$$X = 1.3 \times 10^{-4}\ M$$

d. Ag_2SO_4; $1.1\ 10^{-5}$
e. $Al(OH)_3$; 5×10^{-33}

17.20 The solubility product of PbI_2 at 30 °C is 1×10^{-8}. The solubility product of $BaSO_4$ at 30 °C is also 1×10^{-8}. How does the solubility of PbI_2 in moles per liter compare with the solubility of $BaSO_4$ in moles per liter?

17.21 The solubility product of AgCl is 2.8×10^{-10}. What amount of AgCl will dissolve in 1 L of 0.010 *F* KCl? The KCl is 100% ionized.

Solution:

$$[Ag^+] \times [Cl^-] = 2.8 \times 10^{-10}$$

$$[Ag^+] = \frac{2.8 \times 10^{-10}}{[Cl^-]} = \frac{2.8 \times 10^{-10}}{1 \times 10^{-2}} = 2.8 \times 10^{-8}\ M$$

To produce this 2.8×10^{-8} mol of Ag^+, 2.8×10^{-8} mol of AgCl must have gone into solution. In making this calculation we ignored the Cl^- derived from the AgCl since its concentration is negligible, being about $2.8 \times 10^{-8}\ M$.

17.22 The solubility of $BaSO_4$ in water is $1 \times 10^{-5}\ M$. What is its solubility in 0.1 *F* K_2SO_4?

17.23 What volume of 0.10 F $MgCl_2$ is required to dissolve the same amount of Hg_2Cl_2 that will dissolve in 1.00 L of pure water? K_{sp} of Hg_2Cl_2 is 4.0×10^{-18}.

Solution: Hg_2Cl_2 ionizes as follows: $Hg_2Cl_2(s) \rightleftarrows Hg_2^{2+}(aq) + 2Cl^-(aq)$. K_{sp} for $Hg_2Cl_2 = [Hg_2^{2+}] \times [Cl^-]^2 = 4.0 \times 10^{-18}$.

Its solubility in pure water, calculated as in Prob. 17.19, is 1.0×10^{-6} mol/L. Its solubility in 0.10 F $MgCl_2$, in which $[Cl^-]$ is 0.20 M, calculated as in Prob. 17.21, is 1.0×10^{-16} mol/L.

$$\frac{1.0 \times 10^{-6} \text{ mol/L of water}}{1.0 \times 10^{-16} \text{ mol/L of } 0.10\ F\ MgCl_2} = 1.0 \times 10^{10} \text{ L of } 0.10\ F\ MgCl_2\text{/L of water.}$$

17.24 When excess solid Ag_2CrO_4 is shaken with 1 L of 0.10 F K_2CrO_4, 0.723 mg of Ag_2CrO_4 dissolves. Calculate the solubility product of Ag_2CrO_4.

17.25 Silver oxide is in equilibrium with its saturated solution according to the reaction $Ag_2O(s) + H_2O \rightleftarrows 2Ag^+(aq) + 2OH^-(aq)$. The solubility product for AgOH is $[Ag^+] \times [OH^-] = 2 \times 10^{-8}$. What amount of Ag_2O will dissolve in 1 L of solution whose pH is 11?

17.26 The first ionization of sulfuric acid, $H_2SO_4(aq) \rightarrow H^+(aq) + HSO_4^-(aq)$, is 100% complete. The ionization constant for the second ionization, $HSO_4^-(aq) \rightleftarrows H^+(aq) + SO_4^{2-}(aq)$, is 1.2×10^{-2}. The solubility product of $BaSO_4$ is 1.0×10^{-10}. Excess solid $BaSO_4$ was shaken with a solution of sulfuric acid until a saturated solution of $BaSO_4$ was obtained. The pH of this saturated solution was 2. What amount of $BaSO_4$ dissolved per liter of saturated solution?

Hint: K_{sp} for $BaSO_4$ being 1.0×10^{-10}, the maximum amount of $BaSO_4$ that can dissolve in 1 L of pure water is 1.0×10^{-5}. The amount that will dissolve in a solution of H_2SO_4 whose pH is 2 will be much less than 1.0×10^{-5}.

The amount of Ba^{2+} ions in this solution will be equal to the amount of $BaSO_4$ that dissolves. The amount of SO_4^{2-} ions in this solution will equal the amount initially present (from the H_2SO_4) plus the number derived from the $BaSO_4$ that dissolves. The number derived from the latter source is so small compared with the number initially present that it can be ignored in the calculation.

Therefore, to solve the problem first calculate the concentration of SO_4^{2-} ions in a solution of H_2SO_4 whose pH is 2 in the manner outlined in

Prob. 15.46. This value, divided into K_{sp} for $BaSO_4$, gives the concentration of Ba^{2+} and, hence, the amount of $BaSO_4$ that dissolves. The smallness of the answer justifies the assumption made in solving the problem.

17.27* A solution in equilibrium with solid CaC_2O_4 has a pH of 4.0. The sum of the $C_2O_4^{2-}$, $HC_2O_4^-$, and $H_2C_2O_4$ in the solution is 0.20 *F*. Calculate the concentration of Ca^{2+} ions in the solution. K_{sp} of $CaC_2O_4 = 1.3 \times 10^{-9}$.

Hint: See Prob. 16.52.

17.28* A solution contains 0.010 mol Cl^-/L and 0.0010 mol CrO_4^{2-}/L. The K_{sp} of AgCl is 1.56×10^{-10}, and the K_{sp} of Ag_2CrO_4 is 9.0×10^{-12}. What will be the concentration of Cl^- in moles per liter when Ag_2CrO_4 just begins to precipitate on continued addition of Ag^+, the volume of the solution at this point being exactly 1 L?

Solution: When Ag^+ ion is added to the solution represented by this problem, AgCl will begin to precipitate when the product of $[Ag^+]$ and $[Cl^-]$ equals the solubility product, 1.56×10^{-10}. Since $[Cl^-]$ is 1×10^{-2}, the precipitation of AgCl will begin when $[Ag^+]$ is 1.56×10^{-8}. Since the solubility product of Ag_2CrO_4 is 9.0×10^{-12} and $[CrO_4^{2-}]$ is 1×10^{-3}, precipitation of Ag_2CrO_4 will not begin until $[Ag^+]$ is 9.5×10^{-5}. Therefore, at the start, only AgCl precipitates. As more Ag^+ ion is added after initial precipitation of AgCl, more AgCl will precipitate. As more AgCl precipitates, the concentration of the Cl^- ions remaining in solution decreases, and as the concentration of Cl^- ions decreases, the concentration of Ag^+ ions required to continue precipitation of AgCl increases; during the entire AgCl precipitation process the product of $[Ag^+]$ and $[Cl^-]$ must always be equal to 1.56×10^{-10}. Finally, the concentration of Cl^- ions will be low enough so that an Ag^+ ion concentration of 9.5×10^{-5} will be required to precipitate more AgCl. When that happens, Ag_2CrO_4 will also begin to precipitate. Since $[Ag^+] \times [Cl^-]$ must always equal 1.56×10^{-10}, when $[Ag^+]$ is 9.5×10^{-5}, $[Cl^-]$ will be $1.56 \times 10^{-10} \div 9.5 \times 10^{-5}$ or 1.6×10^{-6} mol/L.

17.29* A solution contains 0.000020 mol of Br^-/L and 0.010 mol of Cl^-/L. K_{sp} of AgCl is 1.56×10^{-10}; the K_{sp} of AgBr is 3.25×10^{-13}. Which of these ions will start precipitating first when Ag^+ is added to the above solution? What will be its concentration when the other ion begins to precipitate?

17.30* The solubility products of $AgIO_3$ and $Ba(IO_3)_2$ are 1.0×10^{-8} and 6.0×10^{-10}, respectively. A solution is 8.6×10^{-4} *M* in Ag^+ and 3.74×10^{-3} *M* in Ba^{2+}. Iodate ion is added to this solution slowly and with constant stirring.

a. Which cation precipitates as the iodate salt first? At what IO_3^- ion concentration does this precipitate just start to form?

b. At what IO_3^- ion concentration does the second cation just start to precipitate as the iodate salt?

c. What is the concentration of the first cation when the second cation just starts to precipitate?

17.31* Calculate the concentration of Cl^- in a solution saturated with both AgCl and Ag_2CrO_4, in which the concentration of chromate ion is 1.0×10^{-3} *M*. For Ag_2CrO_4, $K_{sp} = 1.7 \times 10^{-12}$; for AgCl, $K_{sp} = 1.1 \times 10^{-10}$.

17.32* If solid $AgNO_3$ is added to a certain solution 1.00×10^{-7} *F* in KI and 0.100 *F* in NaCl, what amount of AgI per liter will be precipitated before the solution is saturated with AgCl? Solubility products: AgI, 1.00×10^{-16}; AgCl, 1.10×10^{-10}.

17.33* One liter of solution that was in equilibrium with a solid mixture of AgCl and AgI was found to contain 1×10^{-8} mol of Ag^+, 1×10^{-2} mol of Cl^-, and 1×10^{-8} mol of I^-. Enough Ag^+ ions were added, slowly and with constant stirring, to increase the concentration of Ag^+ to 10^{-6} mol/L; the volume of the solution was kept constant at 1 L. What amount of AgCl was precipitated as a result of this addition of Ag^+ ion? What amount of AgI was precipitated as a result of this addition of Ag^+?

17.34* One commonly used titrimetric method of determining the chloride-ion content of a material utilizes the red color of silver chromate as an end-point indicator. A solution of known volume, to which the chromate-ion indicator has been added and which contains a weighed sample of the material, is titrated with a standard $AgNO_3$ solution until the red of the Ag_2CrO_4 just appears. From the measured and observed quantities and the appropriate solubility product constants, the amount of Cl^- present can be determined within a small error. The following data are given:

A 1.7750-g sample is dissolved in 203 cm^3 of water.
A 1-cm^3 volume of 0.00100 *F* K_2CrO_4 is added.
A 46.00-cm^3 volume of 0.250 N $AgNO_3$ was required to produce the red end point.

$$K_{sp} \text{ of AgCl} = 1.50 \times 10^{-10}$$
$$K_{sp} \text{ of } Ag_2CrO_4 = 9.00 \times 10^{-12}$$

a. What is the $[CrO_4^{2-}]$ just as the red color appears?

b. What is the $[Ag^+]$ at this end point?

c. What is the $[Cl^-]$ at this end point?

d. What amount of Ag^+ was added?

e. What amount of AgCl precipitated?

f. What is the total amount of Cl^- present?

g. What is the mass percent of Cl^- in the original sample?

17.35* A 0.20 *F* solution of Na_3PO_4 has a pH of 10.5. What amount of Ag_3PO_4 will dissolve in 1.0 L of this solution? K_{sp} for Ag_3PO_4 is 1.8×10^{-18}.

17.36* The solubility product of $PbSO_4$ is 1.3×10^{-8}. The solubility product of $Pb(ClO_4)_2$ (lead perchlorate) is 2.4×10^{-15}. Perchloric acid ($HClO_4$) is completely ionized. The first ionization of H_2SO_4 is complete; the ionization constant for HSO_4^- is 1.2×10^{-2}. A solution prepared by dissolving pure $HClO_4$ and pure H_2SO_4 in the same beaker of water has a pH of 1.51. To neutralize 100 cm^3 of this solution requires 90.0 cm^3 of 0.050 *F* KOH. What amount of solid $Pb(NO_3)_2$ must be dissolved in 1.00 L of this solution before a precipitate begins to form? Give the formula of the solid that begins to precipitate.

17.37* Calculate the concentration of I^- in a solution obtained by shaking 0.100 *F* KI with an excess of AgCl. K_{sp} of AgCl is 1.1×10^{-10}; K_{sp} of AgI is 1.0×10^{-16}.

Solution: The two equilibria are

$$AgCl(s) \rightleftarrows Ag^+(aq) + Cl^-(aq) \quad \textbf{(a)}$$

$$Ag^+(aq) + I^-(aq) \rightleftarrows AgI(s) \quad \textbf{(b)}$$

The net equation for the principal reaction is

$$AgCl(s) + I^-(aq) \rightleftarrows AgI(s) + Cl^-(aq) \quad \textbf{(c)}$$

The equilibrium constant for reaction (c) is

$$K = \frac{[Cl^-]}{[I^-]} = \frac{[Cl^-] \times [Ag^+]}{[I^-] \times [Ag^+]} = \frac{1.1 \times 10^{-10}}{1.0 \times 10^{-16}} = 1.1 \times 10^6 \quad \textbf{(d)}$$

Let X = $[I^-]$; then

$$0.100 - X = [Cl^-]$$

Substituting these values in the equilibrium-constant equation gives us

$$\frac{0.100 - X}{X} = 1.1 \times 10^6$$

Since $[I^-]$ is only 10^{-6} as large as $[Cl^-]$, and since the maximum value of $[Cl^-]$ is 0.100 *M*, the value of *X* in the expression $0.100 - X$ is so small that it can be dropped.

$$\frac{0.100}{X} = 1.1 \times 10^6$$

$$X = 9.1 \times 10^{-8}\ M = [I^-]$$

Note: Since 0.100 M I^- is added to solid AgCl, a natural inclination when substituting in the equilibrium-constant equation is to let $X = [Cl^-]$ and $0.100 - X = [I^-]$. If this is done, X, being very large (about 0.100), cannot be dropped from the expression $0.100 - X$. The equation will then be,

$$\frac{X}{0.100 - X} = 1.1 \times 10^6$$

Solving, we have

$$X = 1.1 \times 10^5 - 1.1 \times 10^6 X$$
$$1.1 \times 10^6 X + X = 1.1 \times 10^5$$

If X is dropped from the expression, $1.1 \times 10^6, X + X$, the calculated value of X is 0.100. The value of $[I^-]$, since it is $0.100 - X$, will then be $0.100 - 0.100$, or zero. Obviously, X cannot be dropped in this instance.

The correct procedure in situations of this type is to *let X equal that quantity that we know is very small*; X, being very small, can then be dropped from expressions in which it is subtracted from or added to a number that is very large by comparison with X.

Note: The above equilibrium-constant equation tells us that in a saturated solution in equilibrium with the two solids AgCl and AgI containing the common ion Ag^+, the concentrations of the two dissimilar ions, Cl^- and I^-, are to each other as the solubility products of the parent species. That is

$$\frac{[Cl^-]}{[I^-]} = \frac{K_{sp} \text{ for AgCl}}{K_{sp} \text{ for AgI}}$$

This relationship can also be derived as follows: For AgCl,

$$K_{sp} = [Ag^+] \times [Cl^-] = 1.1 \times 10^{-10}$$
$$[Ag^+] = \frac{1.1 \times 10^{-10}}{[Cl^-]}$$

For AgI,

$$K_{sp} = [Ag^+] \times [I^-] = 1.0 \times 10^{-16}$$
$$[Ag^+] = \frac{1.0 \times 10^{-16}}{[I^-]}$$

Since $[Ag^+]$ is the same for both equilibria

$$\frac{1.0 \times 10^{-16}}{[I^-]} = \frac{1.1 \times 10^{-10}}{[Cl^-]}$$

and

$$\frac{[\text{Cl}^-]}{[\text{I}^-]} = \frac{1.1 \times 10^{-10}}{1.0 \times 10^{-16}} = \frac{K_{sp} \text{ of AgCl}}{K_{sp} \text{ of AgI}}$$

This relationship will be an important one to keep in mind while solving certain problems.

17.38* Calculate the concentration of Ag^+ in a solution prepared by mixing 100 cm^3 of a solution 0.200 *F* in both NaCl and KI with 100 cm^3 of 0.100 *F* $AgNO_3$. $K_{sp} = 1.1 \times 10^{-10}$ of AgCl and 1.0×10^{-16} of AgI.

17.39 A solution is 0.10 M in Cl^-, in Br^-, and in I^-. To 1.00 L of this solution is added 0.15 mol of $AgNO_3$. What are the final concentrations of Cl^-, Br^-, and I^- in the solution? $K_{AgCl} = 1.5 \times 10^{-10}$; $K_{AgBr} = 5.0 \times 10^{-13}$; $K_{AgI} = 8.3 \times 10^{-17}$.

17.40* One liter of solution known to contain Zn^{2+} and Ni^{2+} in equal molar concentrations was kept saturated with H_2S. When precipitation was complete it was found that

a. The volume of the solution in equilibrium with the precipitate was 1 L.
b. The pH of the solution was 4.000.
c. Of the Ni^{2+} originally present, 99.000% was precipitated as NiS.

What percent of the Zn^{2+} originally present was precipitated as ZnS? Solubility products: ZnS, 1.3×10^{-20}; NiS, 1.3×10^{-22}.

17.41* The solubility of $CaCO_3$ in water at 25 °C is 1.3×10^{-4} mol/L. The solubility product calculated from this solubility is 4.8×10^{-9}, not 1.7×10^{-8}. Explain.

Hint: Proton transfer from water to CO_3^{2-} is not ignored.

17.42* To 1 L of a 1 *F* solution of Na_2CO_3 is added 10^{-7} mol of $MgCl_2$. Will a precipitate form and, if so, what is the precipitate? K_{sp} of $MgCO_3 = 4.0 \times 10^{-5}$; K_{sp} of $Mg(OH)_2 = 1.3 \times 10^{-11}$.

Hint: Calculate, from K_2 of H_2CO_3, the concentrations of CO_3^{2-} and OH^- in 1 *F* Na_2CO_3.

17.43* A certain trivalent metal ion forms an insoluble hydroxide and an insoluble carbonate; the solubility products are 1.0×10^{-20} and 5.5×10^{-25}, respectively. If a mixture of a very small amount of these two solids is placed in 0.20 *F* Na_2CO_3, will the conversion hydroxide → carbonate occur, or will the reverse change take place?

SOLUBILITY PRODUCTS AND THE HYDROGEN SULFIDE EQUILIBRIUM

A saturated solution of hydrogen sulfide in water at 18 °C and standard barometric pressure is approximately 0.10 *M* in H_2S. Hydrogen sulfide is a very weak acid; hence, its percent ionization is small. Although H_2S ionizes in two stages,

$$H_2S(aq) \rightleftarrows H^+(aq) + HS^-(aq) \qquad K_1 = 1.0 \times 10^{-7}$$
$$HS^-(aq) \rightleftarrows H^+(aq) + S^{2-}(aq) \qquad K_2 = 1.3 \times 10^{-13}$$

the overall ionization can be represented by one equation,

$$H_2S(aq) \rightleftarrows 2H^+(aq) + S^{2-}(aq)$$

The ionization constant for this overall reaction is 1.3×10^{-20}. That is,

$$\frac{[H^+] \times [S^{2-}]}{[H_2S]} = 1.3 \times 10^{-20}\ M^2$$

When the concentration of H^+ is fixed by addition of acid or base to the solution, this overall ionization constant can justifiably be used to calculate $[S^{2-}]$. Since the solution is saturated, the concentration of H_2S will be constant, namely 0.10 *M*. We can, therefore, combine this constant value with the ionization constant to get the equation

$$\frac{[H^+]^2 \times [S^{2-}]}{0.10} = 1.3 \times 10^{-20}\ M^2$$
$$[H^+]^2 \times [S^{2-}] = 1.3 \times 10^{-21}\ M^3$$

This constant, 1.3×10^{-21}, is the *ion product* for a saturated solution of H_2S. It is a very useful constant in calculations involving reactions in which an acidified or alkalized saturated solution of hydrogen sulfide is either a reactant or a product.

PROBLEMS

17.44* Solutions containing 1×10^{-6} mol/L of Hg^{2+}, Cu^{2+}, Pb^{2+}, Sn^{2+}, Ni^{2+}, Fe^{2+}, and Mn^{2+}, respectively, are saturated with H_2S at 18 °C. The ion product $[H^+]^2 \times [S^{2-}]$ for a saturated (0.10 *M*) solution of H_2S is 1.3×10^{-21}. In each case what is the greatest H^+ concentration that will just allow precipitation of the sulfide to start? The solubility product of each sulfide is given.

a. Hg^{2+}; 1×10^{-50}

Solution:

$$[Hg^{2+}] \times [S^{2-}] = 1 \times 10^{-50}$$
$$[S^{2-}] = \frac{1 \times 10^{-50}}{1 \times 10^{-6}\ \text{mol Hg}^{2+}/\text{L}} = 1 \times 10^{-44}\ M$$

This is the concentration of S^{2-} that must be present for precipitation of HgS to start from a solution containing 1×10^{-6} mol/L of Hg^{2+}.

$$[H^+]^2 \times [S^{2-}] = 1.3 \times 10^{-21}$$

$$[H^+]^2 = \frac{1.3 \times 10^{-21}}{1 \times 10^{-44} \text{ mol } S^{2-}/L} = 1.3 \times 10^{23}$$

$$[H^+] = 3.6 \times 10^{11}\ M$$

This is the concentration of H^+ that will be in equilibrium with 1×10^{-44} mol/L of S^{2-}.

b. Cu^{2+}; 4×10^{-36}
c. Pb^{2+}; 4×10^{-26}
d. Sn^{2+}; 1×10^{-24}
e. Ni^{2+}; 1×10^{-22}
f. Fe^{2+}; 4×10^{-17}
g. Mn^{2+}; 8×10^{-14}

17.45* A solution containing 1×10^{-6} mol/L of Cd^{2+} was kept saturated with H_2S until precipitation was complete. The concentration of H^+ was kept at 0.2 M during the precipitation. What mass of CdS was precipitated per liter of solution? K_{sp} for CdS is 6×10^{-27}.

Solution:

$$[H^+]^2 \times [S^{2-}] = 1.3 \times 10^{-21}$$

$$[S^{2-}] = \frac{1.3 \times 10^{-21}}{[H^+]^2} = \frac{1.3 \times 10^{-21}}{(0.2)^2} = 3.25 \times 10^{-20}\ M$$

This is the final concentration of S^{2-} in the solution.

$$[Cd^{2+}] \times [S^{2-}] = 6 \times 10^{-27}$$

$$[Cd^{2+}] = \frac{6 \times 10^{-27}}{[S^{2-}]} = \frac{6 \times 10^{-27}}{3.25 \times 10^{-20}} = 1.85 \times 10^{-7}$$

This is the concentration of Cd^{2+} ions left in the solution.

$$\text{Moles of CdS precipitated} = 1 \times 10^{-6} - 1.85 \times 10^{-7}$$
$$= 8.15 \times 10^{-7}$$
$$\text{Mass of Cd Sprecipitated} = 8.15 \times 10^{-7} \text{ mol } \times 144.5 \text{ g/mol}$$
$$= 1.18 \times 10^{-4} \text{ g } = 1 \times 10^{-4} \text{ g}$$

17.46* The solubility product of SnS is 1.1×10^{-24}. One liter of 0.00013 M Sn^{2+} was kept saturated with H_2S until precipitation was complete, at which time

0.0135 g of SnS had precipitated. What was the pH of the solution at the end of the precipitation?

17.47* What mass of Ag^+ must be present per liter before Ag_2S will start to precipitate from a saturated solution of H_2S whose pH is 2.0? The solubility product of Ag_2S is 1×10^{-50}.

17.48* The solubility product of Cu_2S is 4.4×10^{-49}. What must be the pH of a saturated solution of H_2S containing 2.0×10^{-18} mol/L of Cu^+ for Cu_2S to just barely start precipitating?

17.49* The solubility product of SnS is 1.0×10^{-24}. If a 5.0×10^{-12} M solution of Sn^{2+} whose pH is maintained at 4.0 is saturated with H_2S, will a precipitate form?

17.50* A solution is 0.0000010 M with respect to Pb^{2+} and 0.0050 M with respect to Cu^{2+}. If the solution is kept saturated with H_2S, what is the hydrogen-ion concentration that will permit the maximum precipitation of CuS but will not allow the precipitation of PbS? K_{sp} for CuS is 3.5×10^{-38} and that for PbS is 1.0×10^{-29}.

17.51* If NiS just begins to precipitate from a 0.0010 F $NiCl_2$ solution saturated with H_2S when the pH is 1.0, what is the solubility product of NiS?

SOLUBILITY PRODUCTS AND THE AMMONIA EQUILIBRIUM

PROBLEMS

(See Tables A.2 and A.4 of the appendix for values of equilibrium constants.)

17.52* When excess solid $Mg(OH)_2$ is shaken with 1.00 L of 1.0 F NH_4Cl, the resulting saturated solution has a pH of 9.0. The net equation for the reaction that occurs is

$$Mg(OH)_2(s) + 2NH_4^+(aq) \rightleftarrows Mg^{2+}(aq) + 2NH_3(aq) + 2H_2O$$

Calculate the solubility product of $Mg(OH)_2$.

Solution:

$$NH_3(aq) + H_2O \rightleftarrows 2NH_4^+(aq) + OH^-(aq)$$

$$K = \frac{[NH_4^+] \times [OH^-]}{[NH_3]} = 1.8 \times 10^{-5}$$

Since pH = 9.0, $[H^+] = 1.0 \times 10^{-9}$ and $[OH^-] = 1.0 \times 10^{-5}$. $K_{sp} = [Mg^{2+}] \times [OH^-]^2$. Since we know $[OH^-]$, all we need is to find $[Mg^{2+}]$. The important fact to notice is that the molar concentration of Mg^{2+} is one-half the molar concentration of NH_3. The molar concentration of the NH_3 can be calculated as follows:

From the ionization constant formula for NH_3 we can determine that

$$\frac{[NH_4^+]}{[NH_3]} = \frac{1.8 \times 10^{-5}}{[OH^-]} = \frac{1.8 \times 10^{-5}}{1.0 \times 10^{-5}} = 1.8$$

Since we started with 1.00 L of 1.0 F NH_4Cl, $[NH_4^+] + [NH_3] = 1.0$. Let $X = [NH_3]$; $1.0 - X = [NH_4^+]$.

$$\frac{1.0 - X}{X} = 1.8$$

Solving, we have

$$X = 0.36 = [NH_3]$$

$$\frac{1}{2}X = 0.18 = [Mg^{2+}]$$

$$K_{sp} = [Mg^{2+}] \times [OH^-]^2 = 0.18 \times (1.0 \times 10^{-5})^2 = 1.8 \times 10^{-11}$$

17.53* Excess $Mg(OH)_2$ is added to a solution that is 0.20 F in NH_4NO_3 and 0.50 F in NH_3. Calculate the concentration of Mg^{2+} ions at equilibrium. K_{sp} of $Mg(OH)_2 = 1.2 \times 10^{-11}$.

17.54* The K_{sp} of $Mg(OH)_2$ is 8.9×10^{-12}. What mass of NH_4^+ must be present in 1.0 L of 0.10 F NH_3 containing 0.30 g of Mg^{2+} to prevent $Mg(OH)_2$ from being precipitated?

Hint: First calculate the maximum $[OH^-]$ that will just fail to produce a precipitate of $Mg(OH)_2$ in the solution. Then calculate the amount of NH_4^+ that must be present in 0.10 F NH_3 to give this $[OH^-]$.

17.55* The K_{sp} of $Mn(OH)_2$ is 2.0×10^{-13}. What mass of NH_4Cl must be present in 100 cm^3 of 0.20 F NH_3 to prevent precipitation of $Mn(OH)_2$ when the solution is added to 100 cm^3 of 0.20 F $MnCl_2$?

17.56* You are given 200 cm^3 of a solution containing Mn^{2+} and Mg^{2+}, each at 0.01 M, and 200 cm^3 of a solution 0.40 F in ammonia. What mass of solid ammonium chloride should be added to the latter so that when the solutions are mixed, the $Mn(OH)_2$ will be precipitated as completely as possible but the $Mg(OH)_2$ will remain unprecipitated? K_{sp} of $Mg(OH)_2$ is 1.4×10^{-11}, that for $Mn(OH)_2$ is 4.5×10^{-14}.

Hint: Note that NH_4^+ is liberated in the precipitation of $Mn(OH)_2$.

17.57* The K_{sp} of $Fe(OH)_3$ is 6×10^{-38}. How much NH_4^+ must be present in order to prevent the precipitation of $Fe(OH)_3$ in a solution 0.1 F in NH_3 and 0.0010 M in Fe^{3+}? Would it be possible to dissolve that much NH_4^+ in a liter?

SOLUBILITY PRODUCT EQUILIBRIA INVOLVING SELECTED WEAK ACIDS

PROBLEMS

(See Table A.2 of the appendix for values of ionization constants.)

17.58* The solubility product of $AgC_2H_3O_2$ is 4.0×10^{-4}. To 1.0 L of a solution 1.0 F in $HC_2H_3O_2$ and 0.10 F in HNO_3 is added just enough solid $AgNO_3$ to start precipitation of $AgC_2H_3O_2$. What amount of $AgNO_3$ is added?

17.59* A solution was prepared by dissolving 1.80 mol of $NaC_2H_3O_2$ and 1.00 mol of $HC_2H_3O_2$ in enough water to give 1.00 L of solution. What is the maximum concentration of Fe^{3+} that can exist in this solution without precipitation of $Fe(OH)_3$? The solubility product of $Fe(OH)_3$ is 6.0×10^{-38}.

17.60* The K_{sp} of $Fe(OH)_3$ is 6.0×10^{-38}.

a. What is the formal solubility of $Fe(NO_3)_3$ in a solution that is 0.20 F in NH_3 and 0.36 F in NH_4NO_3?

b. What volume of this NH_3–NH_4NO_3 solution is needed to dissolve the same amount of $Fe(NO_3)_3$ that will dissolve in 1.00 L of a solution that is 0.10 F in HCOOH and 0.42 F in HCOOK?

17.61* A saturated solution was prepared by shaking excess solid $CaCO_3$ with water containing a small amount of HCl. When equilibrium was established, the pH of the saturated solution was found to be 7.0. What amount of $CaCO_3$ dissolved per liter of solution? The solubility product of $CaCO_3$ is 7×10^{-9}.

Hint: Let $X = [Ca^{2+}]$ = moles of $CaCO_3$ dissolved. X will then equal $[CO_3^{2-}] + [HCO_3^-] + [H_2CO_3]$. Knowing K_1 and K_2 for H_2CO_3 and knowing the pH of the solution the relative amounts of CO_3^{2-}, HCO_3^-, and H_2CO_3 are known. The necessary equations for calculating X can then be set up.

17.62* The solubility of $BaCO_3$ in water saturated with CO_2 at 1 atm is 0.01 mol/L. The concentration of H_2CO_3 in this solution is 0.04 M. The net equation is $BaCO_3(s) + H_2CO_3(aq) \rightleftarrows Ba^{2+}(aq) + 2HCO_3^-(aq)$. Calculate the equilibrium constant for this reaction and calculate the solubility product of $BaCO_3$.

17.63* One liter of water in contact with excess solid $BaCO_3$ was kept saturated with CO_2. When equilibrium was established in the reaction

$$BaCO_3(s) + H_2CO_3(aq) \rightleftarrows Ba^{2+}(aq) + 2HCO_3^-(aq)$$

the concentration of the H_2CO_3 was 0.040 *M*. Calculate the concentration of the CO_3^{2-} ion in the solution. The solubility product of $BaCO_3$ is 1.0×10^{-8}.

17.64* When excess solid $BaSO_3$ is added to 1.0 L of dilute HCl, 4.0×10^{-4} mol of $BaSO_3$ dissolves. No SO_2 gas is evolved in the process, and no complex ions are formed. The pH of the resulting solution is 5.0. Calculate the solubility product of $BaSO_3$.

17.65* In order to just prevent precipitation of BaF_2 in a solution that is 0.10 *F* in $BaCl_2$ and 0.10 *F* in KF, it is necessary to adjust the pH to a value of 1.96 by addition of HCl. Calculate the solubility product of BaF_2.

17.66* A solution containing 0.01 *M* Zn^{2+}, 0.1 *F* acetic acid, and 0.05 *F* NaOAc is saturated with H_2S. What concentration of Zn^{2+} remains in solution? K_{sp} of ZnS is 1.3×10^{-20}.

Hint: Note that 2 mol of H^+ is liberated for each mole of ZnS precipitated.

17.67* A solution contains 0.01 *F* $Ca(NO_3)_2$, 0.01 *F* $Sr(NO_3)_2$, and 0.5 *F* oxalic acid. To what value should the hydrogen-ion concentration be adjusted by addition of HCl or NaOH in order to precipitate as much as possible of the calcium while leaving all the strontium in solution? K_{sp} of calcium oxalate is 2.6×10^{-9}; K_{sp} of strontium oxalate is 7.0×10^{-8}.

Hint: Note that 1 mol of oxalate is removed for each mole of CaC_2O_4 precipitated.

COMPLEX IONS AND SOLUBILITY PRODUCTS

When a cation (called the *central ion*) combines with one or more anions or neutral molecules (called *ligands*) to form a new ion, this new ion is called a *complex ion*. Complex ions are often weak electrolytes and as such dissociate incompletely to form the original species. We can write equations for these processes:

$$Zn^{2+}(aq) + 4NH_3(aq) \rightleftharpoons Zn(NH_3)_4^{2+}(aq)$$

and

$$Zn(NH_3)_4^{2+}(aq) \rightleftharpoons Zn^{2+}(aq) + 4NH_3(aq)$$

The equilibrium constants for ionization of complex ions are referred to as *instability constants*, which may be represented by K_{inst}. (See Table A.3 of the appendix.) Just as H_3PO_4 and other polyprotic acids ionize in stages, each stage having its own ionization constant K_1, K_2, or K_3, with the overall ionization constant being the product of K_1, K_2, and K_3, so complex ions also dissociate in stages, with each stage having its own instability constant.

$$Zn(NH_3)_4^{2+}(aq) \rightleftharpoons Zn(NH_3)_3^{2+}(aq) + NH_3(aq)$$

$$Zn(NH_3)_3^{2+}(aq) \rightleftarrows Zn(NH_3)_2^{2+}(aq) + NH_3(aq)$$
$$Zn(NH_3)_2^{2+}(aq) \rightleftarrows Zn(NH_3)^{2+}(aq) + NH_3(aq)$$
$$Zn(NH_3)^{2+}(aq) \rightleftarrows Zn^{2+}(aq) + NH_3(aq)$$

The overall instability constant is the product of the four individual constants and is represented by the formula

$$K_{inst} = \frac{[Zn^{2+}] \times [NH_3]^4}{[Zn(NH_3)_4^{2+}]}$$

Just as the overall ionization constant of polyprotic acids can be used to calculate the concentration of the anion only if excess H^+ or OH^- is present in the system, so with complex-ion equilibria the overall instability constant can be used only when excess ligand is present. In all problems involving complex ions excess ligand will be present; accordingly, overall instability constants can be used.

PROBLEMS

(See Tables A.2 and A.3 of the appendix for values of equilibrium constants.)

17.68* Calculate the concentration of Cu^{2+} in a solution that is 0.10 F in $CuSO_4$ and 1.40 F in NH_3. The instability constant for $Cu(NH_3)_4^{2+}$ is 4.7×10^{-15} mol^4/L^4.

Solution: The very low value of the instability constant means that essentially all the copper ion in solution will be in the form of the ion complex. Therefore, $[Cu(NH_3)_4^{2+}] = 0.10$ mol/L. Then, $[NH_3] = 1.40 - 4(0.10) = 1.00$ mol/L

$$\frac{[Cu^{2+}]\,[NH_3]^4}{[Cu(NH_3)_4^{2+}]} = 4.7 \times 10^{-15}\ \text{mol}^4/\text{L}^4$$

and

$$[Cu^{2+}] = 4.7 \times 10^{-16}\ \text{mol/L}$$

The low $[Cu^{2+}]$ thus calculated justifies the assumption that $Cu(NH_3)_4^{2+}$ is by far the predominant copper-containing species. It should be noted that, in effect, the NH_3 acts as a *buffer*; it ties up the Cu^{2+} in the form of the weak electrolyte $Cu(NH_3)_4^{2+}$.

17.69* The instability constant for $Ag(NH_3)_2^+$ is 6.0×10^{-8} mol^2/L^2.

a. What is the molar concentration of NH_3 needed to convert exactly 50% of the silver ion to the complex ion in Y F $AgNO_3$?

b. What is the formal concentration of NH_3 in this solution?

Hint: At equilibrium $[Ag^+] = Y/2$ and $[Ag(NH_3)_2^+] = Y/2$. Let $X = [NH_3]$ at equilibrium. The formal concentration of NH_3 will be equal to $[NH_3] + 2 \times [Ag(NH_3)_2^+]$.

17.70* The element Q has stable oxidation states of 2, 3, and 4. A mixture of mass 207 g contains *a* moles of QCl_2, *b* moles of $Q_2(SO_4)_3$, and *c* moles of $Q(CrO_4)_2$.

1. A solution containing 1.35 mol of Ba^{2+} is needed to precipitate all the sulfate and chromate in a 207-g sample.
2. A second identical 207-g sample requires 0.180 mol of MnO_4^- to oxidize all the Q to the +4 state (MnO_4^- is reduced to Mn^{2+}).
3. A third identical 207-g sample requires 2.60 mol of ammonia to completely convert Q^{2+} and Q^{4+} to their very stable ammine complexes, $Q(NH_3)_4^{2+}$ and $Q(NH_3)_6^{4+}$ (Q^{3+} forms neither a stable ammine complex nor an insoluble hydroxide).

a. What amount of each compound was present in a 207-g sample of this mixture?

b. Calculate the atomic weight of Q.

17.71* $Ga(OH)_3$ is practically insoluble in water, its solubility product being about 1.0×10^{-36}. A beaker containing 1 L of a saturated solution of $Ga(OH)_3$ prepared by stirring solid $Ga(OH)_3$ with water contains 4.0×10^{-4} mol of excess solid $Ga(OH)_3$. When 1.2×10^{-3} mol of solid KOH is dissolved in the 1 L of solution, all the solid $Ga(OH)_3$ dissolves, a very stable complex hydroxo ion being formed. The resulting solution has a pH of 10.9. Calculate the formula of the complex ion of Ga^{3+} and OH^-.

17.72* Cobalt(II) hydroxide $Co(OH)_2$ is practically insoluble in water. It dissolves to some extent in NH_3 to form a very stable complex ion with NH_3 ligands. In an effort to determine the composition of the complex ion, a chemist carried out three experiments involving addition of *excess* solid $Co(OH)_2$ to three different solutions containing NaOH and NH_3. *At equilibrium* the following *molar* concentrations of NH_3, OH^-, and complex ion were found in the solutions:

Solutions	Conc. of NH_3	Conc. of OH^-	Conc. of Complex Ion
First experiment	0.5 *M*	0.1 *M*	3.1×10^{-9} *M*
Second experiment	1.0 *M*	0.1 *M*	2.0×10^{-7} *M*
Third experiment	2.0 *M*	0.1 *M*	1.3×10^{-5} *M*

Calculate the formula of the complex ion.

Hint: The formula of the complex ion is $Co(NH_3)_X^{2+}$.

What will be true of the value of $\dfrac{[Co^{2+}] \times [NH_3]^X}{[Co(NH_3)_X^{2+}]}$ in each experiment?

Since $[OH^-]$ is the same in each experiment, what will be true of $[Co^{2+}]$ in each experiment?

17.73* The solubility product of AgCl is 1.2×10^{-10}. The instability constant for $Ag(NH_3)_2^+$ is 6.0×10^{-8}. What must be the formal concentration of a solution of NH_3 in water if 1.0 L of this solution will just barely dissolve 0.020 mol of AgCl?

Solution: The net equation for the principal reaction is

$$AgCl(s) + 2NH_3(aq) \rightleftarrows Ag(NH_3)_2^+(aq) + Cl^-(aq)$$

Since 0.020 mol of AgCl dissolves, $[Cl^-] = 0.020\ M$.

Since K_{sp} for AgCl $= 1.2 \times 10^{-10}$,

$$[Ag^+] = \frac{1.2 \times 10^{-10}}{0.20 \times 10^{-1}} = 6.0 \times 10^{-9}\ M$$

That means that practically all the dissolved AgCl is present as $Ag(NH_3)_2^+$.

Therefore, $[Ag(NH_3)_2^+] = 0.020\ M$.

The equilibrium constant for the principal reaction is

$$K = \frac{[Ag(NH_3)_2^+] \times [Cl^-]}{[NH_3]^2}$$

$$= \frac{[Ag(NH_3)_2^+] \times [Cl^-] \times [Ag^+]}{[NH_3]^2 \times [Ag^+]}$$

$$= \frac{K_{sp}}{K_{inst}} = \frac{1.2 \times 10^{-10}}{6.0 \times 10^{-8}} = 2.0 \times 10^{-3}$$

Let X = the concentration of NH_3 at equilibrium. Substituting the values of the three species in the above equilibrium-constant equation, we have

$$\frac{0.020 \times 0.020}{X^2} = 2.0 \times 10^{-3}$$

Solving, we obtain $X = 0.45\ M = [NH_3]$ at equilibrium. The formation of the 0.020 mol of $Ag(NH_3)_2^+$ required the consumption of 0.040 mol of NH_3. Therefore, the formal concentration of the original solution of NH_3 was $0.45 + 0.040 = 0.49\ F$.

17.74* Excess solid AgCl is treated with 100 cm^3 of 1 F NH_3. What mass of AgCl will dissolve? The solubility product of AgCl is 1.0×10^{-10}. The instability constant for $Ag(NH_3)_2^+$ is 6.0×10^{-8}.

17.75* The solubility product of $AgIO_3$ is 4.5×10^{-8}. When excess solid $AgIO_3$ is treated with 1.0 L of 1 F NH_3, 85 g of $AgIO_3$ dissolves.

$$AgIO_3(s) + 2NH_3(aq) \rightleftarrows Ag(NH_3)_2^+(aq) + IO_3^-(aq)$$

Calculate the equilibrium constant for the reaction

$$Ag(NH_3)_2^+(aq) \rightleftarrows Ag^+(aq) + 2NH_3(aq)$$

17.76* $K_{sp} = 1.8 \times 10^{-10}$ for AgCl; 5.2×10^{-13} for AgBr. The instability constant for $Ag(NH_3)_2^+$ is 6.0×10^{-8}.

a. Exactly 1 mmol of silver chloride is shaken with exactly 1 L of water. What is the concentration of silver ion in the saturated solution?

b. Solid potassium bromide is added to the solution. What amount of potassium bromide has been added to the solution at the point at which the first bit of silver bromide forms?

c. What amount of potassium bromide has been added at the point at which the last bit of silver chloride disappears?

d. At the point at which the solid is converted completely to silver bromide, addition of potassium bromide is terminated and ammonia is added until all the silver bromide dissolves. What is the concentration of ammonia in this solution?

17.77* The cation M^{2+} forms the complex ion MCl_4^{2-}, whose instability constant is 1.0×10^{-21}. The solubility product of MI_2 is 1.0×10^{-15}. Calculate the amount of Cl^- ions that must be present in 1.0 L of aqueous solution in order for 0.010 mol of MI_2 to dissolve when excess solid MI_2 is added to the 1.0 L of solution.

17.78* K_{sp} for $Zn(OH)_2$ is 1×10^{-16}. When excess $Zn(OH)_2$ is treated with 1.0 L of 0.4 *F* KCN, the reaction

$$Zn(OH)_2(s) + 4CN^-(aq) \rightleftarrows Zn(CN)_4^{2-}(aq) + 2OH^-(aq)$$

occurs. When equilibrium is reached, the pH is 13. Calculate the equilibrium constant for the reaction

$$Zn(CN)_4^{2-}(aq) \rightleftarrows Zn^{2+}(aq) + 4CN^-(aq)$$

17.79* The solubility product of $Cd(OH)_2$ is 2.00×10^{-14}. The instability constant of $Cd(CN)_4^{2-}$ is 1.40×10^{-19}. The ionization constant of HCN is 4.00×10^{-10}. Excess solid $Cd(OH)_2$ is added to 1 L of KCN solution. When equilibrium is established, the pH of the solution is 12.4 and the concentration of $Cd(CN)_4^{2-}$ in the solution is 0.0120 *M*. Calculate the concentrations of Cd^{2+}, CN^-, and HCN in the final solution. Calculate the formal concentration of KCN in the original KCN solution.

17.80* **a.** Silver cyanide AgCN is soluble in water to the extent of 1.34×10^{-4} mg/100 cm^3 of water. On the assumption that when AgCN dissolves, the species present are Ag^+ and CN^-, what is the solubility product of AgCN?

b. Actually, a very stable complex is formed; the constant for the reaction $Ag(CN)_2^-(aq) \rightleftarrows Ag^+(aq) + 2CN^-(aq)$ is $K = 9.0 \times 10^{-22}$. Considering this fact also, find the true solubility product of AgCN.

Solution:

a. The solubility product, calculated as in Prob. 17.7, is 1.0×10^{-16}.

b. Assuming the correctness of (a), the equilibrium system would be $AgCN \rightleftarrows Ag^+ + CN^-$, and $[Ag^+]$ and $[CN^-]$ would each be 1.0×10^{-8} *M*. But the following reaction occurs:

$$Ag^+(aq) + 2CN^-(aq) \rightleftarrows Ag(CN)_2^-(aq)$$

Since $K_{inst} = 9.0 \times 10^{-22}$, this reaction is nearly complete to the right. At equilibrium, $[CN^-] = X$, $[Ag^+] = 5.0 \times 10^{-9} + 0.50X$, and $[Ag(CN)_2^-] = 5.0 \times 10^{-9} - 0.50X$. If we substitute these values for $[CN^-]$, $[Ag^+]$, and $[Ag(CN)_2^-]$ in the formula for the instability constant and solve for X, we will find the correct values of $[Ag^+]$ and $[CN^-]$ and hence the correct value of the solubility product.

17.81* Calculate the solubility of AgI in 0.1 *F* $Hg(NO_3)_2$. The main reaction is $AgI(s) + Hg^{2+}(aq) \rightleftarrows HgI^+(aq) + Ag^+(aq)$.

For the equation $HgI^+(aq) \rightleftarrows Hg^{2+}(aq) + I^-(aq)$, $K = 10^{-13}$.

K_{sp} for AgI is 1×10^{-16}.

17.82* The solubility product of AgCN is 2.6×10^{-19}. The instability constant of the dicyanoargentate(I) ion is $9.0 \times 10^{-22}\ M^2$. The ionization constant of HCN is 4.0×10^{-10} *M*. Calculate the molar concentrations of all ionic and molecular species in 0.100 *F* HCN to which has been added sufficient solid AgCN to form a saturated solution.

Hint: The net equation for the principal reaction is

$$AgCN(s) + HCN(aq) \rightleftarrows Ag(CN)_2^-(aq) + H^+(aq)$$

The equilibrium constant for this reaction is

$$K = \frac{[Ag(CN)_2^-] \times [H^+]}{[HCN]}$$

In calculating the numerical value of this equilibrium constant we will, as in previous problems, resolve it into constants whose values we know. We start with the most complex equilibrium, that which involves $Ag(CN)_2^-$. We multiply both numerator and denominator by $[Ag^+] \times [CN^-]^2$. Having done this we find that we have also introduced the terms for K_{sp} of AgCN and K_a for HCN.

$$K = \frac{[Ag(CN)_2^-] \times [H^+] \times [Ag^+] \times [CN^-] \times [CN^-]}{[HCN] \times [Ag^+] \times [CN^-]^2}$$

$$= \frac{K_{AgCN} \times K_{HCN}}{K_{inst}} = \frac{2.6 \times 10^{-19} \times 4.0 \times 10^{-10}}{9.0 \times 10^{-22}} = 1.2 \times 10^{-7}$$

With this information we can then solve for $[H^+]$, $[Ag(CN)_2^-]$, and [HCN]. Then, using the appropriate equilibrium constants, we can calculate the concentrations of the other species in the solution.

17.83* To 1.0 L of 0.1 *F* HCN is added 9.9 g of CuCl. Assuming that there is no change in the volume of the solution when the CuCl is added, calculate the molar concentration for each species in the solution. K_{inst} for $[Cu(CN)_2^-]$ is 5×10^{-28} M^2; K_{sp} of CuCl is 3.2×10^{-7}.

17.84* Calculate the pH of a solution prepared by adding excess solid $Cu(OH)_2$ to 1.0 *F* NH_4Cl. K_{sp} of $Cu(OH)_2$ is 1.6×10^{-19}; K_{inst} of $Cu(NH_3)_4^{2+}$ is 4.7×10^{-15}.

17.85* ZnS will precipitate from 0.010 *F* $Zn(NO_3)_2$ solution on saturation with H_2S only if the pH is greater than 1.00. ZnS will not be precipitated from a solution 0.010 *F* in $Zn(NO_3)_2$ and 1.00 *M* in CN^- unless the pH is greater than 9.00. Calculate *K* for the reaction $Zn(CN)_4^{2-}(aq) \rightleftarrows Zn^{2+}(aq) + 4CN^-(aq)$. A solution saturated with H_2S is 0.10 *M* in H_2S. The overall ionization constant of H_2S is 1.3×10^{-20}.

17.86* At 50 °C the concentration of undissociated H_2S in equilibrium with H_2S gas at a partial pressure of 1.00 atm is 0.075 *M*. In 1.00 L of solution buffered at pH 4.00 that is 0.0045 *F* in $Ni(NO_3)_2$ and 0.500 *F* in NaCl, the partial pressure of H_2S gas required to just begin precipitation of NiS is 0.0333 atm. Assuming the *K*'s of H_2S and the K_{sp} for NiS given in the appendix are applicable at 50 °C and that $NiCl^+$ is the only chloro complex of Ni^{2+} formed, calculate the equilibrium constant for the reaction $NiCl^+(aq) \rightleftarrows Ni^{2+}(aq) + Cl^-(aq)$.

17.87* The following solubility equilibria apply to $Zn(OH)_2$:

$$Zn(OH)_2(s) \rightleftarrows Zn^{2+}(aq) + 2OH^-(aq) \quad K_{sp} = 5.0 \times 10^{-17}\ M$$
$$Zn(OH)_2(s) + 2OH^-(aq) \rightleftarrows Zn(OH)_4^{2-}(aq) \quad K = 0.25\ M$$

Over what pH range can $Zn(OH)_2$ be quantitatively precipitated such that the total concentration of ions containing zinc in equilibrium with solid $Zn(OH)_2$ is less than 10^{-4} *M*?

Chapter 18

Thermodynamics

Thermodynamics is an experimentally derived system for dealing with the macroscopic properties of matter. One of its most important uses is that it enables us to predict whether or not a given process is favorable and to what extent it can occur. It says nothing about the time that may be required for the process to occur.

Thermodynamic systems are described by assigning values to *state functions*, which are properties of the system, and whose values depend only on the particular state we are specifying and not on how the state is reached. Pressure, volume, and temperature are examples of state functions. A *process* is a change in state of the system from an initial to a final state along a certain path. A process is described by specifying the values of the changes in the state functions (symbolized by a Δ, e.g., ΔT is change in temperature) and the values of other thermodynamic quantities associated with the process called *path functions*.

While the values of the changes in state functions are independent of the path followed and depend only on the initial and final states of the system, the values of path functions are determined by the exact nature of the path followed. State functions are designated by capital letters, path functions by lowercase letters.

Generally the initial and final states in processes we will deal with are *equilibrium states*. If the system is in an equilibrium state it will not change with time unless it is disturbed in some manner by something in the surroundings. Although thermodynamics has a very wide range of applicability, we shall restrict ourselves to certain idealized systems that can be described relatively simply.

FIRST LAW OF THERMODYNAMICS

The *first law of thermodynamics* is an experimentally derived generalization about the behavior of matter, which can be stated as: The energy of the universe is constant. A mathematical statement of this law is

$$\Delta E = q + w \tag{18-1}$$

The change in energy ΔE in going from an initial to a final state is equal to the heat q absorbed by the system plus the work w done on the system.

The energy E of a system is a state function that can be defined as the stored-up capacity to do mechanical work. The heat q is a path function; it is the transfer of energy due to a temperature difference. It has a positive sign when the system absorbs heat from the surroundings and a negative sign when the system gives heat to the surroundings. Work w is also a path function; it is the transfer of energy brought about by moving something against an opposing force. It has a negative sign when the system does work on the surroundings and a positive sign when the surroundings do work on the system. Notice that the sign convention for q and w is that they are positive when the direction of energy transfer is to increase the energy of the system.

Since work requires movement, when ordinary chemical systems undergo a change in state, work can be performed only if there is a change in volume of the system. You should remember that $w = 0$ for all constant-volume processes, and therefore $\Delta E = q_V$, where q_V is the heat absorbed at constant volume.

The thermodynamics of ideal gas systems is often relatively simple because the systems are simple. The ideal gas may be in a container that is completely insulated (i.e., the walls of the container do not allow heat flow). All changes in state are then said to be *adiabatic*, $q = 0$. The ideal gas may be in a sealed container that cannot change its volume. All changes in state will then be at constant volume, so $w = 0$. The ideal gas may be in a container that is a cylinder which allows heat flow and is fitted with a frictionless piston, held in place by atmospheric pressure and/or weights.

Since the piston in this system can move, volume changes can occur and work can be performed. This type of work is called pressure–volume (PV) work because it results from the displacement of the piston due to a volume change against an opposing pressure. If the volume of the gas in the cylinder increases ($\Delta V > 0$) the system is doing work on the surroundings ($w < 0$), and if the volume of the gas in the cylinder decreases ($\Delta V < 0$) the surroundings are doing work on the system ($w > 0$).

Calculation of the quantity of work in processes of this type can readily be performed in certain situations. If the movement of the piston occurs against a constant opposing pressure P_{opp} then

$$w = -P_{\text{opp}} \Delta V \tag{18-2}$$

For example, suppose that a gas in a cylinder has $P = 3$ atm and that a piston is held in place by 1 atm of external pressure and enough weights to total 3 atm. If the weights are then removed quickly, the gas in the cylinder will expand against an opposing pressure of 1 atm from its initial state to a final state in which the pressure of the gas is 1 atm. Since the opposing external pressure during the entire expansion has been 1 atm, $w = -(1 \text{ atm})\Delta V$.

Note that the gas was initially in an equilibrium state, but removing the weights from the piston threw it out of equilibrium. Therefore, it spontaneously expanded until it finally reached a new equilibrium state.

It is also theoretically possible to bring about this same change by a path in which the system is always in an equilibrium state. Such a process is said to be *reversible*. A reversible process might occur when the weights are removed from the piston in infinitely small increments. If this entire process occurs at a constant temperature T then

$$w = nRT \ln \frac{V_1}{V_2} \qquad \textbf{(18-3)}$$

where R is the gas constant, n is the amount of gas (in moles), T is the absolute temperature (K), and V_1 and V_2 are the initial and final volumes, respectively.

To summarize, we can calculate w for three types of processes:

1. Constant volume: $w = 0$
2. Volume change against a constant opposing pressure: $w = -P_{\text{opp}} \Delta V$
3. Reversible volume change: $w = nRT \ln V_1/V_2$

The value of q can also be calculated for certain processes of ideal gases. An *isothermal* process is one that occurs at constant temperature. For any isothermal process of an ideal gas, by definition $\Delta E = 0$ and from the first law $q = -w$. And so, if we can calculate w for the isothermal process, we have q as well. It is also possible to calculate q for processes in which the temperature changes, provided that either the volume or the pressure stays constant. For any temperature change at constant volume,

$$q_V = nC_V\Delta T \qquad \textbf{(18-4)}$$

where q_V is the heat absorbed at constant volume, C_V is the heat capacity at constant volume and has the value $\frac{3}{2}R$ for 1 mol of an ideal gas, and n is the amount of gas. For any temperature change at constant pressure

$$q_P = nC_P\Delta T \qquad \textbf{(18-5)}$$

where q_P is the heat absorbed at constant pressure, C_P is the heat capacity at constant pressure and has the value $\frac{5}{2}R$ for 1 mol of an ideal gas, and n is the amount of gas.

We can show from the first law that $\Delta E = q_V$, since $w = 0$ at constant volume. We find it convenient to define another state function H, the enthalpy, such that $\Delta H = q_P$, since most chemical processes occur at constant pressure. It can be shown by use of the first law that

$$\Delta H = \Delta E + \Delta(PV) \quad \textbf{(18-6)}$$

where $\Delta(PV) = P_2V_2 - P_1V_1$ for state 1, the initial state, and for state 2, the final state in the process of interest. For a constant-pressure process $\Delta(PV) = P\Delta V$.

Many of these ideas can also be applied to systems containing only liquids and solids by making the excellent approximation that the volume changes that solids and liquids undergo during ordinary processes are negligible. Therefore, $\Delta H \cong \Delta E$, $C_P \cong C_V$, and $w \cong 0$. For systems containing gases as well as condensed phases, only the volume changes of the gases need be considered. In cases where a gas is produced or consumed during the process, again it is only the volume changes of the gases that need to be considered. For example, if an isothermal process produces a change in the quantity of ideal gas, we can readily show, by substituting from the ideal gas equation into Eq. (18-6), that

$$\Delta H = \Delta E + RT\Delta n \quad \textbf{(18-7)}$$

PROBLEMS

18.1 Calculate the change in energy for a system that absorbs 10 J of heat from the surroundings while performing 8 J of work in the surroundings.

Solution: The first law states that the change in energy is equal to the heat absorbed by the system plus the work done on it.

$$\Delta E = q + w = 10 \text{ kJ} + (-8 \text{ kJ}) = 2 \text{ kJ}$$

18.2 The energy of a system decreases by 17.2 kJ when it evolves 22.3 kJ of heat. How much work is performed?

18.3 A system containing 2.5 mol of an ideal gas absorbs 1700 J/mol of heat. How much work must it perform in order for its energy to remain constant?

18.4 A 20-L volume of gas in a cylinder with a frictionless piston is at a pressure of 3 atm. Enough weight is suddenly removed from the piston to lower the external pressure to 1.5 atm. The gas then expands isothermally until it reaches a new equilibrium state. Calculate ΔE, ΔH, q, and w for this change in state.

Solution: First calculate the final volume using Boyle's law: $P_1V_1 = P_2V_2$, (3 atm)(20 L) = (1.5 atm)(V_2), $V_2 = 40$ L. Since the process is isothermal, $\Delta E = 0$. We can easily show that ΔH must also be zero for an isothermal process of an ideal gas. Since at a given temperature $P_1V_1 = P_2V_2$, $\Delta(PV) = 0$ and, from Eq. (18-6), $\Delta E = \Delta H = 0$.

If $\Delta E = 0$, then from Eq. (18-1), $q = -w$. To calculate the value of w we analyze the path the process follows. The volume is changing against a constant opposing pressure of 1.5 atm. The relationship $w = -P_{opp}\,\Delta V$ applies: $w = -(1.5\text{ atm})(V_2 - V_1) = -(1.5\text{ atm})(40\text{ L} - 20\text{ L}) = -30$ L-atm.

Generally we will wish to express thermodynamic quantities in units of joules. The conversion factor is 1 L-atm = 101 J. Thus $w = -3010$ J and $q = -(-3010\text{ J}) = 3010$ J.

18.5 One mole of an ideal gas occupying a volume of 7 L expands isothermally and reversibly to a volume of 12 L at 300 K. Calculate ΔE, ΔH, q, and w.

Solution: Since the process is isothermal $\Delta E = 0$, $\Delta H = 0$, and $q = -w$. Since the process is also reversible $w = nRT \ln V_1/V_2$. To calculate w, we substitute the data into this relationship taking care to use suitable units. In order to obtain an answer in joules we use $R = 8.314$ J/mol-K. Thus

$$w = (1\text{ mol})\left(8.314\,\frac{1}{\text{mol-K}}\right)(300\text{ K})\left(\ln\frac{7}{12}\right) = -1340\text{ J and}$$
$$q = 1340\text{ J.}$$

18.6 A 5.20-L cylinder with a frictionless piston contains ideal gas at a pressure of 4.10 atm. Weight is quickly removed from the piston so that the external pressure is 1.20 atm. The gas then expands isothermally to a new equilibrium state. Calculate ΔE, ΔH, q, and w for this change in state.

18.7 A 17.8-L cylinder with a frictionless piston contains ideal gas at a pressure of 1.20 atm. Weight is suddenly added to the piston so that the external pressure is 4.10 atm. The gas then contracts isothermally to a new equilibrium state. Calculate ΔE, ΔH, q, and w for this change in state.

18.8 Two moles of ideal gas undergoes reversible isothermal expansion at 298 K from a volume of 3.1 L to a volume of 6.2 L. Calculate ΔE, ΔH, q, and w for this change in state.

18.9 Three moles of ideal gas contracts reversibly and isothermally at 298 K from a volume of 6.2 L to a volume of 3.1 L. Calculate ΔE, ΔH, q, and w for this change in state.

18.10 One mole of ideal gas at 300 K reversibly undergoes a pressure drop from 4.12 atm to 2.06 atm. Calculate ΔE, ΔH, q, and w for this change in state.

18.11 The temperature of 2 mol of ideal gas is raised from 298 K to 348 K at constant volume. Calculate ΔE, ΔH, q, and w for this change in state.

Solution: Since the volume is constant, $w = 0$ and $\Delta E = q$. Substituting into Eq. (18-4), we have

$$q_V = nC_V\Delta T = (2\text{ mol})\left(\frac{3}{2}\right)\left(8.31\frac{\text{J}}{\text{mol-K}}\right)(348\text{ K} - 298\text{ K})$$
$$= 1250\text{ J} = \Delta E$$

In order to calculate ΔH we substitute $\Delta(PV) = \Delta(nRT)$ into Eq. (18-6) to obtain

$$\Delta H = \Delta E + \Delta(nRT) = \Delta E + nR\Delta T \qquad \textbf{(18-8)}$$

since for this change in state the number of moles of gas is constant. Thus, $\Delta H =$ 1250 J + (2 mol)(8.31 J/mol-K)(50 K) = 2080 J.

18.12 The temperature of 2 mol of an ideal gas is raised from 298 K to 348 K at constant pressure. Calculate ΔE, ΔH, q, and w for this change in state.

Solution: Since the pressure is constant, we can substitute into Eq. (18-5):

$$\Delta H = q_P = nC_P\Delta T$$
$$= (2\text{ mol})\left(\frac{5}{2}\right)\left(8.31\frac{\text{J}}{\text{mol-K}}\right)(348\text{ K} - 298\text{ K}) = 2080\text{ J}$$

We can then calculate ΔE by substitution into Eq. (18-8).

$$2080\text{ J} = \Delta E + (2\text{ mol})\left(8.31\frac{\text{J}}{\text{mol-K}}\right)(50\text{ K}) = 1250\text{ J}$$

Since we know q and ΔE, we can use the first law to calculate w:

$$w = \Delta E - q = 1250\text{ J} - 2080\text{ J} = -830\text{ J}$$

18.13 The temperature of 2.50 mol of an ideal gas is raised from 273 K to 373 K at constant volume. Calculate ΔE, ΔH, q, and w.

18.14 The temperature of 2.50 mol of an ideal gas is raised from 373 K to 473 K at constant pressure. Calculate ΔE, ΔH, q, and w.

18.15 Calculate the work for the following four processes. One mole of an ideal gas at 300 K in a cylinder with a frictionless piston is at 4 atm pressure caused by the pressure of the air and three weights of equal mass pushing on the piston. The pressure is dropped to 1 atm by (a) taking off all three weights at once, (b) taking off two weights, waiting, taking off the third, (c) taking off one weight, waiting, taking off the next two at once, or (d) taking off the weights one at a time.

18.16 The temperature of 2.25 mol of an ideal gas is lowered from 350 K to 175 K at constant pressure. Calculate ΔE, ΔH, q, and w.

18.17 The change in state

$$1 \text{ mol ideal gas } (400 \text{ K}, 2 \text{ atm}) \rightarrow 1 \text{ mol ideal gas } (300 \text{ K}, 1 \text{ atm})$$

is brought about by a number of two-step processes. For each two-step process calculate ΔE, ΔH, q, and w.

a. Reversible expansion at 400 K to final volume; cooling at this volume to 300 K

b. Sudden pressure drop to 1 atm and isothermal expansion back to equilibrium; cooling at constant pressure of 1 atm to 300 K

c. Reversible cooling at constant pressure of 2 atm to 300 K; reversible isothermal expansion until pressure is 1 atm

Hint: For processes involving several changes in variables, it is most convenient to treat each change separately in a distinct step and then sum the quantities for each step to get values for the entire process.

18.18 The heat of vaporization of water is 40,670 J/mol at 373 K (100 °C). Calculate ΔE, ΔH, q, and w for the evaporation of 1000 g of water at 373 K and 1 atm.

Hint: Calculate ΔE by using Eq. (18-7) and w by using the first law.

18.19* Using C_P for $H_2O(l) = 74.9$ J/mol and C_P for $H_2O(g) = 25.0$ J/mol and the heat of vaporization given in Prob. 18.18, calculate ΔE, ΔH, q, and w for the change in state 1 mol $H_2O(l)$ at 80 °C and 1 atm $\rightarrow$ 1 mol $H_2O(g)$ at 110 °C and 1 atm.

Hint: Consider the process as three steps: [warming $H_2O(l)$ from 80 °C to 100 °C], [evaporating $H_2O(l)$ to $H_2O(g)$], and [warming $H_2O(g)$ from 100 °C to 110 °C].

18.20* Using the data in the previous two problems and C_P for $H_2O(s) = 36.8$ J/mol and ΔH_{fusion} of $H_2O(s) = 5980$ J/mol, calculate ΔE, ΔH, q, and w for the change in state 1 mol $H_2O(l) \rightarrow$ 1 mol $H_2O(s)$ at -10 °C and 1 atm.

18.21 For the reaction $2NO_2(g) \rightarrow N_2O_4(g)$ at 300 K, ΔE is found to be -54.8 kJ. Calculate ΔH.

18.22 Find $\Delta H - \Delta E$ for each of the following reactions carried out at 400 K:

a. $CaCO_3(s) \rightarrow CaO(s) + CO_2(g)$

b. $4NH_3(g) + 5O_2(g) \rightarrow 4NO(g) + 6H_2O(g)$

c. $4Al(s) + 3MnO_2(s) \rightarrow 2Al_2O_3(s) + 3Mn(s)$

18.23 Find the value of $\Delta H - \Delta E$ for a process in which the temperature of 3.0 mol of ideal gas is lowered from 400 K to 300 K.

SECOND LAW OF THERMODYNAMICS

The first law tells us that any process must satisfy the requirement of conservation of energy. But real processes have another feature, a preferred spontaneous direction. For example, if a hot and a cold object are placed in contact, heat will flow from hot to cold until the temperature of each object is the same. The reverse process, heat flowing from one object to another when both are at the same temperature does not occur, even though it does not violate the first law.

Intuitively we know that systems proceed spontaneously from a condition of nonequilibrium to the equilibrium state. Such a process is called an *irreversible process*. Once at equilibrium, systems change no further unless disturbed by some external agent. The second law of thermodynamics deals with this phenomenon. It can be stated in many ways. One statement is: There is a quantity S called *entropy*, which is a state function. In an irreversible process, the entropy of the universe increases. In a reversible process the entropy of the universe remains constant. At no time does the entropy of the universe decrease.

We can show that this new state function S is a measure of the disorder of the system. Therefore, we can make qualitative predictions about changes in the entropy of a system ΔS when the system undergoes a change in state by evaluating the relative order and disorder of the initial and final states. For example, in an isothermal expansion the increased volume of the final state results in more disorder and thus in a positive ΔS for the process.

We can calculate a value of ΔS for a system undergoing a change in state by using a definition of ΔS:

$$\Delta S = \frac{q_{\text{rev}}}{T} \qquad \textbf{(18-9)}$$

where q_{rev} is the heat that would flow if the change in state were carried out reversibly. This relationship can be used for any isothermal process.

You should note that, like any state function, ΔS for the system does not depend on the path of the process, but only on the initial and final states. However, in order to calculate ΔS we must calculate q for the hypothetical reversible process. For an ideal gas we can use Eq. (18.3) since $q_{\text{rev}} = -w_{\text{rev}}$. Substituting into Eq. (18-9) gives us ΔS for any isothermal volume change of an ideal gas,

$$\Delta S = nR \ln \frac{V_2}{V_1} \qquad \textbf{(18-10)}$$

Calculation of ΔS for processes in which the temperature changes but that occur at constant pressure or constant volume is also possible by using the derived relationships

$$\Delta S = nC_p \ln \frac{T_2}{T_1} \qquad \textbf{(18-11)}$$

or

$$\Delta S = nC_V \ln \frac{T_2}{T_1} \tag{18-12}$$

The second law deals with ΔS for the universe, so in order to judge whether a process can occur, we need to calculate not only ΔS for the system but also ΔS for the surroundings, since the sum of these two quantities gives ΔS for the universe. In order to calculate ΔS_{surr}, we must calculate the value of q for the specific path the system follows during the change in state. Then

$$\Delta S_{surr} = -\frac{q}{T} \tag{18-13}$$

where T is the temperature of the surroundings.

Using the entropy of the universe as a criterion for spontaneity is awkward since we must consider both the system and the surroundings. If we define another state function we can consider only the system. For processes that take place at constant temperature and pressure, as most ordinary chemical processes do, this state function is the *free energy* G, defined by

$$G = H - TS \tag{18-14}$$

or

$$\Delta G = \Delta H - T\Delta S \tag{18-15}$$

The value of ΔG for the system is a criterion of spontaneous change for processes that occur at constant temperature and pressure. If $\Delta G = 0$ for a process, the process is a reversible one; if $\Delta G < 0$, the process is spontaneous. Processes for which $\Delta G > 0$ are not spontaneous, and the system can undergo such a process only under influence from the surroundings.

PROBLEMS

18.24 Calculate ΔS of the universe and ΔG of the system for the change in state that occurs when 2.00 mol of an ideal gas at 300 K occupying a volume of 24.6 L expands isothermally against an opposing pressure of 1.00 atm to reach equilibrium.

Solution: First we calculate the final volume of the ideal gas at equilibrium:

$$V = \frac{nRT}{P} = \frac{(2.00)(0.082)(300)}{1.00} = 49.2 \text{ L}$$

In order to calculate ΔS of the universe, we calculate ΔS_{sys} and ΔS_{surr} separately and sum.

$$\Delta S_{sys} = \frac{q_{rev}}{T} = nR \ln \frac{V_2}{V_1} = (2.00 \text{ mol})\left(8.31 \frac{\text{J}}{\text{mol-K}}\right) \ln \frac{49.2}{24.6}$$

$$= 11.5 \text{ J/K}$$

Since ΔS_{surr} requires the actual heat of the process, we calculate as in Prob. 18.4,

$$q = -w = P\Delta V = (1.00)(24.6) = 24.6 \text{ L-atm} = 2490 \text{ J}$$

$$\Delta S_{surr} = -\frac{q}{T} = -\frac{2490 \text{ J}}{300 \text{ K}} = -8.30 \text{ J/K}$$

Then

$$\Delta S_{univ} = \Delta S_{sys} + \Delta S_{surr} = 11.5 \text{ J/K} + (-8.3 \text{ J/K}) = 3.2 \text{ J/K}$$

As the process is isothermal, ΔG for the system can be calculated using

$$\Delta G = \Delta H - T\Delta S$$

$$\Delta H = 0 \text{ (isothermal process of an ideal gas)}$$

$$\Delta G = 0 - (300 \text{ K})(11.5 \text{ J/K}) = -3450 \text{ J}$$

Note: In accordance with the second law, the entropy of the universe has increased and ΔG for the system is negative since this process is one in which a system not at equilibrium proceeds to equilibrium.

18.25 Calculate ΔS of the universe and ΔG of the system for the change in state that occurs when 2.5 mol of an ideal gas expands isothermally at 300 K against a constant opposing pressure of 0.50 atm from a volume of 15 L to an equilibrium volume.

18.26 Calculate ΔS of the universe and ΔG of the system for the change in state occurring when 2.0 mol of an ideal gas contracts isothermally at 0 °C under a pressure of 1.5 atm from a volume of 40 L to an equilibrium volume. Assume the surroundings are at 0 °C.

18.27 The heat of vaporization of water at its boiling point is 40,900 J/mol. Calculate ΔS of vaporization at the boiling point at 1 atm.

Solution: Since the vaporization is a reversible process at constant T and P, $\Delta G = 0$ and $\Delta H = T\Delta S$. Substituting 40,700 J/mol = (373 K)(ΔS), we have

$$\Delta S = 109 \text{ J/mol-K}$$

18.28 At its boiling point it is found that gold has ΔH of vaporization = 310 kJ/mol and ΔS of vaporization = 124 J/mol-K. Calculate the boiling point of gold.

18.29 Calculate ΔS of the system for the change in state in Prob. 18.19.

18.30* Calculate ΔS of the universe and ΔG of the system for the change in state in Prob. 18.20.

18.31* In the adjoining grid, for each process written on the left put into each box a + if the indicated parameter is positive, a − if it is negative, and a 0 if it is zero (IG = ideal gas).

Process	q	w	ΔE	ΔH	ΔS
1 mol IG (25 °C, 1 atm) → 1 mol IG (25 °C, 0.5 atm)					
$H_2O(l) \rightarrow H_2O(g)$ (25 °C, 1 atm)		−	+		
$H_2O(l) \rightarrow H_2O(s)$ (−10 °C, 1 atm)		0			
1 mol IG (25 °C, 10 L) → 1 mol IG (100 °C, 10 L)					
$H_2(g)$ (1 atm, 3000 K) → $2H(g)$ (equilibrium pressure, 3000 K)		0			

FREE ENERGY AND *K*

We have defined a thermodynamic function G, the free energy, and seen that we can use the sign of ΔG, the difference in free energy between a final and an initial state, as a criterion to predict the direction of spontaneous change for processes carried out at constant T and P. We can use the magnitude of ΔG for a process to evaluate the extent to which a spontaneous change occurs. We can again employ the definition of a standard state outlined in Chapter 8, and tabulate (Table A.7, appendix) experimentally determined values for the standard free energy of formation ΔG_f° of various substances. In a manner analogous to that used for ΔH°, we can calculate ΔG° for a reaction.

$$\Delta G^\circ = \sum \Delta G_f^\circ(\text{products}) - \sum \Delta G_f^\circ(\text{reactants}) \qquad \textbf{(18-16)}$$

where ΔG° represents the free-energy change of a process in which all the components of a system are at a pressure of 1 atm. The value of ΔG° is related to the value of K_p, the equilibrium constant for the process, by the extremely important relationship

$$\Delta G^\circ = -RT \ln K_p \qquad \textbf{(18-17)}$$

Remember that ΔG° is related to K_p, the equilibrium constant that has gases in units of atmospheres and solutes in solution in units of moles/liter.

PROBLEMS

18.32 The standard free energy of formation (ΔG_f°) of NO_2 is 51 kJ/mol. Calculate the value of K_p at 25 °C for the reaction

$$2O_2(g) + N_2(g) \rightleftarrows 2NO_2(g)$$

Solution: Since this reaction is the formation of 2 mol of $NO_2(g)$ from its elements in their standard states, $\Delta G° = 2(\Delta G_f°) = 102$ kJ. Substitution into $\Delta G° = -RT \ln K_p$ using $R = 8.31$ J/mol-K gives

$$\ln K = -\frac{102{,}000}{(8.31)(298)} = -41.2$$

$$K = 1.3 \times 10^{-18}$$

18.33 Use the data in Table A.7 of the appendix to calculate $\Delta G°$ and K_p at 298 K for the reaction $CaCO_3(s) \rightleftarrows CaO(s) + CO_2(g)$

Solution: The value of $\Delta G°$ at 298 K can be calculated from the values of $\Delta G_f°$ given in Table A.7 by using the procedure of Prob. 8.21 and Eq. (18-16).

$$\Delta G° = [(-604 \text{ kJ}) + (-394 \text{ kJ})] - [(-1129 \text{ kJ})] = 131 \text{ kJ}$$

Once we have $\Delta G°$ at 298 K, substitution into Eq. (18-17) gives the value of K_p,

$$\ln K_p = -\frac{131{,}000}{(8.31)(298)} = -52.9$$

$$K_p = 1.1 \times 10^{-23}$$

18.34 Use the data in Table A.7 of the appendix to calculate $\Delta G°$ and K_p for the reaction in which 2 mol of $HBr(g)$ forms from its elements in their standard states at 298 K and 1 atm. Repeat the calculation for the formation from $Br_2(g)$.

18.35 Use the data in Table A.7 of the appendix to calculate $\Delta G°$ and K_p for the combustion of 1 mol of $C_2H_4(g)$ at 298 K to carbon dioxide and liquid water.

18.36 The equilibrium constant K for the reaction $CO_2(g) + H_2(g) \rightleftarrows CO(g) + H_2O(g)$ is 2.0 at $T = 1007$ °C. Calculate the value of $\Delta G°$.

Solution: Substituting into $\Delta G° = -RT \ln K$, we have

$$\Delta G° = (-8.31)(1007 + 273) \ln 2 = -7.4 \times 10^3 \text{ J}$$

18.37 For the reaction $H_2(g) + Cl_2(g) \rightleftarrows 2HCl(g)$, $\Delta G° = -190$ kJ, calculate K_p at 300 K.

18.38 For the reaction $N_2(g) + \frac{1}{2}O_2(g) \rightleftarrows N_2O(g)$, $\Delta G° = 104$ kJ, calculate K_p at 323 K.

18.39 The value of K_p for the reaction $N_2(g) + 3H_2(g) \rightleftarrows 2NH_3(g)$ at 600 K is 7.90×10^2. Calculate the $\Delta G_f°$ of $NH_3(g)$ at this temperature.

18.40 Use the data given in Table A.7 of the appendix to calculate $\Delta G°$ of the following reactions at 298 K:

a. $HCl(g) + NH_3(g) \rightleftarrows NH_4Cl(s)$

b. $2NO_2(g) \rightleftarrows N_2O_4(g)$

c. $C_2H_2(g) + 2H_2(g) \rightleftarrows C_2H_6(g)$

18.41 For the reaction $Zn(s) + Cu^{2+} \rightleftarrows Cu(s) + Zn^{2+}$, $K = 2.00 \times 10^{37}$ at 25 °C, calculate $\Delta G°$.

18.42 For the reaction $AgCl(s) \rightleftarrows Ag^+ + Cl^-$, $K = 2.12 \times 10^{-10}$ at 27 °C, calculate $\Delta G°$.

Chapter 19

Oxidation-Reduction Processes

A chemical change that takes place as the result of an electron transfer is called an *oxidation-reduction*, or *redox*, reaction. Oxidation is the result of the loss of electrons; reduction is the result of the gain of electrons. One never takes place without the other. When a substance is oxidized it causes another substance to be reduced. Thus the substance that is oxidized is called a *reducing agent*. When a substance is reduced it causes another substance to be oxidized. Thus the substance that is reduced is called an *oxidizing agent*.

BALANCING REDOX EQUATIONS

Many redox equations are so easily balanced that they can be done quickly by *simple inspection*. Thus for the unbalanced equation $Al + H^+ \rightleftarrows Al^{3+} + H_2$, it is quite apparent that the balanced equation will be $2Al + 6H^+ \rightleftarrows 2Al^{3+} + 3H_2$. In this equation, as in all balanced equations, *the number of atoms of each element on the left of the equality sign must equal the number of atoms of that element on the right, and the net sum of the charges on all ions on the left must equal the net sum of the charges on all ions on the right*. Whenever possible, equations should be balanced by simple inspection; there is no point in calling upon some relatively cumbersome balancing routine when simple inspection will suffice.

For many redox equations the balancing is too complicated to be accomplished easily by simple inspection. Such equations can be balanced by the *method of half-reactions* (called also the *ion-electron method*) and by the *method of change in oxidation number* (called also *loss and gain of electrons*).

THE METHOD OF HALF-REACTIONS

We can consider every redox reaction as the sum of two half-reactions. Thus the overall reaction represented by the equation $Zn + Cu^{2+} \rightleftarrows Zn^{2+} + Cu$ is the sum of the two half-reactions $Zn \rightleftarrows Zn^{2+} + 2e^-$ and $Cu^{2+} + 2e^- \rightleftarrows Cu$. (You should note that each of the above half-reactions, like all half-reactions, is balanced with respect to both number of atoms and net charge; the charge balance is in each instance achieved by the use of electrons. The presence of electrons in an equation means that it represents a half-reaction.)

In the first of the above half-reactions Zn is *oxidized* to Zn^{2+}. This *oxidation* is attended by the *loss of two electrons* per atom of Zn; the fact that *electrons are lost* is evidence that *oxidation* has occurred. *Oxidation may be defined as a process in which electrons are lost.* In the second half-reaction Cu^{2+} is *reduced* to Cu. This *reduction* is attended by the *gain of two electrons* per atom of Cu; the fact that *electrons are gained* is evidence that *reduction* has occurred. *Reduction may be defined as a process in which electrons are gained.*

When the two half-reactions that constitute the redox equation are added, the final balanced equation is obtained.

$$\begin{array}{r} Zn \rightleftarrows Zn^{2+} + 2e^- \\ \underline{Cu^{2+} + 2e^- \rightleftarrows Cu \qquad} \\ Zn + Cu^{2+} \rightleftarrows Zn^{2+} + Cu \end{array}$$

Note that in this addition process the electrons, being on opposite sides of the equality sign, cancel and hence do not appear in the final balanced equation. The reason they cancel is that the total number of electrons lost in the first half-reaction is equal to the total number of electrons gained in the second half-reaction.

Balancing by the *method of half-reactions* is based upon the concepts and facts that we have cited above, namely:

1. Every redox equation is the sum of two balanced equations for two half-reactions.
2. The two balanced half-reaction equations can be added to give the final balanced redox equation provided the total number of electrons lost in the half-reaction in which oxidation occurs is equal to the total number of electrons gained in the half-reaction in which reduction occurs. This is equivalent to stating that, in any balanced redox equation, the total number of electrons lost by the element or elements oxidized must equal the total number of electrons gained by the element or elements reduced.

In brief, balancing by this method consists of three steps:

1. The unbalanced skeleton equation is broken down into two unbalanced skeleton half-reactions.

2. Each half-reaction is balanced.
3. The two balanced half-reactions are added together to give the final balanced redox equation; to do this, enough multiples of each balanced half-reaction must, if necessary, be taken so that the total number of electrons lost in one half-reaction equals the total number of electrons gained in the other. As a result, all electrons cancel and do not appear in the balanced equation. (If electrons appear in the final equation, it is not correct.) The following examples will illustrate the use of the *method of half-reactions*.

EXAMPLE 1

Balance the equation $H_2SO_3 + MnO_4^- \rightleftarrows SO_4^{2-} + Mn^{2+}$ (in acid solution).

Step 1 *Write the two skeleton half-reactions.*

$H_2SO_3 \rightleftarrows SO_4^{2-}$ (The reducing agent, H_2SO_3, is oxidized to SO_4^{2-}.)

$MnO_4^- \rightleftarrows Mn^{2+}$ (The oxidizing agent, MnO_4^-, is reduced to Mn^{2+}.)

Step 2 *Balance each half-reaction.*

1. To balance the O, add H_2O.

$$H_2SO_3 + H_2O \rightleftarrows SO_4^{2-}$$
$$MnO_4^- \rightleftarrows Mn^{2+} + 4H_2O$$

2. To balance the H, add H^+.

$$H_2SO_3 + H_2O \rightleftarrows SO_4^{2-} + 4H^+$$
$$MnO_4^- + 8H^+ \rightleftarrows Mn^{2+} + 4H_2O$$

3. To balance the charge, add electrons.

$$H_2SO_3 + H_2O \rightleftarrows SO_4^{2-} + 4H^+ + 2e^-$$
$$MnO_4^- + 8H^+ + 5e^- \rightleftarrows Mn^{2+} + 4H_2O$$

Step 3 *Add the two half-reactions.* Since 2 electrons are lost in the first while 5 are gained in the second, and since the lowest common multiple of 2 and 5 is 10, we multiply the first by 5 and the second by 2. When the resulting equations are added, the $10e^-$ on the two sides of the equations will cancel. Also $16H^+$ and $5H_2O$ will cancel from each side.

$$5H_2SO_3 + 5H_2O \rightleftarrows 5SO_4^{2-} + 20H^+ + 10e^-$$
$$2MnO_4^- + 16H^+ + 10e^- \rightleftarrows 2Mn^{2+} + 8H_2O$$
$$\overline{2MnO_4^- + 5H_2SO_3 \rightleftarrows 2Mn^{2+} + 5SO_4^{2-} + 4H^+ + 3H_2O}$$

When you obtain the final balanced equation, you should always check it carefully to be sure it is balanced with respect to both mass and charge and that it is indeed net. If the coefficients are not in lowest terms, or if identical species appear on both sides of the equation then it is not net.

EXAMPLE 2

$$P_4S_3 + NO_3^- \rightleftarrows H_3PO_4 + SO_4^{2-} + NO \quad \text{(in acid solution)}$$

Step 1 *Write the two skeleton half-reactions.*

$$NO_3^- \rightleftarrows NO$$
$$P_4S_3 \rightleftarrows 4H_3PO_4 + 3SO_4^{2-}$$

Note that the skeleton half-reaction contains the minimum number of moles of product per mole of reactant.

Step 2 *Balance each half-reaction.*

1. To balance the O, add H_2O.

$$NO_3^- \rightleftarrows NO + 2H_2O$$
$$P_4S_3 + 28H_2O \rightleftarrows 4H_3PO_4 + 3SO_4^{2-}$$

2. To balance the H, add H^+.

$$NO_3^- + 4H^+ \rightleftarrows NO + 2H_2O$$
$$P_4S_3 + 28H_2O \rightleftarrows 4H_3PO_4 + 3SO_4^{2-} + 44H^+$$

3. To balance the charge, add e^-.

$$NO_3^- + 4H^+ + 3e^- \rightleftarrows NO + 2H_2O$$
$$P_4S_3 + 28H_2O \rightleftarrows 4H_3PO_4 + 3SO_4^{2-} + 44H^+ + 38e^-$$

Step 3 *Add the two half-reactions.* To balance the electrons multiply the first equation by 38 and the second by 3.

$$38NO_3^- + 152H^+ + 114e^- \rightleftarrows 38NO + 76H_2O$$
$$3P_4S_3 + 84H_2O \rightleftarrows 12H_3PO_4 + 9SO_4^{2-} + 132H^+ + 114e^-$$
$$\overline{3P_4S_3 + 38NO_3^- + 20H^+ + 8H_2O \rightleftarrows 12H_3PO_4 + 9SO_4^{2-} + 38NO}$$

EXAMPLE 3

$$Cu_2S + SO_4^{2-} \rightleftarrows Cu^{2+} + H_2SO_3 \quad \text{(in acid solution)}$$

Step 1 *Write the two skeleton half-reactions.*

$$Cu_2S \rightleftarrows 2Cu^{2+} + H_2SO_3$$
$$SO_4^{2-} \rightleftarrows H_2SO_3$$

Note that H_2SO_3 is a product in each half-reaction.

Step 2 *Balance each half-reaction.*

1. To balance O, add H_2O.

$$
\begin{aligned}
Cu_2S + 3H_2O &\rightleftarrows 2Cu^{2+} + H_2SO_3 \\
SO_4^{2-} &\rightleftarrows H_2SO_3 + H_2O
\end{aligned}
$$

2. To balance H, add H^+.

$$
\begin{aligned}
Cu_2S + 3H_2O &\rightleftarrows 2Cu^{2+} + H_2SO_3 + 4H^+ \\
SO_4^{2-} + 4H^+ &\rightleftarrows H_2SO_3 + H_2O
\end{aligned}
$$

3. To balance the charge, add e^-.

$$
\begin{aligned}
Cu_2S + 3H_2O &\rightleftarrows 2Cu^{2+} + H_2SO_3 + 4H^+ + 8e^- \\
SO_4^{2-} + 4H^+ + 2e^- &\rightleftarrows H_2SO_3 + H_2O
\end{aligned}
$$

Step 3 *Add the two half-reactions*. To balance the electrons multiply the second equation by 4.

$$
\begin{aligned}
Cu_2S + 3H_2O &\rightleftarrows 2Cu^{2+} + H_2SO_3 + 4H^+ + 8e^- \\
4SO_4^{2} + 16H^+ + 8e^- &\rightleftarrows 4H_2SO_3 + H_2O \\
\hline
Cu_2S + 4SO_4^{2-} + 12H^+ &\rightleftarrows 2Cu^{2+} + 5H_2SO_3 + H_2O
\end{aligned}
$$

EXAMPLE 4

$$HS_2O_4^- + CrO_4^{2-} \rightleftarrows SO_4^{2-} + Cr(OH)_4^- \quad \text{(in basic solution)}$$

Step 1

$$
\begin{aligned}
HS_2O_4^- &\rightleftarrows 2SO_4^{2-} \\
CrO_4^{2-} &\rightleftarrows Cr(OH)_4^-
\end{aligned}
$$

Step 2

1.

$$
\begin{aligned}
HS_2O_4^- + 4H_2O &\rightleftarrows 2SO_4^{2-} \\
CrO_4^{2-} &\rightleftarrows Cr(OH)_4^- \quad \text{(O is balanced)}
\end{aligned}
$$

2. To balance the H add H^+ as in acid solution. However, since the solution is basic, these H^+ ions will react with OH^- ions to form H_2O. Accordingly, the procedure is first to add H^+ to balance H, then have this H^+ react with OH^- to form H_2O, then add the resulting two equations. Operationally, we convert the H^+ ions to an equal number

of OH^- ions and add that number of H_2O molecules to the other side of the half-reaction.

$$HS_2O_4^- + 4H_2O \rightleftarrows 2SO_4^{2-} + 9H^+$$
$$9H^+ + 9OH^- \rightleftarrows 9H_2O$$
$$\overline{HS_2O_4^- + 9OH^- \rightleftarrows 2SO_4^{2-} + 5H_2O}$$

$$CrO_4^{2-} + 4H^+ \rightleftarrows Cr(OH)_4^-$$
$$4H_2O \rightleftarrows 4H^+ + 4OH^-$$
$$\overline{CrO_4^{2-} + 4H_2O \rightleftarrows Cr(OH)_4^- + 4OH^-}$$

3.

$$HS_2O_4^- + 9OH^- \rightleftarrows 2SO_4^{2-} + 5H_2O + 6e^-$$
$$CrO_4^{2-} + 4H_2O + 3e^- \rightleftarrows Cr(OH)_4^- + 4OH^-$$
$$HS_2O_4^- + 9OH^- \rightleftarrows 2SO_4^{2-} + 5H_2O + 6e^-$$

Step 3

$$2CrO_4^{2-} + 8H_2O + 6e^- \rightleftarrows 2Cr(OH)_4^- + 8OH^-$$
$$\overline{HS_2O_4^- + 2CrO_4^{2-} + 3H_2O + OH^- \rightleftarrows 2SO_4^{2-} + 2Cr(OH)_4^-}$$

EXAMPLE 5

$$P_4 \rightleftarrows H_2PO_2^- + PH_3 \quad \text{(in basic solution)}$$

Step 1

$$P_4 \rightleftarrows 4H_2PO_2^- \ (P_4 \text{ is oxidized})$$
$$P_4 \rightleftarrows 4PH_3 \quad (P_4 \text{ is reduced})$$

Note: P_4 disproportionates.

Step 2

1.

$$8H_2O + P_4 \rightleftarrows 4H_2PO_2^-$$
$$P_4 \rightleftarrows 4PH_3$$

2.

$$8H_2O + P_4 \rightleftarrows 4H_2PO_2^- + 8H^+$$
$$8H^+ + 8OH^- \rightleftarrows 8H_2O$$
$$\overline{P_4 + 8OH^- \rightleftarrows 4H_2PO_2^-}$$
$$P_4 + 12H^+ \rightleftarrows 4PH_3$$
$$12H_2O \rightleftarrows 12H^+ + 12OH^-$$
$$\overline{P_4 + 12H_2O \rightleftarrows 4PH_3 + 12OH^-}$$

3.

$$P_4 + 8OH^- \rightleftarrows 4H_2PO_2^- + 4e^-$$
$$P_4 + 12H_2O + 12e^- \rightleftarrows 4PH_3 + 12OH^-$$

Step 3

$$3P_4 + 24OH^- \rightleftarrows 12H_2PO_2^- + 12e^-$$
$$P_4 + 12H_2O + 12e^- \rightleftarrows 4PH_3 + 12OH^-$$
$$\overline{4P_4 + 12OH^- + 12H_2O \rightleftarrows 12H_2PO_2^- + 4PH_3}$$

or

$$P_4 + 3OH^- + 3H_2O \rightleftarrows 3H_2PO_2^- + PH_3$$

EXAMPLE 6

$$HPO_3^- + H_2O_2 \rightleftarrows PO_4^{3-} \quad \text{(in basic solution)}$$

Step 1

$$HPO_3 \rightleftarrows PO_4^{3-}$$
$$H_2O_2 \rightleftarrows H_2O$$

Note that H^+ and OH^- are generally not included in a skeleton equation.

Step 2

1.
$$H_2O + HPO_3^- \rightleftarrows PO_4^{3-}$$
$$H_2O_2 \rightleftarrows 2H_2O$$

2.
$$H_2O + HPO_3^- \rightleftarrows PO_4^{3-} + 3H^+$$
$$3H^+ + 3OH^- \rightleftarrows 3H_2O$$
$$\overline{HPO_3^- + 3OH^- \rightleftarrows PO_4^{3-} + 2H_2O}$$
$$H_2O_2 + 2H^+ \rightleftarrows 2H_2O$$
$$2H_2O \rightleftarrows 2H^+ + 2OH^-$$
$$\overline{H_2O_2 \rightleftarrows 2OH^-}$$

3.
$$HPO_3^- + 3OH^- \rightleftarrows PO_4^{3-} + 2H_2O + 1e^-$$
$$H_2O_2 + 2e^- \rightleftarrows 2OH^-$$

Step 3

$$2HPO_3^- + 6OH^- \rightleftarrows 2PO_4^{3-} + 4H_2O + 2e^-$$
$$H_2O_2 + 2e^- \rightleftarrows 2OH$$
$$\overline{2HPO_3^- + H_2O_2 + 4OH^- \rightleftarrows 2PO_4^{3-} + 4H_2O}$$

THE METHOD OF CHANGE IN OXIDATION NUMBER

In any redox reaction the oxidation number of at least one element is increased, and the oxidation number of at least one element is decreased. Thus in the reaction

$$Sn^{2+} + 2Fe^{3+} \rightleftarrows Sn^{4+} + 2Fe^{2+}$$

tin is *oxidized* from an oxidation number of +2 in Sn^{2+} to +4 in Sn^{4+}. At the same time, iron is *reduced* from an oxidation number of +3 in Fe^{3+} to +2 in Fe^{2+}.

Here Fe^{3+} is said to be the *oxidizing agent*, while Sn^{2+} is said to be the *reducing agent*.

In the reaction

$$Cr_2O_7^{2-} + 3H_2S + 8H^+ \rightleftarrows 2Cr^{3+} + 3S + 7H_2O$$

chromium is *reduced* from an oxidation number of +6 in $Cr_2O_7^{2-}$ to +3 in Cr^{3+}, while sulfur is *oxidized* from an oxidation number of −2 in H_2S to 0 in S. $Cr_2O_7^{2-}$ is the *oxidizing agent* or *oxidant*; H_2S is the *reducing agent* or *reductant*.

In determining the oxidation number of a specific element in a given molecule or ion the following rules are applied:

1. In an elemental ion, such as Sn^{4+} or S^{2-}, the oxidation number is equal to the charge on the ion. Thus in Sn^{4+} and S^{2-} the oxidation numbers of Sn and S are, respectively, +4 and −2.
2. The oxidation number of any free element, such as O_2 and S, is zero.
3. In its compounds or ions H has an oxidation number of +1. (The exceptions to this rule are the metal hydrides such as CaH_2, where the oxidation number of H is −1.)
4. In its compounds or ions O has an oxidation number of −2. (The common exceptions to this rule are when the O is bonded to F as in OF_2, in which the oxidation number of O is +2, and when the O is bonded to another O as in the peroxides, such as H_2O_2 and Na_2O_2, in which its oxidation number is −1.)
5. In any neutral molecule the total positive oxidation numbers equal the total negative oxidation numbers. In any ion the charge on the ion equals the difference between the positive and negative oxidation numbers. In the compound H_3AsO_4 the total oxidation number for the four atoms of oxygen is −8; since the three hydrogen atoms have a total oxidation number of +3, the oxidation number of arsenic is +5. In the ion $Cr_2O_7^{2-}$, the total oxidation number for the seven atoms of oxygen is −14; since the charge on the ion is −2, the total positive oxidation number is +12, or +6 for each of the two atoms of chromium.

If we examine the two equations given above and any other equation that we wish to select, we will find that *in every balanced redox equation, the total increase in oxidation number of the element (or elements) oxidized equals the total decrease in oxidation number of the element (or elements) reduced.* In the first of the two equations given above, one atom of tin has its oxidation number increased from +2 to +4; the total increase in oxidation number is 2. Each of the two atoms of iron has its oxidation number reduced from +3 to +2; the total decrease in oxidation number is 2. In the second balanced equation each of the three atoms of S is oxidized from an oxidation number of −2 in H_2S to 0 in free S;

the total increase in oxidation number is 6. Each of the two atoms of Cr is reduced from an oxidation number of +6 in $Cr_2O_7^{2-}$ to +3 in Cr^{3+}; the total decrease in oxidation number is 6.

If we examine the two equations above and any other balanced equation, we will notice also that *the sum of the charges of the ions on each of the two sides of the balanced equation is the same*. In the first equation the sum of the charges on the one Sn^{2+} ion and two Fe^{3+} ions on the left is +8; the total net charge on the one Sn^{4+} ion and the two Fe^{2+} ions on the right is also +8. In the second equation the sum of the charges on the one $Cr_2O_7^{2-}$ ion and the eight H^+ ions is +6. The total charge on the two Cr^{3+} ions on the right is also +6.

The above facts (the equality in the total increase and decrease in oxidation number and the equality of the net charge on the two sides of the equation) are the bases for balancing redox equations by the *method of change in oxidation numbers*. The following examples illustrate the use of the method.

EXAMPLE 7

$$Fe^{2+} + MnO_4^- \rightleftarrows Fe^{3+} + Mn^{2+} \quad \text{(in acid solution)}$$

Step 1 Identify the element or elements oxidized and the element or elements reduced. Note the initial and final oxidation number of each of these elements. Note the change in oxidation number of each of these elements.

increase of 1 (Fe: +2 → +3)

$$\overset{+2}{Fe^{2+}} + \overset{+7}{MnO_4^-} \rightleftarrows \overset{+3}{Fe^{3+}} + \overset{+2}{Mn^{2+}}$$

decrease of 5 (Mn: +7 → +2)

Fe is oxidized; Mn is reduced.

Step 2 Select a sufficient number of moles of each reactant so that the total increase in oxidation number of the reductant equals the total decrease in oxidation number of the oxidant.

total increase of 5 (Fe)

$$5Fe^{2+} + MnO_4^- \rightleftarrows 5Fe^{3+} + Mn^{2+}$$

total decrease of 5 (Mn)

Step 3 Balance the charges on each side of the equation by adding the necessary H^+ ions. (If the solution is alkaline, the charges can be balanced by adding

OH^- ions. If the solution is neutral, either H^+ or OH^- ions may be added; H_2O will provide these ions.)

As in the equation written in step 2, the net charge on the left (from the five Fe^{2+} ions and the one MnO_4^- ion) is +9 and the net charge on the right (from the five Fe^{3+} ions and the one Mn^{2+} ion) is +17. By adding eight H^+ ions to the left the charge on each side will be +17.

$$5Fe^{2+} + MnO_4^- + 8H^+ \rightleftarrows 5Fe^{3+} + Mn^{2+}$$

Step 4 Balance the hydrogen by adding H_2O. If the work has been correct up to this point, balancing the H will also balance the oxygen and, thus, balance the equation.

$$5Fe^{2+} + MnO_4^- + 8H^+ \rightleftarrows 5Fe^{3+} + Mn^{2+} + 4H_2O$$

EXAMPLE 8

$$FeS + NO_3^- \rightleftarrows NO + SO_4^{2-} + Fe^{3+} \quad \text{(in acid solution)}$$

If more than one element is oxidized (and/or reduced), the total increase (decrease) in oxidation number is the sum of the increases (decreases) for each element. The oxidation of FeS by HNO_3 illustrates such a reaction.

Step 1

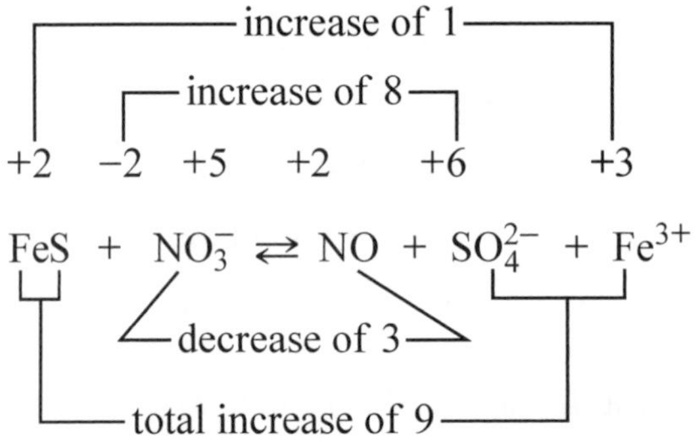

increase of 1 (Fe: +2 → +3)

increase of 8 (S: −2 → +6)

$$\overset{+2\ -2}{FeS} + \overset{+5}{NO_3^-} \rightleftarrows \overset{+2}{NO} + \overset{+6}{SO_4^{2-}} + \overset{+3}{Fe^{3+}}$$

decrease of 3 (N: +5 → +2)

total increase of 9

Step 2

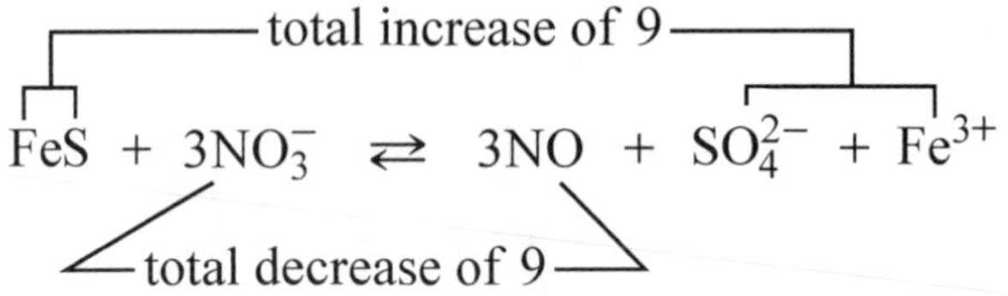

total increase of 9

$$FeS + 3NO_3^- \rightleftarrows 3NO + SO_4^{2-} + Fe^{3+}$$

total decrease of 9

Step 3 $$FeS + 3NO_3^- + 4H^+ \rightleftarrows 3NO + SO_4^{2-} + Fe^{3+}$$

Step 4 $$FeS + 3NO_3^- + 4H^+ \rightleftarrows 3NO + SO_4^{2-} + Fe^{3+} + 2H_2O$$

EXAMPLE 9

$$As_2S_3 + NO_3^- \rightleftarrows NO + SO_4^{2-} + H_3AsO_4 \quad \text{(in acid solution)}$$

If more than 1 mol of the element (or elements) oxidized and/or reduced is present in 1 mol of reactant, the minimum number of moles of product formed per mole of reactant must be given in step 1. Thus when As_2S_3 is oxidized by HNO_3 to yield H_3AsO_4 and SO_4^{2-}, 1 mol of As_2S_3 will yield 2 mol of H_3AsO_4 and 3 mol of SO_4^{2-}. The successive steps in the balancing process will then be

Step 1

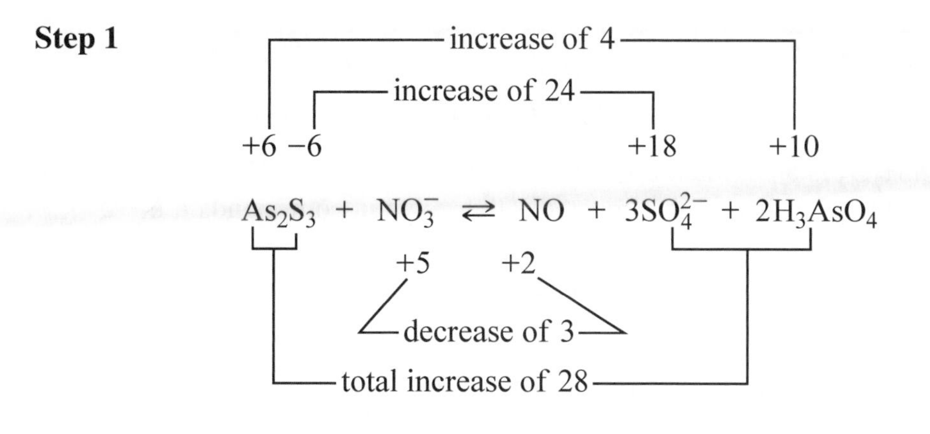

Step 2 $3As_2S_3 + 28NO_3^- \rightleftarrows 28NO + 9SO_4^{2-} + 6H_3AsO_4$

Step 3 $3As_2S_3 + 28NO_3^- + 10H^+ \rightleftarrows 28NO + 9SO_4^{2-} + 6H_3AsO_4$

Step 4 $3As_2S_3 + 28NO_3^- + 10H^+ + 4H_2O \rightleftarrows 28NO + 9SO_4^{2-} + 6H_3AsO_4$

EXAMPLE 10

$$Cu_2S + SO_4^{2-} \rightleftarrows SO_2 + Cu^{2+} \quad \text{(in acid solution)}$$

When a given product (SO_2 in the above equation) is derived from two (or more) separate sources, it should appear twice (or more) in the equation.

Step 1

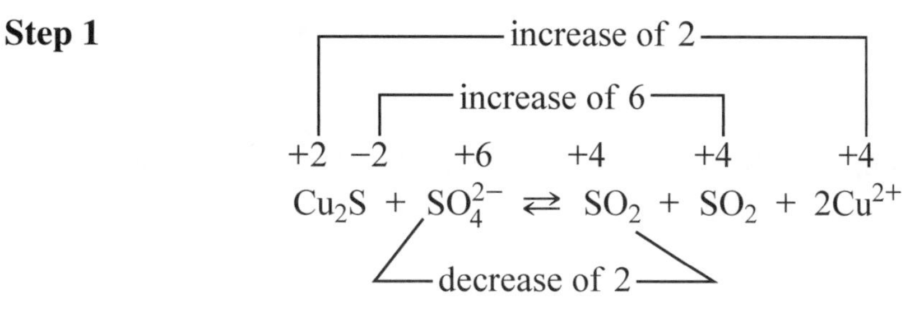

Step 2

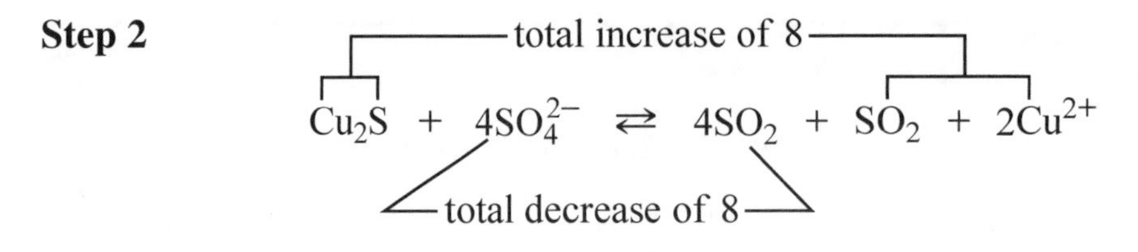

Step 3 $Cu_2S + 4SO_4^{2-} + 12H^+ \rightleftarrows 5SO_2 + 2Cu^{2+}$

Step 4 $Cu_2S + 4SO_4^{2-} + 12H^+ \rightleftarrows 5SO_2 + 2Cu^{2+} + 6H_2O$

EXAMPLE 11

$$H_2O_2 + Cr(OH)_4^- \leftrightarrows CrO_4^{2-} \quad \text{(in basic solution)}$$

When H_2O_2 is an oxidant it must be remembered that in H_2O_2, the oxidation number of O is -1, whereas in almost all other oxygen compounds it is -2.

Step 1

increase of 3

$$\overset{-2}{H_2O_2} + \overset{+3}{Cr(OH)_4^-} \rightleftarrows \overset{+6}{CrO_4^{2-}}$$

(-4 per 2 atoms of O)

decrease of 2

Step 2

increase of 6

$$3H_2O_2 + 2Cr(OH)_4^- \rightleftarrows 2CrO_4^{2-}$$

(-12 for 6 atoms)

decrease of 6

Step 3 $3H_2O_2 + 2Cr(OH)_4^- + 2OH^- \rightleftarrows 2CrO_4^{2-}$

(In basic solution, balance charge by adding OH^-.)

Step 4 $3H_2O_2 + 2Cr(OH)_3^- + 2OH^- \rightleftarrows 2CrO_4^{2-} + 8H_2O$

For most equations, balancing by oxidation number change is generally less time-consuming than balancing by half-reactions. However, in those instances in which the changes in oxidation numbers are not obvious, it may be wiser to use the method of half-reactions.

It should be emphasized that *oxidation number* is a *concept* that has been created by scientists for the purpose of expressing, quantitatively, the relative combining capacities of the constituent elements in a molecule or ion. For most binary species and for most ternary species in which oxygen is a constituent, the calculation of the oxidation number of each constituent element poses no problem when we, by definition, assign H a value of +1 and O a value of −2, and when we recognize that the sum of the oxidation numbers must always equal the net charge on the species. Thus, in MnO_2 the oxidation number of Mn is obviously +4, in MnO_4^- it is +7, and in H_2MnO_4 it is +6.

To determine the oxidation numbers of the two elements in As_2S_3 we can look upon As_2S_3 as having been derived from H_2S, in which the oxidation number of S is −2. Accordingly, we can assume that in As_2S_3 the oxidation number of S is −2; the oxidation number of As will then be +3.

For species such as $CrSCN^{2+}$ in the unbalanced equation

$$CrSCN^{2+} + BrO^- \rightleftarrows Br^- + NO_3^- + CO_3^{2-} + SO_4^{2-} + CrO_4^{2-} \quad \text{(in basic solution)}$$

the determination may be less obvious. As a matter of fact it really makes no great difference what oxidation numbers are assigned to Cr, S, C, and N, respectively, *as long as their sum is equal to +2, the charge on the ion*. We can, if we choose, arbitrarily assign the values Cr = +3, S = −2, and N = −3. The value for C must then be +4. As noted below, the total increase in oxidation number per mole of $CrSCN^{2+}$ is +19.

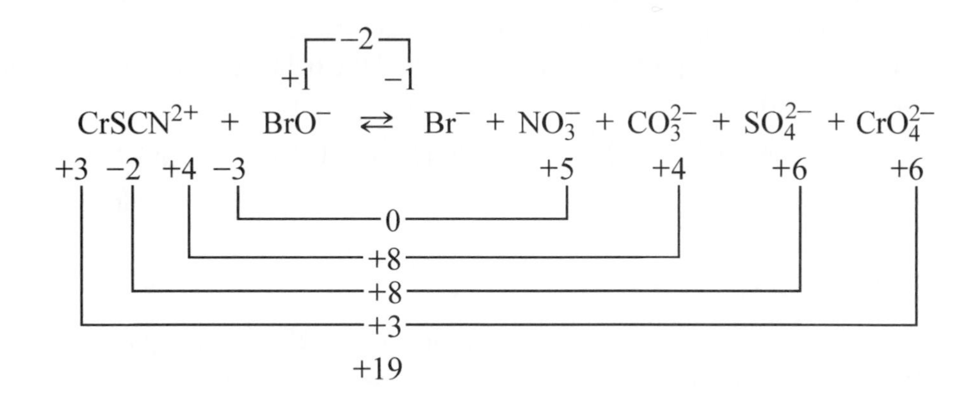

If we assign the values Cr = +3, N = +5, S = −2, and C = −4, the total increase in oxidation number per mole of $CrSCN^{2+}$ is again +19, as noted below.

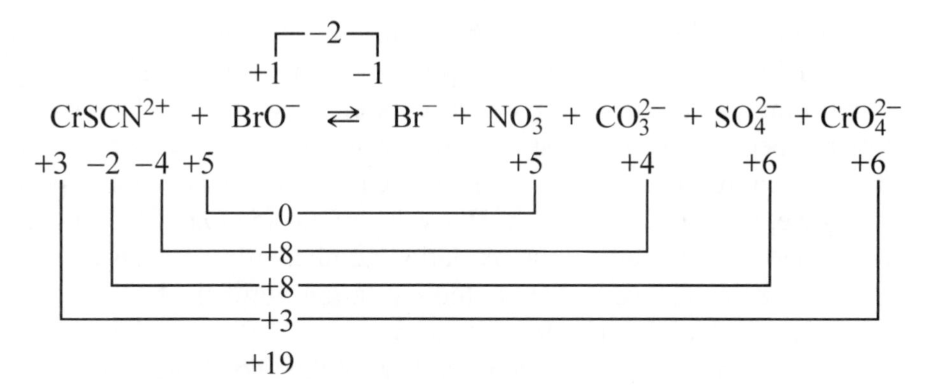

If we assign the values Cr = +6, N = −4, C = −4, and S = +4, the total increase in oxidation number per mole of $CrSCN^{2+}$ is again +19, as noted below

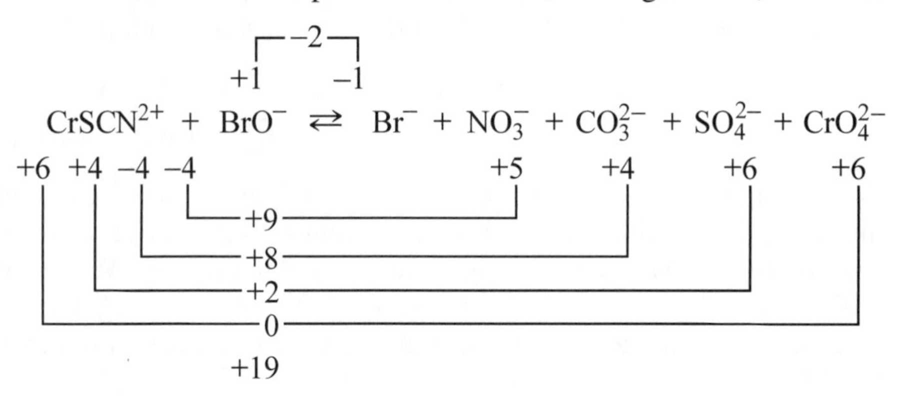

No matter what values we assign to Cr, S, C, and N, as long as the sum equals +2, the total change in oxidation number per mole of $CrSCN^{2+}$ will always be +19 when the indicated products are formed, and the balanced equation, calculated by the four-step process already outlined, will always turn out to be

$$2CrSCN^{2+} + 19BrO^- + 18OH^- \rightleftarrows 19Br^- + 2NO_3^- + 2CO_3^{2-} + 2SO_4^{2-} + 2CrO_4^{2-} + 9H_2O$$

PROBLEMS

Balance each of the following equations. (H^+, OH^-, and H_2O are not included in the unbalanced equation; addition of these species, where necessary, is a part of the balancing process.)

Group A. *In acidic solution.*

19.1 $Sn^{2+} + Ce^{4+} \rightleftarrows Ce^{3+} + Sn^{4+}$

19.2 $H_2S + Fe^{3+} \rightleftarrows Fe^{2+} + S$

19.3 $H_2SO_3 + HNO_2 \rightleftarrows NO + SO_4^{2-}$

19.4 $Br^- + MnO_4^- \rightleftarrows Mn^{2+} + Br_2$

19.5 $Sn^{2+} + H_2O_2 \rightleftarrows Sn^{4+}$

19.6 $I^- + Fe^{3+} \rightleftarrows Fe^{2+} + I_2$

19.7 $Mn^{2+} + HBiO_3 \rightleftarrows Bi^{3+} + MnO_4^-$

19.8 $Mn^{2+} + MnO_4^- \rightleftarrows MnO_2$

19.9 $Sb_2S_3 + NO_3^- \rightleftarrows NO_2 + SO_4^{2-} + Sb_2O_5$

19.10 $SnS_2O_3 + MnO_4^- \rightleftarrows Mn^{2+} + SO_4^{2-} + Sn^{4+}$

19.11 $FeHPO_3 + Cr_2O_7^{2-} \rightleftarrows Cr^{3+} + H_3PO_4 + Fe^{3+}$

19.12 $Hg_4Fe(CN)_6 + ClO_3^- \rightleftarrows Cl^- + NO + CO_2 + Fe^{3+} + Hg^{2+}$

19.13 $Fe_2Fe(CN)_6 + NO_3^- \rightleftarrows NO + CO_2 + Fe^{3+}$

19.14 $FeAsS + NO_3^- \rightleftarrows NO + SO_4^{2-} + H_3AsO_4 + Fe^{3+}$

19.15 $CrSCN^{2+} + Cl_2 \rightleftarrows Cl^- + NO_3^- + CO_2 + SO_4^{2-} + Cr_2O_7^{2-}$

19.16 $Sn(S_2O_3)_2^{2-} + FeS_2O_8^+ \rightleftarrows SO_4^{2-} + Sn^{4+} + Fe^{2+}$

Group B. *In basic solution.*

19.17 $HPO_3^- + OBr^- \rightleftarrows Br^- + PO_4^{3-}$

19.18 $Fe(OH)_2 + O_2 \rightleftarrows Fe(OH)_3$

19.19 $Ni(OH)_2 + OBr^- \rightleftarrows Br^- + NiO_2$

19.20 $Co(OH)_2 + H_2O_2 \rightleftarrows Co(OH)_3$

19.21 $Bi(OH)_3 + Sn(OH)_4^{2-} \rightleftarrows Sn(OH)_6^{2-} + Bi$

19.22 $SO_3^{2-} + Co(OH)_3 \rightleftarrows Co(OH)_2 + SO_4^{2-}$

19.23 $Sn(OH)_4^{2-} + MnO_4^- \rightleftarrows MnO_2 + Sn(OH)_6^{2-}$

19.24 $HS_2O_4^- + AsO_4^{3-} \rightleftarrows AsO_2^- + SO_4^{2-}$

19.25 $PH_3 + CrO_4^{2-} \rightleftarrows Cr(OH)_4^- + P$

19.26 $H_2PO_2^- + CNO^- \rightleftarrows CN^- + HPO_3^-$

19.27 $OCl^- \rightleftarrows Cl^- + ClO_3^-$

19.28 $FeHPO_3 + OCl^- \rightleftarrows Cl^- + PO_4^{3-} + Fe(OH)_3$

19.29 $Cu_2SnS_2 + S_2O_8^{2-} \rightleftarrows SO_4^{2-} + Sn(OH)_6^{2-} + Cu(OH)_2$

19.30 $V \rightleftarrows H_2 + HV_6O_{17}^{3-}$

EQUIVALENTS IN REDOX PROCESSES

In the discussion on neutralization we defined an equivalent of acid or base in such a way that one equivalent of acid would react with one equivalent of base. This approach is also useful in handling the stoichiometry of redox reactions in general and electrolysis in particular. In a redox reaction or an electrolysis there is a transfer of a number of electrons from one chemical species to another. The mass of one equivalent of a species involved in such a process is defined as the mass of 1 mol divided by the number of electrons being transferred per mole of the substance. For example, in the process $Zn(s) + 2Ag^+ \rightleftarrows Zn^{2+} + 2Ag(s)$ there is a transfer of two electrons per mole of $Zn(s)$; the equivalent weight of $Zn(s)$ is $65.4/2 = 32.7$ and 1 mol of $Zn(s)$ represents two equivalents. However, there is a transfer of only one electron per mole of Ag^+ and therefore the mass of one equivalent is the same as the mass of 1 mol, 107.9 g.

Often the number of electrons transferred is not obvious from the inspection of the overall reaction. For example, $2MnO_4^- + 5H_2SO_3 \rightleftarrows 2Mn^{2+} + 5SO_4^{2-} + 4H^+ + 3H_2O$. Here the number of electrons transferred can be determined by dividing the overall process into half-reactions:

$$MnO_4^- + 8H^+ + 5e^- \rightleftarrows 4H_2O + Mn^{2+}$$

$$H_2SO_3 + H_2O \rightleftarrows SO_4^{2-} + 4H^+ + 2e^-$$

Inspection of the half-reactions reveals that 5 electrons are transferred per mole of MnO_4^-, so that 1 mol of MnO_4^- is 5 equivalents. Similarly 2 electrons are transferred per mole of H_2SO_3 so that 1 mol of H_2SO_3 is 2 equivalents. Notice that, in the overall reaction, 2 mol or 10 equivalents of MnO_4^- reacts with 5 mol or 10 equivalents of H_2SO_3.

The relationship between moles and equivalents can also be determined by examination of the oxidation numbers of the atoms in the species involved in the redox process. In any redox reaction the equivalent weight of a substance is its formula weight divided by the change in oxidation number of its component atoms. Thus when MnO_4^- reacts to form Mn^{2+} and H_2O, the oxidation number of Mn decreases from +7 in MnO_4^- to +2 in Mn^{2+}, for a net change of 5. Since there is no change in the oxidation number of O, the equivalent weight of MnO_4^- is one-fifth of its formula weight.

We can express the concentration of a solution of an oxidizing agent or a reducing agent in terms of equivalents, by analogy with molarity. The molarity (M) of a solution is the amount of solute (in moles) in 1 L of solution. The normality (N) of a solution is the amount of solute (in equivalents) in 1 L of solution.

PROBLEMS

19.31 How many equivalents of $KMnO_4$ will be required to react with 30 g of $FeSO_4$ in the following reaction?

$$5Fe^{2+} + MnO_4^- + 8H^+ \rightleftarrows 5Fe^{3+} + Mn^{2+} + 4H_2O$$

Solution:

$$\text{Equivalents } KMnO_4 = \text{equivalents } FeSO_4$$

$$\text{Equivalent weight } FeSO_4 = \frac{\text{formula weight } FeSO_4}{\text{oxidation number change Fe}}$$

$$= \frac{151.9}{1} = \frac{151.9 \text{ g } FeSO_4}{1 \text{ equivalent } FeSO_4}$$

$$\text{Equivalents } FeSO_4 = \frac{30 \text{ g } FeSO_4}{151.9 \text{ g/equivalent } FeSO_4}$$

$$= 0.20 \text{ equivalent } FeSO_4$$

Therefore, 0.20 equivalent of $KMnO_4$ is required.

19.32 What would be the concentration, in grams per liter, of 0.100 *N* $KMnO_4$ when used in the following reaction?

$$2MnO_4^- + 10Cl^- + 16H^+ \rightleftarrows 2Mn^{2+} + 5Cl_2 + 8H_2O$$

Solution: A 0.100 *N* $KMnO_4$ solution contains 0.100 equivalent (equivalent weight) of $KMnO_4$ per liter.

The oxidation number of Mn changes from +7 in MnO_4^- to +2 in Mn^{2+}. This represents an oxidation number change of 5.

$$\text{Redox equivalent weight} = \frac{\text{formula weight}}{\text{oxidation number change}}$$

$$= \frac{158}{5} = 31.6$$

$$0.100 \text{ equivalent} = 3.16 \text{ g} \quad \text{concentration} = 3.16 \text{ g/L}$$

19.33 What mass of $KClO_3$ will be required for the preparation of 400 cm^3 of 0.20 *N* $KClO_3$ for use in the following reaction?

$$ClO_3^- + 3H_2SO_3 \rightleftarrows Cl^- + 3SO_4^{2-} + 6H^+$$

19.34 What volume of 0.50 *N* H_2SO_3 will be required to reduce 120 cm^3 of 0.40 *N* $K_2Cr_2O_7$?

$$Cr_2O_7^{2-} + 3H_2SO_3 + 2H^+ \rightleftarrows 2Cr^{3+} + 3SO_4^{2-} + 4H_2O$$

Hint: Volume × normality = volume × normality.

19.35 What mass of $FeSO_4$ will be oxidized by 24 cm^3 of pH = 1.00 in the following reaction?

$$MnO_4^- + 5Fe^{2+} + 8H^+ \rightleftarrows Mn^{2+} + 5Fe^{3+} + 4H_2O$$

19.36 $Cr_2O_7^{2-}$ will oxidize NO_2^- to NO_3^- in acid solution, the $Cr_2O_7^{2-}$ being reduced to Cr^{3+}.

In one student's experiment 20 cm^3 of 0.100 *F* $K_2Cr_2O_7$ solution reacted with 1.020 g of a mixture of KNO_2 and KNO_3. For this experiment calculate

a. The normality of the $K_2Cr_2O_7$
b. The number of equivalents of $K_2Cr_2O_7$ used
c. The number of equivalents of KNO_2 present in the mixture
d. The equivalent weight of KNO_2
e. The mass of KNO_2 in the mixture
f. The percent of KNO_2 in the mixture

19.37 A 10.1-mL sample of a solution of Cl^- requires 10.8 mL of 0.0834 *M* $KMnO_4$ to reach the equivalence point for its oxidation to ClO^- in alkaline solution. The MnO_4^- forms $MnO_2(s)$. Find the concentration of the Cl^- ion in the original solution.

19.38 A sample of impure zinc metal of mass 2.54 g is analyzed by titration with $KBrO_3$, which oxidizes the zinc metal to the +2 state. The BrO_3^- is reduced to Br_2. The sample requires 50.4 mL of 0.274 *M* potassium bromate solution. Find the percentage of zinc metal in the sample, assuming that it does not contain any other reducing agents.

19.39 Sodium oxalate $Na_2C_2O_4$ in solution is oxidized to $CO_2(g)$ by MnO_4^-, which is reduced to Mn^{2+}. A 50.1 mL volume of a solution of MnO_4^- is required to titrate a 0.339-g sample of sodium oxalate. Find the concentration of the MnO_4^- solution.

19.40* A volume of 32.5 mL of the MnO_4^- solution of Prob. 19.39 is used to titrate a 4.62-g sample of a uranium-containing material. The oxidation of the uranium can be represented by the change $UO^{2+} \rightarrow UO_2^{2+}$. Calculate the percentage of uranium in the sample.

Hint: The answer to Prob. 19.39 is not necessary for this calculation; the stoichiometric relationships between oxalate and UO^{2+} are enough.

FARADAY'S LAW AND ELECTROCHEMICAL EQUIVALENCE

One faraday of electricity, when passed through a solution of an electrolyte, will cause one equivalent weight of substance (in grams) to react, be deposited, or be liberated at each electrode. This important generalization is a part of a broader generalization known as *Faraday's law.*

You should note that since 1 equivalent weight of a substance such as Ag^+ contains 6.022×10^{23} ions, 6.022×10^{23} electrons will be required to electro-deposit 1 equivalent weight of silver according to the reaction $Ag^+ + e^- \rightleftarrows Ag$. Since 1 faraday of electricity will deposit 1 equivalent weight of silver, 1 faraday must represent 6.022×10^{23} electrons. This is the basis for stating that 1 faraday is a *mole of electrons.*

One faraday of electricity is 96,500 coulombs. (The more exact value, to five significant figures, is 96,489.) One coulomb is the charge that is carried when one ampere of current flows for one second. Therefore, 1 faraday = 96,500 A-s = 26.8 A-h. We will assume in the problems that follow that the efficiency of the process (the current efficiency) is 100% unless stated otherwise.

PROBLEMS

19.41 Electricity was allowed to flow until 20 g of copper was deposited from a solution of $CuSO_4$. What charge in coulombs passed through the solution?

Solution: Since the change in oxidation number when Cu^{2+} is reduced to Cu is 2, the equivalent weight of copper is half its atomic weight, namely 31.8.

$$\frac{20.0 \text{ g Cu}}{31.8 \text{ g Cu/equivalent}} \times \frac{96{,}500 \text{ C}}{1 \text{ equivalent}} = 60{,}700 \text{ C}$$

19.42 A current of 2.00 A was allowed to flow through a solution of $AgNO_3$ for 6.00 h. What mass of silver was deposited?

Solution: 6.00 h = 21,600 s. Two amperes of current flowing for 21,600 s is 43,200 C of electricity. The equivalent weight of Ag in $AgNO_3$ is 107.9.

$$\frac{43{,}200 \text{ C}}{96{,}500 \text{ C/equivalent}} \times \frac{107.9 \text{ g Ag}}{1 \text{ equivalent}} = 48.3 \text{ g Ag}$$

19.43 A certain amount of electricity deposited 50.0 g of silver from a solution of $AgNO_3$. What mass of copper will this same amount of current deposit from a solution of $CuSO_4$?

Solution: Since 1 faraday of electricity will deposit 1 equivalent of any element, it follows that the mass of one element deposited by a given amount of electricity will be to the mass of another element deposited by the same amount of electricity as the equivalent weight of the first element is to the equivalent weight of the second element.

$$\frac{\text{Mass of copper deposited}}{\text{Mass of silver deposited}} = \frac{\text{equivalent weight of copper in } CuSO_4}{\text{equivalent weight of silver in } AgNO_3}$$

The equivalent weight of Ag in $AgNO_3$ is 107.9. The equivalent weight of Cu in $CuSO_4$ is 31.8. Therefore,

$$\frac{\text{mass of copper}}{50.0 \text{ g Ag}} = \frac{31.8}{107.9}$$

$$\text{mass of copper} = \frac{31.8}{107.9} \times 50.0 \text{ g} = 14.7 \text{ g}$$

19.44 What mass of cobalt will be deposited from a solution of $CoCl_2$ by 40,000 C of electricity?

19.45 How many coulombs of electricity will be required to deposit 100 g of chromium from a solution of $CrCl_3$?

19.46 What mass of zinc will be deposited from a solution of $ZnCl_2$ by a current of 3.00 A for 20.0 h?

19.47 A quantity of electricity that deposited 70 g of nickel from a solution of $NiCl_2$ will liberate what mass of hydrogen from a solution of HCl?

19.48 A solution of $CuSO_4$ contains Cu^{2+}. A solution of $Na_2Cu(CN)_3$ contains Cu^+ ions in equilibrium with $Cu(CN)_3^{2-}$ ions and CN^- ions. A quantity of electricity that deposits 12 g of copper from a solution of copper sulfate will deposit what mass of copper from a solution of $Na_2Cu(CN)_3$?

19.49 What mass of nickel will be deposited from a solution of $NiCl_2$ by 4.00 A of current for 24.0 h if the current efficiency of the process is 96.0%?

19.50 Calculate the charge, in coulombs, on an electron.

19.51 A current of 2.0 A was passed through a solution of H_2SO_4 for 20 min. What volume of O_2 gas, measured at STP, was liberated?

Hint: 1 mol of O_2 gas is how many equivalents?

19.52 A solution of $CuSO_4$ was electrolyzed, using platinum electrodes. A volume of 6.0 L of O_2 gas, measured at STP, was liberated at the positive electrode. What mass of copper was deposited at the negative electrode?

Solution: The number of equivalents of O_2 liberated at the positive electrode equals the number of equivalents of Cu deposited at the negative electrode.

$$1 \text{ mol } O_2 = 4 \text{ equivalents}$$

$$6 \text{ L } O_2 = \frac{6 \text{ mol}}{22.4} = \frac{6 \times 4}{22.4} \text{ equivalents}$$

$$\frac{6 \times 4}{22.4} \text{ equivalents Cu} \times 31.8 \text{ g Cu/equivalent} = 34 \text{ g Cu deposited}$$

19.53* To electrodeposit all the Cu and Cd from a solution of $CuSO_4$ and $CdSO_4$ in water required 1.20 faradays of electricity. The mixture of Cu and Cd that was deposited had a mass of 50.36 g. What mass of $CuSO_4$ was present in the solution?

19.54* A certain solution of K_2SO_4 was electrolyzed by using platinum electrodes. The combined volume of the dry gases that were evolved was 67.2 cm^3 at STP. If we assume 100% current efficiency and no loss of gases during measurement, how many coulombs of electricity were consumed?

19.55* A current of 2.0 A was passed through a cell containing 800 cm^3 of 1.0 F H_2SO_4. The following electrode reactions took place:

$$\text{anode: } 2H_2O \rightleftarrows O_2(g) + 4H^+ + 4e^-$$
$$\text{cathode: } 2H^+ + 2e^- \rightleftarrows H_2(g)$$

a. What time (in seconds) was required to liberate 0.050 mol of O_2?
b. What volume of H_2 gas, measured over water at 25 °C at a barometric pressure of 740 mmHg, was produced at the same time?
c. How many electrons were involved in the anode reaction in this experiment?
d. What volume of 2 F NaOH would be required to neutralize all the acid remaining in the cell at the end of the electrolysis?

19.56* When 1.00 g of yttrium metal (at. wt. 88.92) is treated with excess H_2SO_4, 378 cm^3 of H_2 gas is liberated at STP. When the resulting solution of yttrium sulfate is electrolyzed with platinum electrodes by using a steady current of 2.00 A, 1.00 g of pure yttrium is deposited on the negative electrode and O_2 gas is liberated at the positive electrode. Calculate
a. The formula for yttrium sulfate
b. The number of minutes the electrolysis had to proceed to deposit the 1.00 g of Y
c. The volume, measured at STP, of the oxygen gas liberated

19.57* The atomic weight of metal M is 52.01. When the melted chloride of M is electrolyzed, for every gram of metal deposited on the cathode, 725 cm^3 of dry chlorine gas, measured at 25 °C and 740 mmHg, is liberated at the anode. Calculate the formula of the chloride of M.

Hint: Equivalents of Cl_2 equals equivalents of M. The atomic weight of M divided by the equivalent weight of M equals the valence of M. This is equivalent to stating that the atomic weight of M divided by the equivalent weight of M equals the change in oxidation number of M when it is electro-deposited.

19.58* A metal M is known to form the fluoride MF_2. When 3300 coulombs of electricity is passed through the molten fluoride, 1.950 g of M is plated out. What is the atomic weight of M?

19.59* When a solution of KI is electrolyzed, using porous silver electrodes, H_2 gas is liberated at the negative electrode (cathode) and insoluble AgI is deposited in the pores of the positive electrode (anode). All the AgI that is formed remains in the pores of the anode. At the conclusion of the experiment the anode has increased in mass by 5.076 g and 530 cm^3 dry H_2 gas, measured at 27 °C and 720 mmHg, has been liberated. Calculate the atomic weight of iodine from these data.

19.60* An aqueous solution of the soluble salt MSO_4 is electrolyzed between inert (platinum) electrodes until 0.0327 g of metal M is deposited on the negative electrode (cathode). To neutralize the solution that was formed in the electrolytic cell required 50 cm^3 of 0.020 *F* KOH. Calculate the atomic weight of metal M.

19.61* A potassium salt of a ternary acid of molybdenum (Mo, at. wt. 95.95) has the formula K_2MoO_x. When an acidified solution of K_2MoO_x is electrolyzed between platinum electrodes, only oxygen gas is liberated at the positive electrode and only molybdenum metal is deposited at the negative electrode. When electrolysis is continued until 0.3454 g of molybdenum is deposited, 121.0 cm^3 of O_2 gas, measured at STP, is liberated. Calculate the formula of the salt.

Solution: If we know the oxidation number of Mo in MoO_x^{2-}, we can calculate the value of x. Since MoO_x^{2-}, is reduced to Mo metal, the *change* in the oxidation number of Mo in the course of the reaction is equal to the oxidation number of Mo in MoO_x^{2-}. The equivalent weight of Mo in this reaction is its atomic weight divided by the change in oxidation number when it is formed from MoO_x^{2-}. Therefore, if we know the equivalent weight of Mo in the above reaction, we can, following the argument outlined above, calculate the value of x.

Equivalents Mo deposited = equivalents O_2 liberated.

Equivalents O_2 = 4 × moles O_2 = 4 × 121 cm^3/22,400 cm^3
= 0.0216 equivalents Mo.

Eq. wt. Mo = 0.3454 g of Mo/0.0216 equivalents Mo = 16.0 g/equivalent.

Change in oxidation number of Mo = at. wt./eq. wt. = 95.95/16.0 = 6.

Therefore, the oxidation number of Mo in MoO_x^{2-} is 6, $x = 4$, and the formula of the compound is K_2MoO_4.

19.62* When 0.20 faraday of electricity is passed through a solution of $Pb(NO_3)_2$, a compound containing 20.7 g of Pb is deposited at the positive electrode. What is the oxidation number of lead in the compound that is deposited?

19.63* A thin layer of gold can be applied to another material by an electrolytic process. The surface area of an object to be gold-plated is 49.8 cm^2 and the density of the gold is 19.3 g/cm^3. A current of 3.25 A is applied to a solution that contains gold in the +3 oxidation state. Calculate the time required to deposit an even layer of gold 1.00×10^{-3} cm thick on the object.

POTENTIALS

The relative tendency of substances to accept or donate electrons in redox reactions is a chemical property that is of great interest to us. We can conveniently express this tendency as a potential (measured in volts) associated with a given half-reaction. For example, we can measure the potential of the half-reaction $Zn^{2+} + 2e^- \rightleftarrows Zn(s)$. Under carefully defined conditions and relative to an appropriate standard, we find that the potential of this half-reaction is −0.763 V. This value gives us information about the tendency of Zn^{2+} to accept electrons and become reduced. It also tells us about the reverse process, the tendency of Zn metal to donate electrons and become oxidized.

Although we can never carry out a single half-reaction, we can still assign values to the potentials of half-reactions by arbitrarily choosing a standard half-reaction and assigning it a potential of exactly 0. This standard half-reaction is $2H^+ \leftrightharpoons H_2(g) + 2e^-$ and is called the *standard hydrogen electrode*. We can then construct a *galvanic cell*, an apparatus that separates the reduction and the oxidation half-reactions from each other and forces the electron transfer to take place through a wire. Such a cell is shown schematically in the figure below. The voltmeter in the line allows us to measure the potential difference between the half-reactions in the two cell compartments.

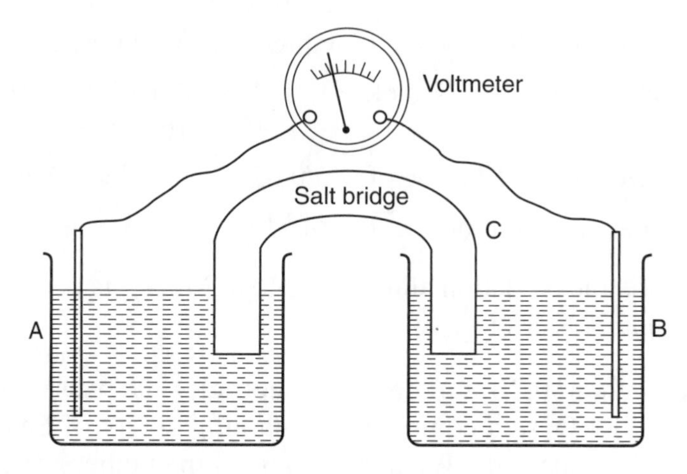

If one of the cell compartments contains the standard hydrogen electrode, the measured potential difference is equal to the potential of the other half-reaction. In general, if one of the compartments contains the components of a half-reaction of known potential, a measurement of the potential difference allows us to find the potential of the other half-reaction. The potentials of a great many half-reactions have been measured in this way.

A typical galvanic cell can be constructed by using zinc and copper. Beaker A in the figure contains a strip of zinc metal in a 1 *M* solution of $ZnSO_4$. Beaker B contains a strip of copper metal in a 1 *M* solution of $CuSO_4$. The U-shaped tube C contains a gel prepared with a solution of K_2SO_4. It is called a *salt bridge* and serves as a conducting medium between the two solutions. The strips of Cu and Zn are connected by copper wires through the voltmeter.

A cell can be described by using a cell notation in which a vertical line separates each phase and a double vertical line represents a salt bridge. The cell shown here is: $Zn|Zn^{2+}(1\ M)\|Cu^{2+}(1\ M)|Cu$. We show only the substances that undergo chemical change or are part of the electric circuit.

A group of galvanic cells operating together is what we commonly call a *battery*. Galvanic cells are important sources of electrical energy. They can be viewed as devices for the conversion of stored chemical energy into electrical energy. It is the difference in potential between the two halves of the galvanic cell that causes the electrons to flow from one to the other through the wire connection.

The greater the potential difference, the greater the extent to which the cell reaction will proceed. Since each half of a galvanic cell corresponds to a half-reaction, and since the potential difference can be found from the potentials of these half-reactions, we can use the values of the potentials of the half-reactions to correlate and predict the redox behavior of many substances.

For convenience, we can make a list of half-reactions and their standard potentials in order of increasing potential. Such a list can be found in Tables A.5 and A.6

of the appendix. By convention all the half-reactions in such a list are written as reductions, so we call the potentials in such a table *reduction potentials*.

It should be emphasized that the voltages given in Tables A.5 and A.6 are for systems in which the concentrations of all substances in solution are 1 *M*, the temperature is 25 °C, and all gases are at a pressure of 1 atm. Under these conditions the potentials are referred to as *standard electrode potentials* and are designated by the symbol $E°$.

If a cell is made up with a strip of Cu in a solution of $CuSO_4$ in one half-cell and a strip of Ag in Ag_2SO_4 in the other half-cell, its voltage is 0.462 V, calculated as follows:

$$\begin{array}{lr} Cu \rightleftarrows Cu^{2+} + 2e^- & -0.337\text{ V} \\ 2Ag^+ + 2e^- \rightleftarrows 2Ag & +0.799\text{ V} \\ \hline Cu + 2Ag^+ \rightleftarrows Cu^{2+} + 2Ag & +0.462\text{ V} \end{array}$$

This system is referred to as a $Cu|Cu^{2+}\|Ag^+|Ag$ cell.

Note, in this example, that doubling the amount of Ag^+ in the reaction $2Ag^+ + 2e^- \rightleftarrows 2Ag$ does not alter the voltage. The reason is that the voltage is a measure of the *work per electron*.

Note, also, that the substance with the highest (most positive) potential, Ag^+ in this case, functions as the oxidizing agent. The Cu^{2+} half-reaction has accordingly been reversed, resulting in its voltage being changed from +0.337 V to −0.337 V. As a result, the sum of the voltages is positive (+0.462). This operation illustrates how one can determine whether or not a certain oxidizing agent will react with a certain reducing agent. The rule is this: If the sum of the two potentials, written with the proper signs (sign of the potential of the reducing agent is reversed), is *positive*, the two substances *will* react with each other; if the sum is negative, they will not react.

Will Cu reduce Sn^{4+} to Sn^{2+}, being itself oxidized to Cu^{2+}?

$$\begin{array}{lr} Cu \rightleftarrows Cu^{2+} + 2e^- & -0.337\text{ V} \\ Sn^{4+} + 2e^- \leftrightarrows Sn^{2+} & +0.15\text{ V} \\ \hline Cu + Sn^{4+} \leftrightarrows Cu^{2+} + Sn^{2+} & -0.19\text{ V} \end{array}$$

The sum is negative. Reaction will not occur.

Application of the above rule to Tables A.5 and A.6 reveals these simple and useful relationships:

- Any reducing agent of higher potential will reduce any oxidizing agent of lower potential.

- Any oxidizing agent of lower potential will oxidize any reducing agent of higher potential.
- Any reducing agent on the right *will* reduce (be oxidized by) any oxidizing agent *below* itself and on the left. It *will not* be oxidized by an oxidizing agent above itself and on the left.
- Any oxidizing agent on the left will oxidize (be reduced by) any reducing agent *above* itself and on the right. It *will not* oxidize a reducing agent *below* itself on the right.

In the reaction $Zn + Cu^{2+} \rightleftarrows Zn^{2+} + Cu$, the Zn functions as the reducing agent and the half-reaction $Zn^{2+} + 2e^- \rightleftarrows Zn$ in Table A.5 proceeds in the reverse direction. In the reaction $Zn^{2+} + Mn \rightleftarrows Zn + Mn^{2+}$, the Zn^{2+} functions as the oxidizing agent and the zinc half-reaction proceeds as written in the table. This half-reaction is reversible and the system can reach an equilibrium state.

What is true of the reaction $Zn^{2+} + 2e^- \rightleftarrows Zn$ is true of all redox (oxidation-reduction) half-reactions. They are reversible and can reach a state of equilibrium as part of an overall redox reaction. For reactions with a high positive potential, indicating that the substance is a strong oxidizing agent, the equilibrium is far to the right. For reactions with a low potential (a large negative potential), meaning that the substance is a strong reducing agent, the equilibrium is far to the left.

Since the reactions are all reversible and can reach a state of equilibrium, they are written with conventional double arrows, $Zn^{2+} + 2e^- \rightleftarrows Zn$.

PROBLEMS

19.64 Calculate the theoretical voltage of each of the following cells, assembled in the manner shown in the figure on page 290.

a. $Zn|1\ M\ Zn^{2+}||1\ M\ Ni^{2+}|Ni$

b. $Al|1\ M\ Al^{3+}||1\ M\ Cu^{2+}|Cu$

c. $Cu|1\ M\ Cu^{2+}||1\ M\ Hg^{2+}|Hg$

d. $Pt|1\ M\ Fe^{3+}, 1\ M\ Fe^{2+}||1\ M\ MnO_4^-, 1\ M\ Mn^{2+}|Pt$

19.65 Calculate the theoretical voltage of each of the following cells with alkaline solutions.

a. $Zn(s)|ZnS(s)||1\ M\ S^{2-}|S(s)|Pt(s)$

b. $Pt(s)|H_2\ (1\ atm)||1\ M\ OH^-, 1\ M\ Zn(OH)_4^{2-}|Zn(s)$

c. $Ni(s)|Ni(OH)_2(s)||1\ M\ OH^-|O_2\ (1\ atm)|Pt(s)$

19.66 State whether or not a reaction will occur when the following are brought together in acidic 1 *M* solution at 25 °C.

a. MnO_4^- and I^-
b. Cr^{2+} and Cu^{2+}
c. Sn^{2+} and H_3PO_4
d. H_2S (1 atm) and Fe^{3+}
e. Sn and Sn^{4+}

19.67 State whether or not a reaction will occur when the following are brought together in alkaline 1 *M* solution at 25 °C.
a. MnO_4^- and ClO^-
b. Cl^- and BrO^-
c. PH_3 and $Sn(OH)_4^{2-}$
d. I^- and Br^-
e. H_2O_2 and MnO_4^-

19.68 Two cells are constructed in which the standard hydrogen electrode is one of the half-cells. When the other half-cell is the $Be(s)|Be^{2+}$ electrode, the $E°$ is measured as 1.70 V with the hydrogen electrode as the cathode. When the $Au(s)|Au^{3+}$ electrode is the other half-cell, the $E°$ is measured as 1.29 V with the hydrogen electrode as the anode. Calculate $E°$ for the cell made of the two metal half-cells and designate the anode and the cathode.

19.69 Two cells are constructed in which the $Ni(s)|Ni^{2+}$ electrode is one of the half-cells. When the other half-cell is $Cu(s)|Cu^{2+}$ the $E°$ is measured as 0.570 V with the nickel electrode as the anode. When the $Mg(s)|Mg^{2+}$ electrode is the other half-cell, the $E°$ is measured as 2.145 V with the magnesium electrode as the anode. Calculate $E°$ for the cell made of the magnesium and copper electrodes and designate the anode and the cathode.

EFFECT OF CHANGE OF CONCENTRATION

Since each half-reaction reaches a state of equilibrium, it follows that the equilibrium composition, and hence the voltage, will change when the concentration of a reactant or product is changed. Thus for the reaction

$$Fe^{3+} + e^- \rightleftarrows Fe^{2+} \quad (0.771 \text{ V})$$

an increase in concentration of Fe^{3+} will shift the equilibrium to the right, thereby increasing the positive potential. If the concentration of Fe^{3+} is lowered, the equilibrium will shift to the left and the potential will decrease. If the concentration of Fe^{2+} is increased, the equilibrium will shift to the left and the potential will decrease, while if the concentration of Fe^{2+} is lowered, the equilibrium will shift to the right and the potential will increase.

The quantitative effect of change in concentration can be calculated by the use of the *Nernst equation*,

$$E = E^\circ - \frac{RT}{nF} \ln \frac{[\text{product}]}{[\text{reactant}]} \quad \textbf{(19-1)}$$

In this formula, E° is the standard potential of the half-reaction, E is its potential under the particular conditions of the experiment, R is the gas *constant* (with a value of 8.31 J/mol-K), F is the *faraday* (with a value of 96,500 J/V), n is the number of electrons transferred in the half-reaction, and T is the absolute temperature. Therefore, at constant temperature RT/F will be constant for all reactions; its value at 25 °C is 0.0257 V/mol. Thus Eq. (19-1) takes the simplified form

$$E = E^\circ - \frac{0.0257}{n} \ln \frac{[\text{product}]}{[\text{reactant}]}$$

Since the product, Fe^{2+} in the above example, is in the reduced state and the reactant, Fe^{3+}, is in the oxidized state, the formula is commonly written

$$E = E^\circ - \frac{0.0257}{n} \ln \frac{[\text{reduced state}]}{[\text{oxidized state}]}$$

Suppose the concentration of Fe^{3+} is 0.1 M and the concentration of Fe^{2+} is 1.0 M:

$$E = +0.771 - \frac{0.0257}{1} \ln \frac{1.0}{0.1} = +0.771 - 0.0592 = 0.712 \text{ V}$$

If more than one reactant or more than one product is involved, each species is included in the concentration term and, as in the standard equilibrium formula, each is raised to a power equal to the amount that reacts. Solids or liquids are not included in the formulation and the concentration of H_2O is constant at 55.6 mol/L. Thus for the half-reaction

$$ClO_3^- + 6H^+ + 6e^- \rightleftarrows Cl^- + 3H_2O \qquad (1.45 \text{ V})$$

$$E = 1.45 - \frac{0.0257}{6} \ln \frac{[Cl^-]}{[ClO_3^-] \times [H^+]^6}$$

and for the half-reaction

$$Zn^{2+} + 2e^- \leftrightarrows Zn(s) \qquad (-0.763 \text{ V})$$

$$E = -0.763 - \frac{0.0257}{2} \ln \frac{1}{[Zn^{2+}]}$$

Note that if all concentrations are 1 M, the term [red]/[oxid] = 1. Since ln 1 = 0, the entire term

$$-\frac{0.0257}{n}\ln\frac{[\text{red}]}{[\text{oxid}]}$$

equals zero, and $E = E°$.

If [oxid] is less than 1 or [red] is greater than 1, making [red]/[oxid] greater than 1, the ln will be positive. As a result, the term

$$-\frac{0.0257}{n}\ln\frac{[\text{red}]}{[\text{oxid}]}$$

will be negative and E will be more negative than $E°$. This is exactly the conclusion that was reached in our qualitative inspection. If the ratio of [red] to [oxid] is less than 1, the ln will be negative, the term

$$-\frac{0.0257}{n}\ln\frac{[\text{red}]}{[\text{oxid}]}$$

will be positive, and E will be less negative then $E°$.

For the following problems, assume that the temperature is 25 °C unless otherwise noted.

PROBLEMS

19.70 Calculate the potential of the half-reaction $Zn \rightleftarrows Zn^{2+} + 2e^-$ when the concentration of the Zn^{2+} ion is 0.10 M.

19.71 Calculate the potential of the half-reaction $Fe^{2+} \rightleftarrows Fe^{3+} + e^-$ when Fe^{2+} and Fe^{3+} are, respectively, 0.40 M and 1.60 M.

19.72 Calculate the potential of the following cells:

a. $Zn(s)|0.200\ M\ Zn^{2+}||0.100\ M\ Cu^{2+}|Cu(s)$

b. $Zn(s)|0.200\ M\ Zn^{2+}||1.00 \times 10^{-4}\ M\ Cu^{2+}|Cu(s)$

c. $Zn(s)|0.200\ M\ Zn^{2+}||1.00 \times 10^{-8}\ M\ Cu^{2+}|Cu(s)$

d. $Zn(s)|1.00 \times 10^{-8}\ M\ Zn^{2+}||0.200\ M\ Cu^{2+}|Cu(s)$

19.73 Calculate the potential of the following cells:

a. $Zn(s)|ZnS(s)|0.0400\ M\ S^{2-}||2.00\ M\ CN^-,\ 1\ M\ Zn(CN)_4^{2-}|Zn(s)$

b. $Zn(s)|ZnS(s)|2.00\ M\ S^{2-}||0.04\ M\ CN^-,\ 1\ M\ Zn(CN)_4^{2-}|Zn(s)$

19.74 Calculate the potential of the cell

$Pt(s)|Cl_2(g)(1\ atm)|1.00\ M\ Cl^-||0.0100\ M\ Mn^{2+},\ 2.00\ M\ MnO_4^-|Pt(s)$

a. at pH = 1.00

b. at pH = 4.00

19.75 The potential of the cell $In(s)|In^{3+}(0.0085\ M)||H^+(1\ M)|H_2(g)(1\ atm)|Pt(s)$ is found to be 0.28 V. Calculate $E°$ of the cell and of the $In(s)$, In^{3+} electrode.

19.76 Calculate the reduction potential of the half-cell $Pt(s)$, $H_2(g)(1\ atm)$, H^+ at pH = 2, pH = 4, and pH = 7.

19.77 A concentration cell is a galvanic cell in which both half-cells are composed of the same substances, but in different concentrations. Since E is determined by concentrations, a potential difference and a current are established. Consider a cell in which both half-cells contain zinc metal electrodes and Zn^{2+} in solution. The $[Zn^{2+}]$ is 1.0 M in one half-cell and 0.10 M in the other. Identify the anode and the cathode and calculate E for the cell.

19.78 Determine E in a concentration cell in which the electrodes are $Sn(s)$ and the $[Sn^{2+}]$ in the two half-cells is 0.34 M and 0.0088 M.

19.79 Find the $[Fe^{3+}]$ in a half-cell of a concentration cell with $Fe(s)$ electrodes whose $E = 0.020$ V and whose other half-cell is the standard electrode.

19.80* Calculate the $[Ni^{2+}]$ required for the potential of the cell $Ni(s)|Ni^{2+}||0.001\ M$ $Sn^{2+}|Sn(s)$ to be 0.0 V.

CALCULATION OF EQUILIBRIUM CONSTANTS

The Nernst equation enables us to calculate the equilibrium constant for the overall oxidation-reduction reaction. Thus suppose we set up a cell like that shown in the figure on page 290 with Br_2 and Br^- in one beaker and I_2 and I^- in the other, close the circuit, and allow the system to stand until the voltage drops to zero. The reaction $Br_2 + 2I^- \rightleftarrows I_2 + 2Br^-$ will then have reached a state of equilibrium. When this point is reached, the potential generated by the reaction $Br_2 + 2e^- \rightleftarrows 2Br^-$ will exactly equal the potential generated by the reaction $I_2 + 2e^- \rightleftarrows 2I^-$.

Since

$$E_{Br} = 1.065 - \frac{0.0257}{2} \ln \frac{[Br^-]^2}{[Br_2]}$$

and

$$E_I = 0.536 - \frac{0.0257}{2} \ln \frac{[I^-]^2}{[I_2]}$$

Then

$$1.065 - \frac{0.0257}{2} \ln \frac{[Br^-]^2}{[Br_2]} = 0.536 - \frac{0.0257}{2} \ln \frac{[I^-]^2}{[I_2]}$$

$$\frac{0.0257}{2}\left(\ln\frac{[Br^-]^2}{[Br_2]} - \ln\frac{[I^-]^2}{[I_2]}\right) = 1.065 - 0.536 = 0.529$$

$$\frac{0.0257}{2}\ln\frac{[Br^-]^2/[Br_2]}{[I^-]^2/[I_2]} = 0.529$$

$$\ln\frac{[I_2] \times [Br^-]^2}{[I^-]^2 \times [Br_2]} = \frac{0.529 \times 2}{0.0257} = 41.2$$

But $\frac{[I_2] \times [Br^-]^2}{[I^-]^2 \times [Br_2]}$ is the equilibrium constant K for the reaction $Br_2 + 2I^- \rightleftarrows 2Br^- + I_2$.

$$\ln K = 41.2$$
$$K = 8 \times 10^{17}$$

This means that the above reaction is practically complete to the right.

If we examine the calculations given above we will find that for a reaction at equilibrium,

$$\ln K = \frac{n(\text{difference in the standard potentials of the two half-reactions})}{0.0257}$$

and

$$K = e^{n\Delta E°/0.0257}$$

where n is the number of electrons gained or lost in each balanced half-reaction and $\Delta E°$ is the difference between the two standard potentials.

PROBLEMS

19.81 Calculate the equilibrium constant for each of the following reactions:

a. $Cl_2(aq) + 2Br^-(aq) \rightleftarrows 2Cl^-(aq) + Br_2(aq)$
b. $2Fe^{3+}(aq) + 2I^-(aq) \rightleftarrows 2Fe^{2+}(aq) + I_2(aq)$
c. $Cr_2O_7^{2-}(aq) + 3Sn^{2+}(aq) + 14H^+(aq) \rightleftarrows 2Cr^{3+}(aq) + 3Sn^{4+}(aq) + 7H_2O$
d. $Zn(s) + Cu^{2+}(aq) \leftrightarrows Zn^{2+}(aq) + Cu(s)$

19.82 Using data from Table A.5 in the appendix, calculate the solubility product of AgBr.

FREE ENERGY AND POTENTIAL

The relationship between $E°$ of a galvanic cell and ΔG is simple and important. It is

$$\Delta G° = -nFE° \quad \textbf{(19-2)}$$

where $\Delta G°$ is measured in joules, $E°$ is measured in volts, F is the Faraday, 96,500 C, and n is the number of moles of electrons transferred by the cell reaction. Notice the negative sign in this equation, which means that the direction of spontaneous change corresponds to a negative standard free energy change and a positive potential.

PROBLEMS

19.83 The $E°$ for the cell $Cu(s)|Cu^{2+}\ (1\ M)||Ag^{+}\ (1\ M)|Ag(s)$ is 0.4594 V. Find $\Delta G°$ for the cell reaction.

19.84 Determine $\Delta G°$ for the cell $Ni(s)|Ni^{2+}\ (1\ M)||Fe^{3+}\ (1\ M)|Fe(s)$ for which $E° = 0.19$ V.

19.85 The $\Delta G°$ of the reaction $2I^{-}(aq) + Cl_2(g) \rightarrow I_2(s) + 2Cl^{-}(aq)$ in aqueous solution is -159 kJ.
a. Calculate $E°$ of the corresponding cell
b. Find $\Delta G°$ and $E°$ for the reaction: $I^{-}(aq) + \frac{1}{2}Cl_2(g) \rightarrow Cl^{-}(aq) + \frac{1}{2}I_2(s)$

19.86 The $\Delta G°$ of the reaction $Cu^{2+} + Au^{+} \rightarrow Cu(s) + Au^{3+}$ in aqueous solution is 183 kJ.
a. Calculate the $E°$ of the reaction
b. Write the cell made from the components in the reaction
c. Calculate $E°$ for the cell

CALCULATION OF THE POTENTIAL OF A HALF-REACTION FROM THE POTENTIALS OF OTHER HALF-REACTIONS

We have learned that when two half-reactions are combined to give a *complete chemical reaction*, the voltage of the couple is the sum of the two voltages, due regard being given to the sign of each voltage.

When the potential of a *half-reaction* is calculated from the potentials of two other half-reactions, we must keep in mind (1) that the expulsion of electrons requires energy, (2) that since the potential of a reaction is a measure of the work per unit charge, the total energy will be the product of volts × electrons, and (3) that in any series of reactions the net change in energy is the algebraic sum of the energy changes in all steps.

Thus, suppose we wish to calculate the potential of the half-reaction $BrO^{-} + 4OH^{-} \rightleftarrows BrO_3^{-} + 2H_2O + 4e^{-}$ from the two half-reactions 1 and 2 shown in Table 19.1.

TABLE 19.1

Reaction	*N*	*E*°	Total Energy
(1) $6OH^- + Br^- \rightleftarrows BrO_3^- + 3H_2O$	$+6e^-$	−0.61 V	$6 \times -0.61 = -3.66$ eV
(2) $2OH^- + Br^- \rightleftarrows BrO^- + H_2O$	$+2e^-$	−0.76 V	$2 \times -0.76 = -1.52$ eV
(3) $4OH^- + BrO^- \rightleftarrows BrO_3^- + 2H_2O$	$+4e^-$	−0.535 V	$4 \times -5.35 = -2.14$ eV

If we subtract reaction 2 from reaction 1, taking proper note of signs, we obtain reaction 3. Note that in reaction 3, the total energy is −2.14 electron volts. Since four electrons are given off per molecule of BrO^-, the potential $E°$ will be −2.14/4 or −0.535.

We can summarize this procedure with the formula

$$E_T^\circ = \frac{n_1E_1^\circ + n_2E_2^\circ + \cdots}{n_T} \tag{19-3}$$

where E_T° is the potential of the new half-reaction, n_T is the number of electrons in this half-reaction, $n_1, n_2, \ldots$ are the numbers of electrons, and $E_1^\circ, E_2^\circ, \ldots$ are the potentials of each half-reaction being combined to make the new half-reaction.

When a balanced equation for a chemical reaction is obtained by combining two half-reactions, the amount of each reactant must be chosen so that the total number of electrons lost by the oxidized substance equals the total number of electrons gained by the reduced substance; as a result, the electrons cancel and no electrons appear in the final balanced equation. In contrast, when a *half-reaction is calculated from two half-reactions*, the electrons do not cancel. Instead, a chemical species common to the two parent half-reactions is canceled. (In the example given above Br^- ions are canceled.) Every half-reaction must include electrons; if electrons are not included in the balanced equation, it is not a half-reaction. Rather it is a balanced equation for a chemical reaction. If electrons appear in the balanced equation, it is a half-reaction, not a chemical reaction. Like all half-reactions, it will not occur unless it is combined with another appropriate half-reaction.

PROBLEMS

19.87 Given the half-reactions,

$$H_2SO_3 + 4H^+ + 4e^- \rightleftarrows S + 3H_2O, \quad E^\circ = 0.45\text{ V}$$

$$SO_4^{2-} + 4H^+ + 2e^- \rightleftarrows H_2SO_3 + H_2O, \quad E^\circ = 0.17\text{ V}$$

calculate $E°$ for the half-reaction

$$SO_4^{2-} + 8H^+ + 6e^- \rightleftarrows S + 4H_2O$$

Hint: Add the two half-reactions, including the total energies, cancel the H_2SO_3, and divide the total energy in electron volts by the number of electrons in the final half-reaction, or substitute in Eq. (19-3).

19.88 Given the half-reactions,

$$ClO_3^- + 6H^+ + 6e^- \rightleftarrows Cl^- + 3H_2O, \quad E° = 1.45\ V$$
$$Cl_2 + 2e^- \rightleftarrows 2Cl^-, \quad E° = 1.36\ V$$

calculate $E°$ for the half-reaction

$$ClO_3^- + 6H^+ + 5e^- \rightleftarrows \tfrac{1}{2}Cl_2 + 3H_2O$$

Hint: Divide the second equation by 2 to give

$$\tfrac{1}{2}Cl_2 + 1e^- \rightleftarrows Cl^-, \quad E° = 1.36\ V$$

Then subtract this equation from the first, including the total energies. Finally, divide the total final energy in electron volts by the number of electrons in the final half-reaction, or substitute in Eq. (19-3).

19.89 Given the standard potentials ($E°$) of the following half-reactions,

Half-Reaction	$E°$ (V)
(1) $M \rightleftarrows M^{2+} + 2e^-$	+0.80
(2) $M \rightleftarrows M^{3+} + 3e^-$	+0.70
(3) $M^{3+} + H_2O \rightleftarrows MO^{2+} + 2H^+ + e^-$	+0.30
(4) $M^{2+} + 2H_2O \rightleftarrows MO_2^+ + 4H^+ + 3e^-$	+0.20

calculate the standard potential of the half-reaction in which M^{2+} is oxidized to MO^{2+}.

Hint: In this instance there are no two equations of which one involves M^{2+} and the other MO^{2+}, each containing the same species. So we must try to create two such equations. Inspection tells us that if we subtract reaction (1) from reaction (2), we will obtain the half-reaction in which M^{2+} is converted to M^{3+}. We note that in reaction (3) M^{3+} is converted to MO^{2+}. Therefore, if we add our newly created equation to that for reaction (3) we will obtain the desired half-reaction.

19.90 Given the following standard potentials in basic solution:

$$BrO^- + 4OH^- \rightleftarrows BrO_3^- + 2H_2O + 4e^-, \quad E° = -0.54\ V$$
$$Br^- + 2OH^- \rightleftarrows BrO^- + H_2O + 2e^-, \quad E° = -0.76\ V$$

Will BrO^- disproportionate to give Br^- and BrO_3^-? If the answer is yes, write the equation for the disproportionation reaction.

Solution: Disproportionation occurs when a substance undergoes a redox reaction with itself. Some of the substance acts as a reducing agent, some as an oxidizing agent. For disproportionation to occur the potential of the half-reaction in which the substance functions as a reducing agent must be higher than the potential of the half-reaction in which it functions as an oxidizing agent. Inspection of the two equations shows that this requirement is satisfied; therefore, disproportionation will occur.

The balanced equation for the reaction is obtained in the conventional manner by multiplying the lower equation by 2 (to give 4 electrons), then subtracting it from the upper equation to give

$$3BrO^- \rightleftarrows 2Br^- + BrO_3^-$$

19.91 Given the following half-reactions in acid solution:

Half-Reaction	$E°$ (V)
$H_2S \rightleftarrows S + 2H^+ + 2e^-$	−0.14
$H_2SO_2 + H_2O \rightleftarrows H_2SO_3 + 2H^+ + 2e^-$	−0.40
$2S + 3H_2O \rightleftarrows S_2O_3^{2-} + 6H^+ + 4e^-$	−0.50
$S + 4H_2O \rightleftarrows SO_4^{2-} + 8H^+ + 6e^-$	−0.36
$S_2O_3^{2-} + 3H_2O \rightleftarrows 2H_2SO_3 + 2H^+ + 4e^-$	−0.4025

Will H_2SO_3 disproportionate to give S and SO_4^{2-}? Give the data upon which your answer is based. If the answer is yes, write the equation for the disproportionation reaction.

19.92 You are given the following half-reactions in alkaline solution:

Half-Reaction	$E°$ (V)
$ClO_2^- + 2OH^- \rightleftarrows ClO_3^- + H_2O + 2e^-$	−0.33
$ClO^- + 2OH^- \rightleftarrows ClO_2^- + H_2O + 2e^-$	−0.66
$Cl^- + 2OH^- \rightleftarrows ClO^- + H_2O + 2e^-$	−0.89
$Cl^- \rightleftarrows \frac{1}{2}Cl_2 + 1e^-$	−1.36

a. Calculate the potential $E°$ of the half-reaction $\frac{1}{2}Cl_2 + 6OH^- \rightleftarrows ClO_3^- + 3H_2O + 5e^-$

b. Will ClO^- disproportionate to give Cl^- and ClO_3^-? Show the calculations that lead to your conclusion. If the answer is yes, write the equation for the disproportionation reaction.

19.93 In acid solution will H_3PO_2 disproportionate to give PH_3 and H_3PO_4? (See Table A.5 of the appendix.)

19.94 Calculate the potential of the half-reaction,

$$H_2PO_3 + 7H^+ + 7e^- \rightleftarrows PH_3 + 3H_2O$$

19.95 What oxidation state of chlorine will be formed when 1.0 *M* ClO_4^- is reduced with excess solid $Mn(OH)_2$ in alkaline solution? (See Table A.6 of the appendix.)

19.96 Calculate the potential of the half-reaction,

$$ClO_4^- + 4H_2O + 8e^- \rightleftarrows Cl^- + 8OH^-$$

19.97 The metal M has five oxidation states, 0, 2, 3, 4, and 5, which are related by the potentials given below; the potentials for zinc, iron, and tin are also included. Assume that all possible reactions are very rapid.

	Half-Reaction	*E*° (V)
A	$M \rightleftarrows M^{2+} + 2e^-$	+0.80
	$Zn \rightleftarrows Zn^{2+} + 2e^-$	+0.76
B	$M \rightleftarrows M^{3+} + 3e^-$	+0.70
C	$M + H_2O \rightleftarrows MO^{2+} + 2H^+ + 4e^-$	+0.60
D	$M^{2+} \rightleftarrows M^{3+} + e^-$	+0.50
	$Fe \rightleftarrows Fe^{2+} + 2e^-$	+0.45
E	$M + 2H_2O \rightleftarrows MO_2^+ + 4H^+ + 5e^-$	+0.44
F	$M^{2+} + H_2O \rightleftarrows MO^{2+} + 2H^+ + 2e^-$	+0.40
G	$M^{2+} + H_2O \rightleftarrows MO^{2+} + 2H^+ + e^-$	+0.30
H	$M^{2+} + 2H_2O \rightleftarrows MO_2^+ + 4H^+ + 3e^-$	+0.20
	$Sn \rightleftarrows Sn^{2+} + 2e^-$	+0.14
I	$M^{3+} + 2H_2O \rightleftarrows MO_2^+ + 4H^+ + 2e^-$	+0.05
J	$MO^{2+} + H_2O \rightleftarrows MO_2^+ + 2H^+ + e^-$	−0.20

a. Determine the potential of the reaction

$$3Sn^{2+} + 2M^{2+} + 4H_2O \rightleftarrows 3Sn + 2MO_2^+ + 8H^+$$

b. Will excess metallic Zn reduce M^{2+} to M?
c. Will excess metallic Zn reduce M^{3+} to M^{2+}?
d. Will excess metallic Zn reduce M^{3+} to M?

e. Give the oxidation state of M produced when 1 *M* MO_2^+ is treated with

1. excess metallic zinc
2. excess metallic tin
3. excess metallic iron

f. Will oxidation state +3 disproportionate to +2 and +5?

g. Show how $E°$ for half-reaction H can be calculated by appropriate combination of $E°$ for

1. half-reactions A and E
2. half-reactions D, G, and J
3. half-reactions J and F
4. half-reactions A, C, and J

19.98* Calculate K for the dissociation of $Zn(NH_3)_4^{2+}$ ion using the data of Tables A.5 and A.6 of the appendix.

Hint: Use reaction 6 of Table A.5 and reaction 13 of Table A. 6.

APPENDIX

TABLE A.1 Vapor Pressure of Water

Temperature (°C)	Pressure (mmHg)	Temperature (°C)	Pressure (mmHg)
0	4.6	21	18.5
1	4.9	22	19.8
2	5.3	23	20.9
3	5.6	24	22.2
4	6.1	25	23.6
5	6.5	26	25.1
6	7.0	27	26.5
7	7.5	28	28.1
8	8.0	29	29.8
9	8.6	30	31.5
10	9.2	31	33.4
11	9.8	32	35.4
12	10.5	33	37.4
13	11.2	34	39.6
14	11.9	35	41.9
15	12.7	36	44.2
16	13.5	37	46.7
17	14.4	38	49.4
18	15.4	39	52.1
19	16.3	40	55.0
20	17.4	100	760.0

TABLE A.2 Ionization Constants of Acids and Bases

Compound	Formula	K_i
Acetic	$HC_2H_3O_2(HOAc)$	1.8×10^{-5}
Arsenic	H_3AsO_4	$K_1 = 2.5 \times 10^{-4}$
		$K_2 = 5.6 \times 10^{-8}$
		$K_3 = 3.0 \times 10^{-13}$
Arsenious	H_3AsO_3	$K_1 = 6 \times 10^{-10}$
Boric	H_3BO_3	$K_1 = 6.0 \times 10^{-10}$
Carbonic	H_2CO_3	$K_1 = 4.2 \times 10^{-7}$
		$K_2 = 4.8 \times 10^{-11}$
Chromic	H_2CrO_4	$K_1 = 1.8 \times 10^{-1}$
		$K_2 = 3.2 \times 10^{-7}$
Formic	$HCHO_2$	1.77×10^{-4}
Hydrocyanic	HCN	4.0×10^{-10}
Hydrofluoric	HF	6.9×10^{-4}
Hydrogen sulfide	H_2S	$K_1 = 1.0 \times 10^{-7}$
		$K_2 = 1.3 \times 10^{-13}$
Hypochlorous	$HClO$	3.2×10^{-8}
Lactic	$HC_3H_5O_3$	1.4×10^{-4}
Nitrous	HNO_2	4.5×10^{-4}
Oxalic	$H_2C_2O_4$	$K_1 = 3.8 \times 10^{-2}$
		$K_2 = 5.0 \times 10^{-5}$
Phenol	HC_6H_5O	1.3×10^{-10}
Phosphoric	H_3PO_4	$K_1 = 7.5 \times 10^{-3}$
		$K_2 = 6.2 \times 10^{-8}$
		$K_3 = 1.0 \times 10^{-12}$
Sulfuric	H_2SO_4	$K_2 = 1.2 \times 10^{-2}$
Sulfurous	H_2SO_3	$K_1 = 1.3 \times 10^{-2}$
		$K_2 = 5.6 \times 10^{-8}$
Ammonia	NH_3	$K_b = 1.8 \times 10^{-5}$
Water	H_2O	$K_i = 1.8 \times 10^{-16}$
		$K_{H_2O} = 1.0 \times 10^{-14}$

TABLE A.3 Complex Ion Equilibria

Ligand	Equation	Instability Constant
Ammonia	$Cd(NH_3)_4^{2+} \rightleftarrows Cd^{2+} + 4NH_3$	7.5×10^{-8}
	$Cu(NH_3)_4^{2+} \rightleftarrows Cu^{2+} + 4NH_3$	4.7×10^{-15}
	$Co(NH_3)_6^{2+} \rightleftarrows Co^{2+} + 6NH_3$	1.3×10^{-5}
	$Co(NH_3)_6^{3+} \rightleftarrows Co^{3+} + 6NH_3$	2.2×10^{-34}
	$Ni(NH_3)_6^{2+} \rightleftarrows Ni^{2+} + 6NH_3$	1.8×10^{-9}
	$Ag(NH_3)_2^{+} \rightleftarrows Ag^{+} + 2NH_3$	5.9×10^{-8}
	$Zn(NH_3)_4^{2+} \rightleftarrows Zn^{2+} + 4NH_3$	3.4×10^{-10}
Cyanide	$Cd(CN)_4^{2-} \rightleftarrows Cd^{2} + 4CN^{-}$	1.4×10^{-19}
	$Cu(CN)_2^{-} \rightleftarrows Cu^{+} + 2CN^{-}$	5.0×10^{-28}
	$Fe(CN)_6^{4-} \rightleftarrows Fe^{2+} + 6CN^{-}$	1.0×10^{-35}
	$Hg(CN)_4^{2-} \rightleftarrows Hg^{2+} + 4CN^{-}$	4.0×10^{-42}
	$Ni(CN)_4^{2-} \rightleftarrows Ni^{2+} + 4CN^{-}$	1.0×10^{-22}
	$Ag(CN)_2^{-} \rightleftarrows Ag^{+} + 2CN^{-}$	1.8×10^{-19}
	$Zn(CN)_4^{2-} \rightleftarrows Zn^{2+} + 4CN^{-}$	1.3×10^{-17}
Hydroxide	$Al(OH)_4^{-} \rightleftarrows Al^{3+} + 4OH^{-}$	1.0×10^{-34}
	$Zn(OH)_4^{2-} \rightleftarrows Zn^{2+} + 4OH^{-}$	3.3×10^{-16}
Chloride	$HgCl_4^{2-} \rightleftarrows Hg^{2+} + 4Cl^{-}$	1.1×10^{-16}
Bromide	$HgBr_4^{2-} \rightleftarrows Hg^{2+} + 4Br^{-}$	2.3×10^{-22}
Iodide	$HgI_4^{2-} \rightleftarrows Hg^{2+} + 4I^{-}$	5.3×10^{-31}
Thiosulfate	$Ag(S_2O_3)_2^{3-} \rightleftarrows Ag^{+} + 2S_2O_3^{2-}$	3.5×10^{-14}

TABLE A.4 Solubility Products at 20 °C

Compound	Product	K_{sp}
Aluminum hydroxide	$[Al^{3+}] \times [OH^{-}]^3$	5×10^{-33}
Barium carbonate	$[Ba^{2+}] \times [CO_3^{2-}]$	1.6×10^{-9}
Barium chromate	$[Ba^{2+}] \times [CrO_4^{2-}]$	8.5×10^{-11}
Barium sulfate	$[Ba^{2+}] \times [SO_4^{2-}]$	1.5×10^{-9}
Barium oxalate	$[Ba^{2+}] \times [C_2O_4^{2-}]$	1.5×10^{-8}
Bismuth sulfide	$[Bi^{3+}]^2 \times [S^{2-}]^3$	1×10^{-70}
Cadmium hydroxide	$[Cd^{2+}] \times [OH^{-}]^2$	2×10^{-14}
Cadmium sulfide	$[Cd^{2+}] \times [S^{2-}]$	6×10^{-27}
Calcium carbonate	$[Ca^{2+}] \times [CO_3^{2-}]$	6.9×10^{-9}
Calcium oxalate	$[Ca^{2+}] \times [C_2O_4^{2-}]$	1.3×10^{-9}

TABLE A.4 (continued)

Compound	Product	K_{sp}
Calcium sulfate	$[Ca^{2+}] \times [SO_4^{2-}]$	2.4×10^{-5}
Chromium hydroxide	$[Cr^{3+}] \times [OH^-]^3$	7×10^{-31}
Cobalt sulfide	$[Co^{2+}] \times [S^{2-}]$	5×10^{-22}
Cupric hydroxide	$[Cu^{2+}] \times [OH^-]^2$	1.6×10^{-19}
Cupric sulfide	$[Cu^{2+}] \times [S^{2-}]$	4×10^{-36}
Ferric hydroxide	$[Fe^{3+}] \times [OH^-]^3$	6×10^{-38}
Ferrous hydroxide	$[Fe^{2+}] \times [OH^-]^2$	2×10^{-15}
Ferrous sulfide	$[Fe^{2+}] \times [S^{2-}]$	4×10^{-17}
Lead carbonate	$[Pb^{2+}] \times [CO_3^{2-}]$	1.5×10^{-13}
Lead chromate	$[Pb^{2+}] \times [CrO_4^{2-}]$	2×10^{-16}
Lead iodide	$[Pb^{2+}] \times [I^-]^2$	8.3×10^{-9}
Lead sulfate	$[Pb^{2+}] \times [SO_4^{2-}]$	1.3×10^{-8}
Lead sulfide	$[Pb^{2+}] \times [S^{2-}]$	4×10^{-26}
Magnesium carbonate	$[Mg^{2+}] \times [CO_3^{2-}]$	4×10^{-5}
Magnesium hydroxide	$[Mg^{2+}] \times [OH^-]^2$	8.9×10^{-12}
Magnesium oxalate	$[Mg^{2+}] \times [C_2O_4^{2-}]$	8.6×10^{-5}
Manganese hydroxide	$[Mn^{2+}] \times [OH^-]^2$	2×10^{-13}
Manganese sulfide	$[Mn^{2+}] \times [S^{2-}]$	8×10^{-14}
Mercurous chloride	$[Hg_2^{2+}] \times [Cl^-]^2$	1.1×10^{-18}
Mercuric sulfide	$[Hg^{2+}] \times [S^{2-}]$	1×10^{-50}
Nickel hydroxide	$[Ni^{2+}] \times [OH^-]^2$	1.6×10^{-16}
Nickel sulfide	$[Ni^{2+}] \times [S^{2-}]$	1×10^{-22}
Silver arsenate	$[Ag^+]^3 \times [AsO_4^{3-}]$	1×10^{-23}
Silver bromide	$[Ag^+] \times [Br^-]$	5×10^{-13}
Silver carbonate	$[Ag^+]^2 \times [CO_3^{2-}]$	8.2×10^{-12}
Silver chloride	$[Ag^+] \times [Cl^-]$	2.8×10^{-10}
Silver chromate	$[Ag^+]^2 \times [CrO_4^{2-}]$	1.9×10^{-12}
Silver iodate	$[Ag^+] \times [IO_3^-]$	3×10^{-8}
Silver iodide	$[Ag^+] \times [I^-]$	8.5×10^{-17}
Silver phosphate	$[Ag^+]^3 \times [PO_4^{3-}]$	1.8×10^{-18}
Silver sulfide	$[Ag^+]^2 \times [S^{2-}]$	1×10^{-50}
Silver thiocyanate	$[Ag^+] \times [CNS^-]$	1×10^{-12}
Stannous sulfide	$[Sn^{2+}] \times [S^{2-}]$	1×10^{-24}
Zinc hydroxide	$[Zn^{2+}] \times [OH^-]^2$	5×10^{-17}
Zinc sulfide	$[Zn^{2+}] \times [S^{2-}]$	1×10^{-20}

TABLE A.5 Some Standard Reduction Potentials in Acid Solution

	Half-Reaction	$E°$ (V)
1	$K^+ + e^- \rightleftharpoons K(s)$	−2.925
2	$Ca^{2+} + 2e^- \rightleftharpoons Ca(s)$	−2.87
3	$Al^{3+} + 3e^- \rightleftharpoons Al(s)$	−1.66
4	$Mn^{2+} + 2e^- \rightleftharpoons Mn(s)$	−1.18
5	$H_3PO_4 + H^+ + e^- \rightleftharpoons H_2O + H_2PO_3$	−0.9
6	$Zn^{2+} + 2e^- \rightleftharpoons Zn(s)$	−0.763
7	$H_3PO_2 + H^+ + e^- \rightleftharpoons P(s) + 2H_2O$	−0.51
8	$H_3PO_3 + 2H^+ + 2e^- \rightleftharpoons H_3PO_2 + H_2O$	−0.50
9	$Cr^{3+} + e^- \rightleftharpoons Cr^{2+}$	−0.41
10	$H_3PO_4 + 2H^+ + 2e^- \rightleftharpoons H_3PO_3 + H_2O$	−0.276
11	$Ni^{2+} + 2e^- \rightleftharpoons Ni(s)$	−0.250
12	$Sn^{2+} + 2e^- \rightleftharpoons Sn(s)$	−0.136
13	$2H_2SO_3 + H^+ + 2e^- \rightleftharpoons HS_2O_4^- + 2H_2O$	−0.08
14	$2H^+ + 2e^- \rightleftharpoons H_2(g)$	0.000
15	$P(s) + 3H^+ + 3e^- \rightleftharpoons PH_3$	0.06
16	$AgBr(s) + e^- \rightleftharpoons Ag(s) + Br^-$	0.0713
17	$2H^+ + S + 2e^- \rightleftharpoons H_2S$	0.141
18	$Sn^{4+} + 2e^- \rightleftharpoons Sn^{2+}$	0.15
19	$SO_4^{2-} + 4H^+ + 2e^- \rightleftharpoons H_2SO_3 + H_2O$	0.17
20	$Cu^{2+} + 2e^- \rightleftharpoons Cu(s)$	0.337
21	$H_2SO_3 + 4H^+ + 4e^- \rightleftharpoons S(s) + 3H_2O$	0.45
22	$I_2 + 2e^- \rightleftharpoons 2I^-$	0.5355
23	$MnO_4^- + e^- \rightleftharpoons MnO_4^{2-}$	0.564
24	$O_2 + 2H^+ + 2e^- \rightleftharpoons H_2O_2$	0.682
25	$Fe^{3+} + e^- \rightleftharpoons Fe^{2+}$	0.771
26	$Ag^+ + e^- \rightleftharpoons Ag(s)$	0.799
27	$NO_3^- + 2H^+ + e^- \rightleftharpoons NO_2 + H_2O$	0.80
28	$Hg^{2+} + 2e^- \rightleftharpoons Hg(l)$	0.854
29	$NO_3^- + 4H^+ + 3e^- \rightleftharpoons NO + 2H_2O$	0.96
30	$HNO_2 + H^+ + e^- \rightleftharpoons NO + H_2O$	1.00
31	$Br_2 + 2e^- \rightleftharpoons 2Br^-$	1.065
32	$MnO_2(s) + 4H^+ + 2e^- \rightleftharpoons Mn^{2+} + 2H_2O$	1.23
33	$Cr_2O_7^{2-} + 14H^+ + 6e^- \rightleftharpoons 2Cr^{3+} + 7H_2O$	1.33
34	$Cl_2 + 2e^- \rightleftharpoons 2Cl^-$	1.3595
35	$ClO_3^- + 6H^+ + 6e^- \rightleftharpoons Cl^- + 3H_2O$	1.45

TABLE A.5 (continued)

	Half-Reaction	$E°$ (V)
36	$MnO_4^- + 8H^+ + 5e^- \rightleftarrows Mn^{2+} + 4H_2O$	1.51
37	$Mn^{3+} + e^- \rightleftarrows Mn^{2+}$	1.55
38	$HBiO_3 + 5H^+ + 2e^- \rightleftarrows Bi^{3+} + 3H_2O$	1.70
39	$H_2O_2 + 2H^+ + 2e^- \rightleftarrows 2H_2O$	1.77
40	$F_2 + 2e^- \rightleftarrows 2F^-$	2.65
41	$F_2 + 2H^+ + 2e^- \rightleftarrows 2HF$	3.06

TABLE A.6 Some Standard Reduction Potentials in Alkaline Solution

	Half-Reaction	$E°$ (V)
1	$Ca(OH)_2(s) + 2e^- \rightleftarrows Ca(s) + 2OH^-$	−3.03
2	$2H_2O + 2e^- \rightleftarrows H_2 + 2OH^-$	−2.93
3	$K^+ + e^- \rightleftarrows K(s)$	−2.925
4	$Al(OH)_4^- + 3e^- \rightleftarrows Al(s) + 4OH^-$	−2.35
5	$H_2PO_2^- + e^- \rightleftarrows P(s) + 2OH^-$	−2.05
6	$HPO_3^{2-} + 2H_2O + 2e^- \rightleftarrows H_2PO_2^- + 3OH^-$	−1.57
7	$Mn(OH)_2(s) + 2e^- \rightleftarrows Mn(s) + 2OH^-$	−1.55
8	$ZnS(s) + 2e^- \rightleftarrows Zn(s) + S^{2-}$	−1.44
9	$Zn(CN)_4^{2-} + 2e^- \rightleftarrows Zn(s) + 4CN^-$	−1.26
10	$Zn(OH)_4^{2-} + 2e^- \rightleftarrows Zn(s) + 4OH^-$	−1.216
11	$PO_4^{3-} + 2H_2O + 2e^- \rightleftarrows HPO_3^{2-} + 3OH^-$	−1.12
12	$2SO_3^{2-} + 2H_2O + 2e^- \rightleftarrows S_2O_4^{2-} + 4OH^-$	−1.12
13	$Zn(NH_3)_4^{2+} + 2e^- \rightleftarrows Zn(s) + 4NH_3$	−1.03
14	$CNO^- + H_2O + 2e^- \rightleftarrows CN^- + 2OH^-$	−0.97
15	$SO_4^{2-} + H_2O + 2e^- \rightleftarrows SO_3^{2-} + 2OH^-$	−0.93
16	$Sn(OH)_6^{2-} + 2e^- \rightleftarrows Sn(OH)_4^{2-} + 2OH^-$	−0.90
17	$P(s) + 3H_2O + 3e^- \rightleftarrows PH_3 + 3OH^-$	−0.89
18	$Sn(OH)_4^{2-} + 2e^- \rightleftarrows Sn(s) + 4OH^-$	−0.76
19	$Ni(OH)_2(s) + 2e^- \rightleftarrows Ni(s) + 2OH^-$	−0.72
20	$Fe(OH)_3(s) + e^- \rightleftarrows Fe(OH)_2(s) + OH^-$	−0.56

(continued)

TABLE A.6 (continued)

	Half-Reaction	$E°$ (V)
21	$S + 2e^- \rightleftarrows S^{2-}$	−0.48
22	$CrO_4^{2-} + 4H_2O + 3e^- \rightleftarrows Cr(OH)_4^- + 4OH^-$	−0.13
23	$O_2 + 2H_2O + 2e^- \rightleftarrows H_2O_2 + 2OH^-$	−0.076
24	$MnO_2(s) + 2H_2O + 2e^- \rightleftarrows Mn(OH)_2(s) + 2OH^-$	−0.05
25	$Cu(NH_3)_4^{2+} + e^- \rightleftarrows Cu(NH_3)_2^+ + 2NH_3$	0.0
26	$Mn(OH)_3(s) + e^- \rightleftarrows Mn(OH)_2(s) + OH^-$	0.1
27	$Co(NH_3)_6^{3+} + e^- \rightleftarrows Co(NH_3)_6^{2+}$	0.1
28	$Co(OH)_3(s) + e^- \rightleftarrows Co(OH)_2(s) + OH^-$	0.17
29	$ClO_3^- + H_2O + 2e^- \rightleftarrows ClO_2^- + 2OH^-$	0.33
30	$ClO_4^- + H_2O + 2e^- \rightleftarrows ClO_3^- + 2OH^-$	0.36
31	$O_2 + 2H_2O + 4e^- \rightleftarrows 4OH^-$	0.401
32	$IO^- + H_2O + 2e^- \rightleftarrows I^- + 2OH^-$	0.49
33	$NiO_2(s) + 2H_2O + 2e^- \rightleftarrows Ni(OH)_2(s) + 2OH^-$	0.49
34	$MnO_4^- + e^- \rightleftarrows MnO_4^{2-}$	0.564
35	$MnO_4^- + 2H_2O + 3e^- \rightleftarrows MnO_2(s) + 4OH^-$	0.588
36	$MnO_4^{2-} + 2H_2O + 2e^- \rightleftarrows MnO_2(s) + 4OH^-$	0.60
37	$ClO_2^- + H_2O + 2e^- \rightleftarrows ClO^- + 2OH^-$	0.66
38	$BrO^- + H_2O + 2e^- \rightleftarrows Br^- + 2OH^-$	0.76
39	$H_2O_2 + 2e^- \rightleftarrows 2OH^-$	0.88
40	$ClO^- + H_2O + 2e^- \rightleftarrows Cl^- + 2OH^-$	0.89

TABLE A.7 Standard Heats of Formation and Free Energies of Formation at 298 K (25 °C)

Compound	$\Delta H°_f$ (kJ/mol)	$\Delta G°_f$ (kJ/mol)
$AgCl(s)$	−127	−110
$AlCl_3(s)$	−704	−629
$Al_2O_3(s)$	−1670	−1576
$BCl_3(g)$	−395	−380
$BaCl_2(s)$	−860	−811

TABLE A.7 (continued)

Compound	ΔH_f° (kJ/mol)	ΔG_f° (kJ/mol)
$BaO(s)$	−558	−529
$Br_2(g)$	31	3
$C(g)$	718	671
$CCl_4(g)$	−107	−58
$CH_4(g)$	−75	−51
$C_2H_2(g)$	227	209
$C_2H_4(g)$	52	68
$C_2H_6(g)$	−84.7	−33
$CHCl_3(g)$	−100	−69
$CO(g)$	−110.5	−137.3
$CO_2(g)$	−393.5	−394.4
$CaCl_2(s)$	−795	−750
$CaCO_3(s)$	−1207	−1129
$CaO(s)$	−636	−604
$CsCl(s)$	−447	−419
$Cs_2O(s)$	−318	−290
$CuCl_2(s)$	−220	−176
$CuO(s)$	−155	−127
$FeCl_2(s)$	−342	−302
$FeCl_3(s)$	−399	−334
$FeO(s)$	−272	−251
$Fe_2O_3(s)$	−822	−742
$HBr(g)$	−36	−53
$HCl(g)$	−92	−95
$HF(g)$	−269	−273
$HI(g)$	26	2
$H_2O(g)$	−242	−229
$H_2O(l)$	−286	−237
$H_2S(g)$	−20	−34
$I_2(g)$	62	19
$KCl(s)$	−437	−408
$LiCl(s)$	−402	−377
$Li_2O(s)$	−596	−560
$MgCl_2(s)$	−642	−592
$MgO(s)$	−601	−570
$NH_3(g)$	−46	−16
$NH_4Cl(s)$	−315	−203

(continued)

TABLE A.7 (continued)

Compound	ΔH°_f (kJ/mol)	ΔG°_f (kJ/mol)
$NO(g)$	90	87
$NO_2(g)$	34	51
$N_2O(g)$	82	104
$N_2O_4(g)$	10	98
$NaCl(s)$	−411	−384
$Na_2O(s)$	−416	−377
$O_3(g)$	142	163
$PCl_3(l)$	−300	−260
$PCl_5(s)$	−400	–
$P_4O_{10}(s)$	−2940	−2676
$PbCl_2(s)$	−359	−314
$PbO(s)$	−217	−188
$PbO_2(s)$	−277	−217
$RbCl(s)$	−433	−404
$Rb_2O(s)$	−330	−297
$SO_2(g)$	−297	−300
$SO_3(g)$	−395	−371
$SiO_2(s)$	−911	−856
$SrCl_2(s)$	−829	−781
$SrO(s)$	−592	−562
$ZnCl_2(s)$	−415	−369
$ZnO(s)$	−348	−318

TABLE A.8 Bond Energies

Bond	Energy (kJ/mol)	Bond	Energy (kJ/mol)
C—C	346	H—H	436
C=C	610	H—Br	366
C≡C	835	H—Cl	432
C—Cl	335	H—I	299
C—H	413	H—N	391
Br—Br	193	H—O	463
Cl—Cl	243	N≡N	946
F—F	158	N=O	628
I—I	151	O=O	498

TABLE A.9 Electronegativities

Element	Electronegativity
H	2.2
Li	1.0
Be	1.6
B	2.0
C	2.6
N	3.0
O	3.4
F	4.0
Na	0.9
Mg	1.3
P	2.2
S	2.6
Cl	3.2
Br	3.0
I	2.7
Cs	0.7

ANSWERS TO PROBLEMS

Chapter 2

2.2. 5.9 ft, 1.8 m, 180 cm. **2.3.** 25.4 oz, 0.795 qt, 0.750 L. **2.4.** 1000 L.
2.5. 89 km/h, 25 m/s. **2.6.** 16,000 ft^2. **2.8(a).** 1×10^{-3} kg/cm^3; **(b).** 1×10^6 g/m^3; **(c).** 1×10^3 kg/m^3; **(d).** 1000 g/L. **2.9.** 2.7×10^{19} g, 2.7×10^{22} mg, 2.7×10^{25} μg.
2.10. 5000 mmHg, 5000 torr, 6×10^5 Pa. **2.11.** 3.37×10^6 J.
2.12. For example, 305 K (89 °F), atmospheric pressure 1.00×10^5 Pa, relative humidity 85%, winds 4.4 m/s. **2.13(a).** 22.2 °C; **(b).** −28.9 °C. **2.14(a).** 53.6 °F; **(b).** −58.0 °F.
2.15(a). 25 °C, 77 °F; **(b).** −273 °C, −459 °F. **2.16.** −40°. **2.17.** −34°. **2.18.** 73 °J.
2.19(a). 2.1×10^{10}; **(b).** 7.6×10^2; **(c).** 2.7×10^{-3}; **(d).** 1.8×10^{-6}; **(e).** 1.0×10^{-1}.
2.20(a). 2.7×10^{-1}; **(b).** 1.8×10^2. **2.21(a).** 2.2×10^{-6}; **(b).** 6.5×10^{-2}.
2.22(a). 13.58; **(b).** 10.00; **(c).** 0.46; **(d).** 1000; **(e).** 399. **2.23.** 3.1557×10^7 s/solar year.

Chapter 3

3.1. 200 g/mol Au. **3.2.** 32 g/mol S. **3.3.** 79.91. **3.4.** 87.15. **3.5.** ^{107}Ag = 51.75%, ^{109}Ag = 48.25%. **3.7.** 99.98% ^1H, 0.02% D. **3.8.** 10.01. **3.10.** 0.9103 mol-K.
3.11. 6.958 mol Ne. **3.12(a).** 9.34 mol B; **(b).** 0.518 mol Pt; **(c).** 0.424 mol U.
3.13. 0.250 mol S_8, 1.00 mol S_2, 2.00 mol S. **3.15.** 196 g Cl. **3.16.** 0.972 g Li.
3.17. 56.6 g Co. **3.18(a).** 81.9 g Xe; **(b).** 25.0 g Ca; **(c).** 4.33 g Li. **3.19.** 269 g.
3.23. 1.53×10^{24} atoms Mn. **3.24.** 67.7 g V. **3.25.** 52.8 g Cr. **3.26.** 2.35×10^{24} atoms C.
3.27. 7.99×10^{-4} mol U. **3.28.** 5.82×20^{26} atoms Ni. **3.29.** 2.73×10^{26} atoms Cr.
3.30. 8.91×10^{25} atoms.

Chapter 4

4.7. 0.65 mol. **4.9.** 0.942 mol. **4.10.** 229 g/mol. **4.11.** 79.8, bromine.
4.13. 3.04 mol. **4.14.** 22 mol. **4.15(a).** 1.7 mol O; **(b).** 3.4 mol O; **(c).** 10 mol O; **(d).** 7.7 mol O. **4.17.** 0.0063 mol penicillin. **4.18.** 3.9 mol. **4.19.** 8.3×10^{-5} mol SiO_2.
4.21. 3.43×10^{22} molecules. **4.23.** 1.31×10^{-3} mol. **4.25.** 3.0×10^8 molecules.
4.26. 7.19×10^{22} molecules. **4.28.** 7.21×10^7 g. **4.29.** 6.69 g. **4.30.** 156 g/mol.
4.31. 5.86×10^{23} atoms. **4.33.** 95.0. **4.35.** 161 g. **4.36.** 1.83 mol. **4.38.** 6.67×10^{-4} mol.
4.39. 356 g. **4.41.** 0.956 g. **4.42.** 5.97 g. **4.43.** 17.2 tons. **4.45(a).** 40.5%; **(b).** 45.3%. **4.46(a).** 1.06 mol; **(b).** 235 g; **(c).** 76.7 g; **(d).** 51.4 g; **(e).** 3.74 mol; **(f).** 0.36 mol; **(g).** 10.3 g; **(h).** 2.25×10^{22} molecules; **(i).** 8.81×10^{22} atoms; **(j).** 58.5%; **(k).** $\frac{1}{6}$; **(l).** 0.0699 g; **(m).** 14.4 g. **4.47.** zincblende, 67.10%. **4.48.** Dealer B.
4.49. 76.5. **4.52.** $KClO_4$. **4.53.** SnS_2. **4.54.** $Ni(CO)_4$. **4.55.** no. **4.57.** Cu_2O.
4.59. Na_2SO_3. **4.60.** $Mg_3P_2O_8$. **4.61.** C_2H_5N. **4.62.** X_2Y_3. **4.64.** $C_{12}H_{20}N_4$.
4.65. $C_6H_{12}O_6$. **4.66.** C_3H_8. **4.67.** C_4H_8O. **4.68.** C_3H_4. **4.69.** C_2H_3.
4.70. C_5H_5N. **4.71.** C_3H_8. **4.72.** C_4H_8O. **4.73.** $C_{15}H_{15}N_3$. **4.74.** C_3H_3ClN.
4.75. CH_3NO_2. **4.76.** $C_9H_{11}NO_2$. **4.77.** $C_6H_7NO_2S$. **4.78.** C_6H_9BrO. **4.79.** XY_3.
4.80. XZ_2, X_2Z_5, 3.18 times as great.

Chapter 5

5.6(a). 1.5 mol N_2, 4.5 mol F_2; **(b).** 0.134 mol N_2, 0.402 mol F_2. **5.7.** 5.25 mol O_2, 3 mol CO_2, 4.5 mol H_2O. **5.8.** 1.31 mol $BaSO_4$, 0.524 mol Ag_3AsO_4. **5.9.** 1.90 mol As_2O_5. **5.10.** 6.48 mol $FeCl_3$. **5.11.** 1.34 mol HBr. **5.12.** 0.98 mol CCl_4. **5.13.** 0.60 mol B_5H_9. **5.15.** 249 g Zn. **5.17.** 479 g Cl_2. **5.18.** 16.1 g $AsCl_3$. **5.20.** 190 lb Zn. **5.21.** 41.9 g CuO. **5.22.** 2600 kg Fe_2O_3, 24,400 mol CO_2. **5.23.** 2.54 g MgO. **5.24.** 101 g Ag_3PO_4. **5.25.** 80.0%. **5.26.** 4.00%. **5.27.** 224 tons Pb. **5.28.** 11.8 g O_2, 5.41 g CO_2, 15.7 g SO_2. **5.29.** Pb_3O_4. **5.30.** $3MnO_2 \rightarrow Mn_3O_4 + O_2$. **5.31.** $Pb_3O_4 + 4H_2 \rightarrow 3Pb + 4H_2O$. **5.32.** $2NaNO_3 \rightarrow 2NaNO_2 + O_2$. **5.34.** 84.2 g AgCl. **5.35.** $Ca_3(PO_4)_2$. **5.36.** 61.9 g Bi_2S_3. **5.37.** 34.2 g HCl. **5.38.** 0.85 g W remaining, 25.3 g WCl_6. **5.39.** limiting reagent is gold, 13 g $KAuF_4$. **5.40.** limiting reagent is MnO_2, 96 g Mn. **5.41.** $NaNO_3$. **5.42.** 89.3%. **5.43.** 55.0%. **5.44.** 100 g CO_2. **5.45.** 128. **5.47.** 108. **5.48.** 14.0, 107.9. **5.49.** 67.4 g Mn. **5.50.** $C_4H_8O_2$. **5.51.** 188. **5.52.** 92.93. **5.53.** 15.2 g CS_2. **5.54.** $2CrCl_3 + 3H_2 \rightarrow 2Cr + 6HCl$.

Chapter 6

6.2. 42.9%. **6.3.** 2.73 g C. **6.11.** 46.1%. **6.12.** 58.0%. **6.13.** 10.7 g. **6.14.** 29.7 g. **6.15.** 0.88. **6.16.** 2.49 g NaCl. **6.17.** 24.0% $CaCO_3$. **6.18.** 20.8 g S_2Cl_2. **6.19.** 1.7 g C_2H_2. **6.20.** 2.04 g. **6.21.** 33.9 g. **6.22.** 21.6 g Al. **6.23.** 75%. **6.24.** 16.8%.

Chapter 7

7.2. 550 L. **7.3.** 59 L. **7.4.** 660 atm. **7.5.** 1.1 atm. **7.6.** 17 L. **7.7.** 4.3 atm. **7.9.** 710 cm^3. **7.10.** 15.2 ft^3. **7.11.** 606 °C. **7.12.** 16.4 m^3. **7.13.** 38 °C. **7.15.** 45.5 L. **7.16.** 97 °C. **7.17.** 13.8 atm. **7.18.** 552 mmHg. **7.19.** 0.84 L. **7.20(a).** decreases; **(b).** decreases; **(c).** increases. **7.21.** 64 °C. **7.22.** 99.6 L. **7.23.** 26 atm. **7.24.** 44.9 L. **7.26.** 9.50 L. **7.28.** 0.817 mol. **7.30.** 24.5 L. **7.32.** 236 g. **7.34.** 109 g. **7.35.** 34.9 L. **7.37.** 30.1 g/mol. **7.38.** 0.50:0.57. **7.39.** 19 g C_2H_2. **7.40.** 9.27 mol. **7.41.** 21.8 L. **7.42.** 30.0 g. **7.48.** 359 g. **7.49.** 208 L. **7.50.** 14.6 atm. **7.51.** 46 L. **7.52.** 1.25 g C. **7.53.** 2.28 atm. **7.54.** 2.6×10^{-3} g Ne. **7.55.** 1.204×10^{24}. **7.56.** 170 g/mol. **7.57.** 10.4 L. **7.58.** N_2. **7.59.** less than 460 g. **7.60.** 0.019 atm. **7.61.** 146 g/mol. **7.62.** 119 g/mol. **7.63.** 2.2 L. **7.64.** 0.193 g. **7.66.** 1.25 g/L. **7.68.** 0.298 g/L. **7.69.** O_2, 1.07. **7.70.** 0.13 g/L. **7.71.** 0 °C. **7.72.** 2.2 g/L. **7.73.** 0.38 g/L. **7.74.** 0.089 g/L. **7.75.** 20 mmHg. **7.76.** 0.50. **7.77.** 0.25. **7.78.** 148 mmHg. **7.79.** 178 mmHg. **7.80.** 1400 mmHg. **7.81.** 159 g. **7.82.** 0.53 atm. **7.83.** 0.39 atm. **7.86.** 15 mmHg. **7.87.** 70 g. **7.88.** 0.321 g. **7.89(a).** 67.2 L; **(b).** 17.6 L; **(c).** 6.25 mol; **(d).** 168 g; **(e).** 1.96 g/L; **(f).** 81.8%; **(g).** 200 mmHg; **(h).** 1.0 g/L. **7.92.** 3.70 g. **7.94.** 3000 ft^3 O_2, 6000 ft^3 SO_3. **7.95.** $Zr + 4HCl \rightarrow ZrCl_4 + 2H_2$. **7.97.** 2.33:1. **7.98.** $2H_2S \rightarrow 2H_2 + S_2$. **7.99.** 0.068 atm. **7.100.** 7.3×10^{-3} mol CO_2. **7.101.** $\frac{3}{4}$. **7.102.** 1.1 atm. **7.103.** 25%. **7.104.** 0.58. **7.105.** 21%. **7.106.** 28.054. **7.107.** 13.6%. **7.108.** 43.8%. **7.109.** 44.5%. **7.110.** 13.6%. **7.111.** 128 mmHg.

7.112. 3.36 g. **7.113.** 83 mmHg. **7.114.** 0.42. **7.116.** 109 ft^3. **7.117.** 23.5 mmHg. **7.118.** 0.047 mol. **7.119.** 239 L. **7.121.** 0.019 mol H_2, 0.45 g NaH. **7.122(a).** 0.665 atm; **(b).** 1.39 atm. **7.123.** 11.2 mmHg. **7.125.** 0.020. **7.126.** 11.7 mmHg. **7.128.** 1.02 ft/min. **7.129.** 8:9. **7.130.** 33.8 cm^3. **7.131.** 36. **7.132.** 2.4 km/s. **7.133.** 30.069. **7.134.** 16.043. **7.135.** 1.04.

Chapter 8

8.2. 122 g H_2O. **8.3.** 131 kJ. **8.4.** 75.4 kJ. **8.5.** 78.90 kJ/K. **8.6.** −1370 kJ/mol. **8.7.** 5.7 cal/g. **8.8.** 6.01 kJ/mol. **8.10.** 25 kJ/g. **8.11.** −290 kJ/mol. **8.12.** 3060 kJ/mol. **8.13.** 89.6 g CH_3OH. **8.14.** 4.7×10^7 L. **8.15.** 74.9 g CH_4. **8.16.** 96.3 J. **8.17.** 850 J. **8.18.** 20.7 K. **8.19.** 7.79 g octane. **8.20.** 357 K. This calculation neglects the heat required to warm the melted ice to the final temperature. **8.24(a).** −177 kJ; **(b).** −58 kJ; **(c).** −1411 kJ; **(d).** −312 kJ. **8.25.** −2220 kJ/mol. **8.26.** −395 kJ/mol. **8.27(a).** 103 kJ; **(b).** −96 kJ. **8.28.** 62 kJ/mol. **8.30.** 277 kJ. **8.31.** −293.7 kJ/mol. **8.32.** −324 kJ. **8.35(a).** 103 kJ; **(b).** 436 kJ; **(c).** −111 kJ; **(d).** 92 kJ; **(e).** −886 kJ. **8.36.** −163 kJ. **8.37.** 566 kJ. **8.38.** 705 kJ. **8.39.** 173 kJ/mol. **8.40.** 151 kJ/mol. The bond is weaker in ClF_5 due to crowding of the 5 F atoms around the central Cl. **8.41.** 328 kJ/mol.

Chapter 9

9.1. $5.8 \times 10^{14}\ s^{-1}$. **9.2.** 0.05 cm. **9.3.** 3.3×10^4 cm/s. **9.4.** 2.7×10^{-14} J. **9.5.** 6.6×10^{-19} J. **9.6.** 1.5×10^{18}. **9.7.** 1.2×10^{20}. **9.8.** 3.3×10^{-19} J, 200 kJ/mol. **9.9.** 9.9×10^{-19} J. **9.10.** $6.7 \times 10^4\ s^{-1}$. **9.11.** 3.7×10^7 cm/s. **9.12.** 4.04×10^{-4} cm, 1.28×10^{-4} cm, 4.33×10^{-5} cm, 9.47×10^{-6} cm. **9.13.** 1.2×10^{-5} cm. **9.14.** 3.04×10^{-19} J/atom. **9.15.** 2.19×10^{-18} J/atom. **9.16.** 1.2×10^7 cm/s. **9.17.** 1.3×10^{-9} cm. **9.18.** 1.6×10^{-32} cm. **9.19.** $2d$, $1p$, $3f$. **9.20(c).** $[Ar]3d^{10}4s^24p^4$; **(d).** $[Xe]6s^2$. **9.21(a).** $[Ar]3d^{10}4s^24p^6$; **(b).** $[Kr]4d^{10}5s^25p^6$; **(c).** $[Xe]4f^{14}5d^{10}6s^26p^6$. **9.22(b).** $[Ar]3d^{10}4s^24p^2$; **(c).** [Kr]; **(d).** $[Kr]4d^{10}5s^25p^2$. **9.23(a).** N; **(b).** Ar; **(c).** Ni; **(d).** Si, S. **9.24(a).** $[Ar]3d^{10}4s^24p^3$; **(b).** $[Kr]4d^{10}5s^25p^3$; **(c).** $[Xe]4f^{14}5d^{10}6s^26p^3$. All have the same valence electronic configuration. **9.25(a).** $[Kr]5s^1$; **(b).** $[Kr]5s^24d^1$; **(c).** $[Kr]5s^24d^3$; **(d).** $[Kr]5s^24d^5$. **9.26(a).** [Ar]; **(b).** [Ar]; **(c).** [Ar]. **9.27.** I^-, Te^{2-}, Sb^{3-}, Cs^+, Ba^{2+}, etc. **9.28.** 17, 35, 53. **9.29.** 34. **9.30(a).** Be; **(b).** Ti; **(c).** Cr; **(d).** Fe. **9.31(a).** 5; **(b).** 1; **(c).** 4; **(d).** 0. **9.32.** $S^+[Ne]3s^23p^3$; $S^{2+}[Ne]3s^23p^2$; $S^{3+}[Ne]3s^23p^1$; $S^{4+}[Ne]3s^2$; $S^{5+}[Ne]3s^1$; S^{6+} [Ne]. Only S^{4+} and S^{6+} do not have unpaired electrons. **9.33.** $[Xe]4f^{14}5d^{10}6s^1$. **9.34(a).** $[He]2s^22p_x \uparrow 2p_y \downarrow$; **(b).** $[He]2s^22p^53s^1$; **(c).** $[He]2p^1$. **9.35.** They are filling the $4f$ shell. The outer valence shell is $6s^2$. **9.36(a).** $1s^22s^22p^43s^1$; **(b).** $1s^22s^22p^43s^23p^44s^1$; **(c).** He 2, O 8, Si 14. **9.37.** O. **9.38(a).** $1s^32s^32p^1$; **(b).** $1s^32s^32p^93s^33p^1$; **(c).** 3, 15, 27.

Chapter 10

10.1. (e), (d), (a), (b), (c). **10.2.** (b), (g), (a), (f), (e), (c), (d). **10.3(a).** 9; **(b).** 19; **(c).** 35; **(d).** 22, 32; **(e).** 34; **(f).** 15; **(g).** 15, 16; **(h).** 17. **10.4.** Bonds = valence electrons when electrons < 4, and bonds = 8 − valence electrons when electrons ≥ 4. **10.5.** (b), (d). **10.6.** (c), (e).

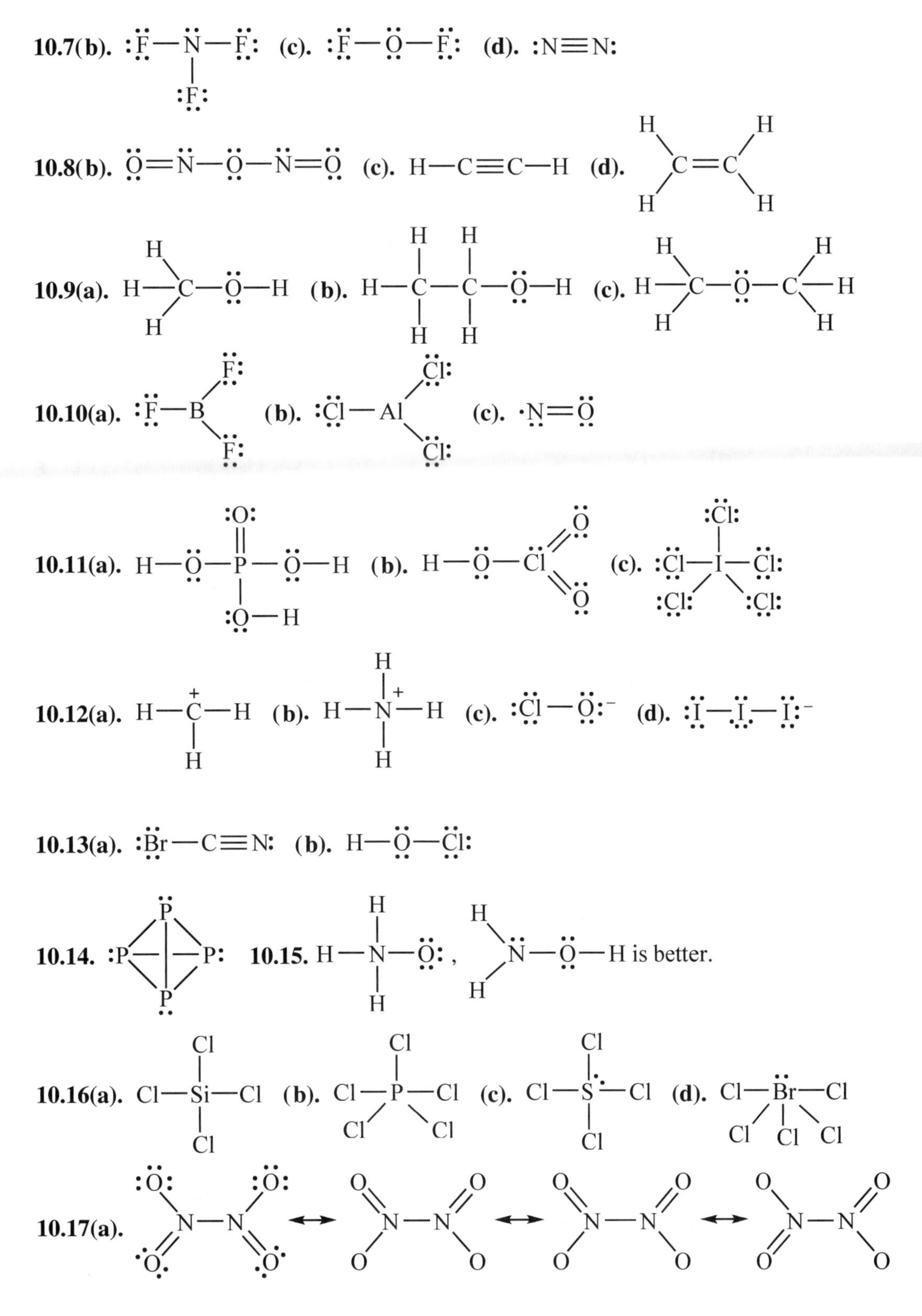
10.7(b).
(c).
(d).
10.8(b).
(c).
(d).
10.9(a).
(b).
(c).
10.10(a).
(b).
(c).
10.11(a).
(b).
(c).
10.12(a).
(b).
(c).
(d).
10.13(a).
(b).
10.14.
10.15.
is better.
10.16(a).
(b).
(c).
(d).
10.17(a).

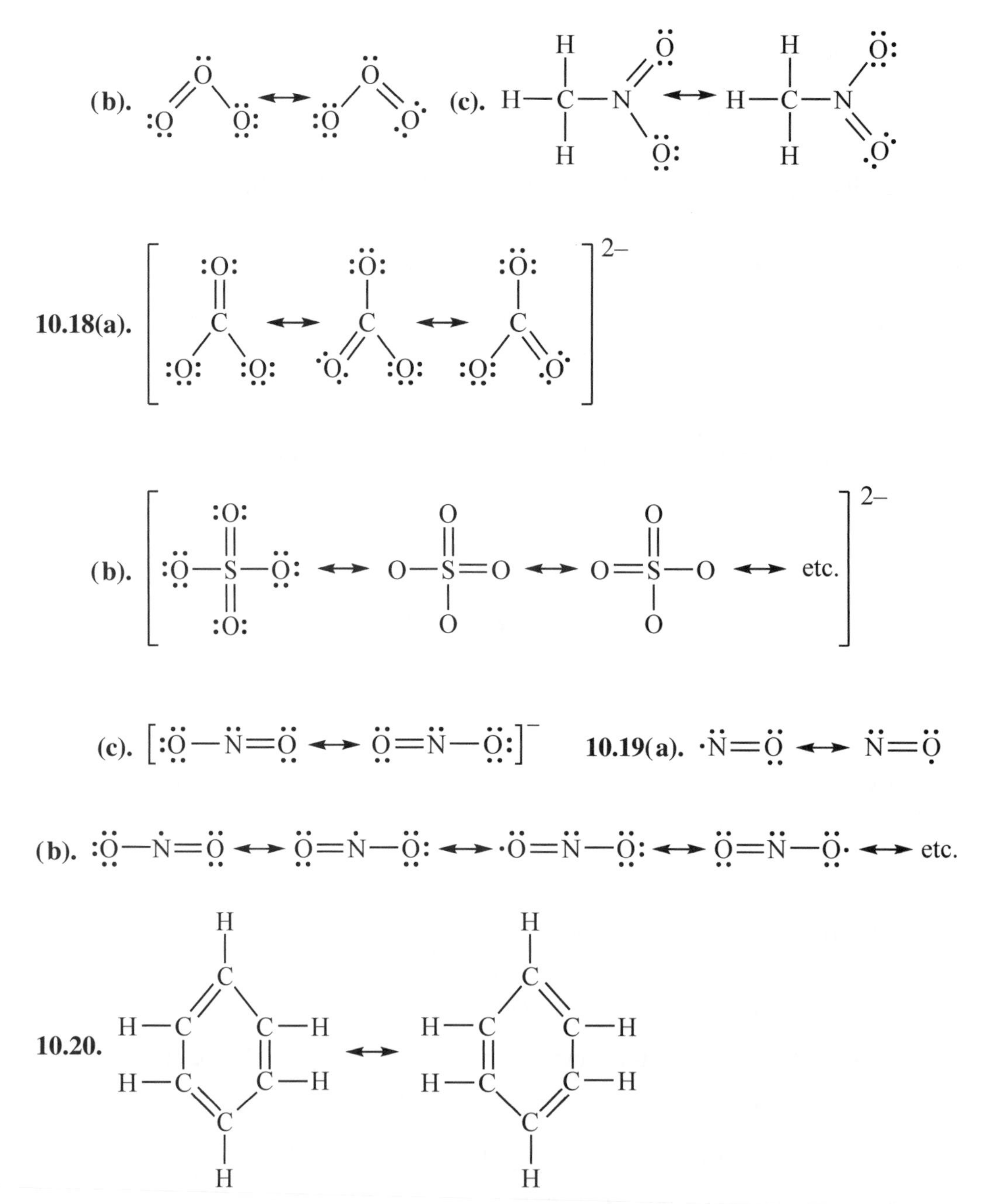

10.21. Resonance in the allyl cation cation ↔ $^+CH_2—CH{=}CH_2$ which is not possible without the double bond. **10.22(a).** $2sp^3$; **(b).** $3sp^2$; **(c).** $2sp$; **(d).** $2sp$; **(e).** $2sp^2$; **(f).** $2sp^3$. **10.23(a).** $2sp$, $2sp$; **(b).** $2sp^2$, $2sp^2$; **(c).** $2sp^3$, $2sp^2$; **(d).** $2sp$, $2sp$. **10.25(a).** C—H (σ, $sp^3 - 1s$), C—Cl (σ, $sp^3 - 3p$); **(b).** C—H (σ, $sp - 1s$), C—C (σ, $sp - sp$) (π, $2p_x - 2p_x$) (π, $2p_y - 2p_y$);

(c). C—H (σ, $sp - 1s$), C—N (σ, $sp - 2p_x$) (π, $2p_y - 2p_y$) (π, $2p_z - 2p_z$); **(d).** N—H (σ, $sp^3 - 1s$), N—O (σ, $sp^3 - sp^3$), O—H (σ, $sp^3 - 1s$). **10.26.** C—O (σ, $sp - 2p_x$) (π, $2p_y - 2p_y$) (π, $2p_z - 2p_z$). **10.27.** N═N═O, N—N (σ, $2p_x - sp$) (π, $2p_y - 2p_y$), N—O (σ, $sp - 2p_x$) (π, $2p_z - 2p_z$); N≡N—O, N—N (σ, $2p_x - sp$) (π, $2p_y - 2p_y$) (π, $2p_z - 2p_z$) N—O (σ, $sp - 2p_x$). **10.28.** The unhybridized *p* orbital of the N overlaps with each of the unhybridized *p* orbitals of the 3 O atoms to form a bonding orbital spread over all 4 atoms, which holds 2 electrons. **10.29(a).** linear; **(b).** trigonal planar; **(c).** tetrahedral; **(d).** bent. **10.30(a).** tetrahedral; **(b).** pyramidal; **(c).** bent; **(d).** bent. **10.31(a).** trigonal planar; **(b).** trigonal planar; **(c).** bent; **(d).** bent. **10.32.** :Ö═C═C═C═Ö:. **10.33.** $BeBr_2$, $HgBr_2$. **10.34(a).** trigonal planar; **(b).** trigonal planar; **(c).** pyramidal; **(d).** tetrahedral. **10.35(a).** trigonal bipyramidal; **(b).** octahedral; **(c).** octahedral; **(d).** tetrahedral. **10.36(b).** at 5 of the 6 corners of a distorted octahedron; **(c).** linear (the 2 apices of a trigonal bipyramid); **(d).** at 4 of the corners of a trigonal bipyramid (the electron pair is in one of the equatorial positions).

Chapter 11

11.2. 18.7%, 0.187. **11.3.** 10.0 g sugar. **11.4.** 28.9 g gold. **11.5.** 66%. **11.10.** 6.46%. **11.11.** 20 g. **11.12.** 18%. **11.13.** 12.5 g. **11.14.** 10%. **11.19.** 0.298. **11.20.** 0.819. **11.21.** 0.459. **11.22.** 0.0106. **11.23.** 280 g glucose. **11.25.** 1.44 *m*. **11.27.** $f = 0.0151$, $m = 0.853$. **11.28.** 9.2 g. **11.29.** 55.6 *m*. **11.30.** 19.0 *m*. **11.31.** 14 g acetone, 106 g H_2O. **11.32.** 0.0061. **11.33.** 1.7 *m*. **11.34.** 0.0176. **11.35.** 2.50 *m*, 0.0431. **11.41.** 4.0 g. **11.42.** 5.06 L. **11.43.** 0.32 *M*. **11.44.** 9.8 g. **11.45.** 1.30×10^{-3} *M*. **11.46.** 6.3 g/100 cm^3. **11.47.** 1.08×10^{-4} *M*. **11.48.** 6.62 *M*. **11.49.** 0.641 *M*. **11.50.** 9.391 *M*. **11.51.** 2.51 *M*. **11.52.** 4.60 *M*. **11.53.** 1.21 g/cm^3. **11.54.** 1271 cm^3. **11.56.** 0.0227, 13.3%, 1.29 *m*. **11.57.** 0.635 *M*. **11.60(a).** 0.48 mol; **(b).** 0.24 mol. **11.61.** 7.6 g. **11.63.** 4.40 *M*. **11.64.** 2.00 *M*. **11.66.** 11.7 L. **11.68.** 1.83 *M*. **11.69.** 0.32 *M*. **11.70.** 128 cm^3, 4.59 g. **11.71.** 1500 cm^3. **11.72.** 4.0 g. **11.73.** 0.250 *M*, 0.125 *M*. **11.74.** 0.69 *M*.

Chapter 12

12.2. 128. **12.4.** 116. **12.5.** −0.67 °C. **12.6.** 90.6 g. **12.7.** 46. **12.8.** 5. **12.9.** 32.042. **12.10.** 351.4 K. **12.11.** 148 g of sugar. **12.12.** 165. **12.13.** 265.7 K. **12.14.** 358. **12.15.** 245 g. **12.16.** 100.42 °C. **12.18.** 101.12 °C. **12.19.** 11.2 g. **12.20.** 3 mol. **12.21.** 272.3 K. **12.22.** 374.9 K. **12.23.** 5.65 *m*. **12.24.** MF_3. **12.25.** dimerized. **12.27.** 0.0498. **12.28.** −1.62 °C. **12.31.** 42.4 mmHg. **12.32.** 31.3 mmHg. **12.33.** 0.2438 atm. **12.34.** 0.02863 atm. **12.35.** 4.3 g C_6Cl_6. **12.36.** 267. **12.37.** 0.0536 atm. **12.38.** 0.0306 atm. **12.40.** $P_{C_2H_6O} = 21.9$ mmHg, $P_{H_2O} = 22.6$ mmHg, $f_{H_2O} = 0.509$. **12.41.** 0.294, 0.144. **12.42.** $P_{C_7H_8} = 37.1$ mmHg, $P_{C_6H_6} = 117.9$ mmHg. **12.43.** $P_T = 0.353$ atm, $f_{C_7H_8} = 0.261$. **12.44.** 0.131 atm of isopropyl alcohol, 0.068 atm of propyl alcohol. **12.45.** 0.740. **12.47.** 251,000.

12.48. 54.8 atm. **12.49.** 1.52 g. **12.50.** 25.7 atm. **12.51(a).** 2.45 atm; **(b).** 0.0245 atm; **(c).** 0.000245 atm. **12.52.** 13.2 atm. **12.53.** 1.2×10^{-3} g.

Chapter 13

13.3. 0.010 mol/L-min, 0.0096 mol/L-min, 0.0090 mol/L-min. **13.4.** 0.12 mmHg/s, 0.093 mmHg/s, 0.070 mmHg/s. **13.5.** 0.06 mmHg/s, 0.047 mmHg/s, 0.035 mmHg/s. **13.6.** NO = 14.2 mmHg/min, NH_3 = 14.2 mmHg/min, O_2 = 17.8 mmHg/min. **13.7.** $-\Delta P_{NO}/\Delta t = 4.8 \times 10^{-4}$ atm/s , $-\Delta P_{O_2}/\Delta t = 2.4 \times 10^{-4}$ atm/s. **13.8.** 1.3×10^{-2} atm/s. **13.9.** 8.0×10^{-5} atm/s, 6.0×10^{-5} atm/s, 4.8×10^{-5} atm/s, $\Delta P_{CO_2}/\Delta t = 3.0 \times 10^{-5}$ atm/s. **13.10.** 7.8×10^{-4} atm/s, 5.9×10^{-4} atm/s, 3.6×10^{-4} atm/s, 2.0×10^{-4} atm/s, total = 4.1×10^{-4} atm/s. **13.11.** 1.34 mmHg/min. **13.12.** 3rd order; 1st order in O_2, 2nd order in NO. **13.13.** $\frac{3}{2}$ order; 1st order in H_2, $\frac{1}{2}$ order in Br_2. **13.15.** 0.0103 min^{-1}. **13.16.** 2.08×10^{-6} mmHg/s. **13.17.** 2nd order; 3.2×10^{-4} L/mol-s. **13.18.** rate = $kP_{N_2O_3}$. **13.20.** 2.28×10^{-3} h^{-1}. **13.21.** 3.05×10^{-2} mmHg/min. **13.22.** $k = 2.20 \times 10^{-5}$ s^{-1}. **13.23.** $k = 3.20 \times 10^{-4}$ s^{-1}. **13.24(a).** 224,000 s; **(b).** 470,000 s; **(c).** 0.487 *M*. **13.26(a).** 47.7 s; **(b).** 147 s; **(c).** 0.180 atm. **13.27.** 73,200 s. **13.28.** 48.2 min. **13.29.** 337 mmHg; 663 mmHg. **13.31.** 3.24×10^{-3} L/mol-min. **13.32.** 2.10×10^{-6} mmHg/s. **13.33.** 2500 s. **13.34.** 27.3 mmHg. **13.35(a).** 4.5×10^{-3} L/mol-s; **(b).** 8.9×10^{-3} M; **(c).** 8400 s. **13.36(a).** Because of the stoichiometry, if $[I^-] = [ClO^-]$ at the beginning, it will be equal throughout the course of the reaction and the rate equation can be treated as one term second order; **(b).** $k = 0.34$ L/mol-s; **(c).** $[IO^-] = 0.013$ M. **13.39.** 0.27 h^{-1}, 2.6 h, 0.83 h. **13.40(a).** 0%; **(b).** 25%; **(c).** 33.3%. **13.41.** 224 s^{-1}. **13.42.** 44.4 kJ. **13.43.** 1.0×10^5 J/mol. **13.44.** 96,700 J/mol. **13.45.** 3.62×10^{-5} L/mol-s. **13.46.** $K = k_1/k_{-1}$.

Chapter 14

14.5. 0.2 mol PCl_3, 0.2 mol Cl_2, 0.8 mol PCl_5. **14.8.** 0.041. **14.9.** 267. **14.10.** 13.3. **14.11.** 30. **14.12.** 1.6. **14.13.** 16 mol. **14.14.** 33.4 mol. **14.17.** 3.3 mol. **14.18.** 9.3 g. **14.19.** 3.04×10^{-3}. **14.20.** 1.6 mol NO_2. **14.23(a).** $K_1 = 6.9$ L^2/mol^2; **(b).** $K_2 = 0.38$ mol/L, $K_2 = 1/\sqrt{K_1}$. **14.24.** $[N_2O_5] = 0.94$, $[N_2O_3] = 1.62$, $[N_2O] = 1.44$. **14.25.** 0.77 mol. **14.26.** 0.23 mol. **14.27.** 1.92. **14.28.** 0.96 atm. **14.29.** 0.1. **14.30.** 0.68. **14.31.** 66. **14.33.** 0.47 atm. **14.36(a).** $P_{N_2} = 0.016$ atm, $P_{H_2} = 0.048$ atm, $P_{NH_3} = 0.97$ atm; **(b).** 0.032 is fraction decomposed. **14.37.** $P_{N_2} = 6.6 \times 10^{-6}$ atm. **14.38.** $P_{H_2} = 0.19$ atm, $P_{CH_4} = 0.34$ atm. **14.42.** 0.12 atm. **14.43.** 0.314 atm. **14.44.** $K_p = 0.0534$, $K = 1.33$. **14.45.** 1.0 atm. **14.46.** $K = 1.33 \times 10^{-2}$, $K_p = 0.665$. **14.47.** 0.0343 mol. **14.48.** 1.15×10^4 mmHg. **14.49(a).** 0.814 atm, 0.372 atm; **(b).** 1.78 atm.

Chapter 15

15.3(a). 1.7×10^{-7}; **(b).** 5.6×10^{-10}. **15.4(a).** 5.0×10^{-10}; **(b).** 5.0×10^{-12}. **15.7(a).** 8; **(b).** 2.7; **(c).** 2.5; **(d).** 10; **(e).** 11.3. **15.10(a).** 3.2×10^{-2}; **(b).** 2.5×10^{-14}. **15.11(a).** 4.0×10^{-11}; **(b).** 1.6×10^{-8}. **15.12.** choice (b). **15.14(a).** $1.7 \times 10^{-5}\,M$; **(b).** $1.8 \times 10^{-5}\,M$; **(c).** $4 \times 10^{-10}\,M$. **15.15.** 0.0015%, 0.015%. **15.17.** 0.10 F; $1.4 \times 10^{-3}\,M$. **15.19.** $1.3 \times 10^{-3}\,M$. **15.20.** 10%. **15.21.** pH = 2.30. **15.22.** $1.3 \times 10^{-12}\,M$. **15.23.** 9.7×10^{-11}. **15.24(a).** 0.063; **(b).** 0.062; **(c).** 0.0045. **15.26.** 8.4. **15.27.** $7.5 \times 10^{-6}\,M$. **15.28.** 0.11 M; $2.2 \times 10^{-5}\,M$. **15.29.** 6×10^{-8}. **15.30.** pH = 4.93, pOH = 9.07. **15.31.** 7.7×10^{-5}. **15.32.** 0.028 mol NaOAc. **15.33.** 0.24 mol NH_4Cl. **15.34.** 0.072 mol NH_3. **15.36.** 1.8×10^{-5} mol OH^-. **15.37.** $5.4 \times 10^{-6}\,M$. **15.38.** 8×10^{-5}. **15.41.** $1.3 \times 10^{-13}\,M$. **15.42.** $3.2 \times 10^{-7}\,M$. **15.44.** $1.3 \times 10^{-19}\,M$. **15.45.** 5.0×10^{-4}; 2.8×10^{-10}; $8.4 \times 10^{-22}\,M$. **15.47.** 0.30 mol. **15.48.** 1.14×10^{-4}, 8.33.

Chapter 16

16.7. 0.04 N. **16.8.** 800 cm^3. **16.9.** 1.6 N. **16.10.** 29 cm^3. **16.11.** 0.83 g. **16.12.** 83.0. **16.13.** 218 g. **16.14.** 37.5 cm^3. **16.15.** 2.24 L. **16.16.** 0.5 N. **16.17.** 8.72. **16.18.** 5.22. **16.19.** 11.15. **16.20.** 0.00154 mol available protons. **16.21.** 8.65%. **16.22.** 0.172 M. **16.23.** 3.19. **16.24(a).** 7; **(b).** 5; **(c).** 11. **16.25.** 13.0, 12.5, 11.7, 10.7, 9.7, 7.0. **16.26.** 11.2, 9.4, 8.4, 7.4, 6.4, 5.3. **16.28.** 5.2. **16.29.** F of NaOAc = 18 × F of HOAc. **16.30.** HF. **16.31.** 7.2. **16.32.** 0.1 mol. **16.33.** 1.39 mol, − 0.14. **16.35.** 12.5 cm^3. **16.36.** 132 cm^3 NaOH, 868 cm^3 HOAc. **16.37.** 116 g sodium phenoxide, 94 g phenol. **16.38.** pH = 3.85 before, pH = 3.77 after. **16.39.** pH = 3.85 before, pH = 2.28 after. This buffer solution has less capacity because it was prepared from smaller amounts of acid and base. **16.41.** $[OH^-] = [HSO_3^-] = 1.3 \times 10^{-4}\,M$; $[K^+] = 0.20\,M$; $[H^+] = 7.5 \times 10^{-11}\,M$; $[H_2SO_3] = 7.7 \times 10^{-13}\,M$. **16.42.** 11.8. **16.43.** $5 \times 10^{-2}\,M$. **16.44.** $5 \times 10^{-4}\,M$. **16.45(a).** 4.6; **(b).** 9.6; **(c).** 3.7. **16.46.** $4 \times 10^{-8}\,M$. **16.47.** $[Na^+] = [NH_4^+] = 1.2 \times 10^{-3}\,M$; $[HCN] = 1.7 \times 10^{-3}\,M$. **16.48.** $[Na^+] = [C_2H_3O_2^-] = 0.20\,M$; $[NH_3] = 0.20\,M$; $[NH_4^+] = [OH^-] = 1.9 \times 10^{-3}\,M$; $[HOAc] = 5.9 \times 10^{-8}\,M$; $[H^+] = 5.3 \times 10^{-12}\,M$. **16.49.** $2.8 \times 10^{-10}\,M$. **16.50.** 5.00 F. **16.51.** 0.36 mol. **16.52.** $[H^+] = 1.0 \times 10^{-4}\,M$; $[OH^-] = 1.0 \times 10^{-10}\,M$; $[H_2C_2O_4] = 1.9 \times 10^{-5}\,M$; $[HC_2O_4^-] = 1.2 \times 10^{-2}\,M$; $[C_2O_4^{2-}] = 7.6 \times 10^{-3}\,M$. **16.53.** 42.5 g. **16.55.** 0.70 F. **16.56(a).** 11.2; **(b).** 7.1. **16.57.** 0.33. **16.58(a).** 4.0; **(b).** 0.050 mol; **(c).** 8.0; **(d).** 0.26 mol. **16.59.** 0.134 mol. **16.60.** 12; $3.3 \times 10^{-5}\,M$; 2.14 M; 2.25 M; $4.2 \times 10^{-15}\,M$.

Chapter 17

17.3. 1.8×10^{-28}. **17.4.** 1.5×10^{-32}. **17.5.** 1.78×10^{-18}. **17.6.** 4.92×10^{-9}. **17.7(b).** 8.8×10^{-17}; **(c).** 3.3×10^{-13}; **(d).** 2.0×10^{-10}; **(f).** 8.38×10^{-9}. **17.9.** $3 \times 10^{-11}\,M$. **17.10(a).** 4×10^{-36}; **(b).** 2×10^{-16}; **(c).** 6×10^{-24}; **(d).** 2×10^{-23}. **17.11.** 0.008 M. **17.12.** 1.7×10^{-7}. **17.13.** $Fe(OH)_3$ ppts; $BaSO_4$ does not. **17.14.** 9.3×10^{-8} mol. **17.15.** 4×10^{-6}. **17.17.** 4.0×10^{-5}. **17.18.** 1.6×10^{-5}.

17.19(b). 1.7×10^{-5}; **(d).** 1.4×10^{-2}; **(e).** 4×10^{-9}. **17.20.** 13.6 times as great. **17.22.** $1 \times 10^{-9}\,M$. **17.24.** 1.9×10^{-12}. **17.25.** 1×10^{-5}. **17.26.** 2.8×10^{-8}. **17.27.** $2.0 \times 10^{-8}\,M$. **17.29.** Cl^-; $0.0096\,M$. **17.30(a).** Ag^+; $1.2 \times 10^{-5}\,M$; **(b).** $4.0 \times 10^{-4}\,M$; **(c).** $2.5 \times 10^{-5}\,M$. **17.31.** $2.7 \times 10^{-6}\,M$. **17.32.** 1.00×10^{-8}. **17.33.** 9.9×10^{-3}; 9.9×10^{-9}. **17.34(a).** $4 \times 10^{-6}\,M$; **(b).** $1.5 \times 10^{-3}\,M$; **(c).** $1.0 \times 10^{-7}\,M$; **(d).** 1.15×10^{-2}; **(e).** 1.11×10^{-2}; **(f).** $(1.11 \times 10^{-2}) + (2.5 \times 10^{-8})$ mol; **(g).** 22.2. **17.35.** 2.2×10^{-6}. **17.36.** 6.5×10^{-11}; $Pb(ClO_4)_2$. **17.38.** $2.0 \times 10^{-15}\,M$. **17.39.** $0.10\,M$, $0.050\,M$, and 8.3×10^{-6}. **17.40.** 0%. **17.42.** yes; $Mg(OH)_2$. **17.43.** reverse. **17.44(b).** $1.8 \times 10^{4}\,M$; **(c).** $0.18\,M$; **(d).** $3.6 \times 10^{-2}\,M$; **(e).** $3.6 \times 10^{-3}\,M$; **(f).** 5.7×10^{-6}; **(g).** $1.3 \times 10^{-7}\,M$. **17.46.** 0.66. **17.47.** 3×10^{-15} g. **17.48.** 4.0. **17.49.** no. **17.50.** $11\,M$. **17.51.** 1.3×10^{-22}. **17.53.** $5.9 \times 10^{-3}\,M$. **17.54.** 1.2 g. **17.55.** 14 g. **17.56.** 1.2 g. **17.57.** $4.6 \times 10^{5}\,M$; no. **17.58.** 2.2 mol. **17.59.** $6 \times 10^{-11}\,M$. **17.60(a).** 6.0×10^{-23}; **(b).** 1.2×10^{-14} L. **17.61.** 4×10^{-3} mol. **17.62.** 1×10^{-4}; 1×10^{-8}. **17.63.** $1.0 \times 10^{-6}\,M$. **17.64.** 8.4×10^{-10}. **17.65.** 3.2×10^{-6}. **17.66.** $5 \times 10^{-9}\,M$. **17.67.** $0.35\,M$. **17.69(a).** $2.5 \times 10^{-4}\,M$; **(b).** $(2.5 \times 10^{-4} + Y)F$. **17.70(a).** $a = 0.20$, $b = 0.25$, $c = 0.30$; **(b).** 51.0. **17.71.** $Ga(OH)_4^-$. **17.72.** $Co(NH_3)_6^{2+}$. **17.74.** 0.55 g. **17.75.** 8×10^{-8}. **17.76(a).** $1.34 \times 10^{-5}\,M$; **(b).** 3.9×10^{-8} mol; **(c).** 1.0×10^{-3} mol; **(d).** $0.35\,M$. **17.77.** 4.1×10^{-2} mol. **17.78.** 3×10^{-16}. **17.79.** $[Cd^{2+}] = 3.2 \times 10^{-11}\,M$, $[CN^-] = 2.7 \times 10^{-3}\,M$, $[HCN] = 2.7 \times 10^{-6}\,M$; $0.051\,F$. **17.80(a).** 1.0×10^{-16}; **(b).** 1.5×10^{-19}. **17.81.** $1 \times 10^{-2}\,M$. **17.82.** $[H^+] = [Ag(CN)_2^-] = 1.1 \times 10^{-4}\,M$; $[OH^-] = 9.3 \times 10^{-11}\,M$; $[Ag^+] = 7.0 \times 10^{-13}\,M$; $[HCN] = 0.10\,M$; $[CN^-] = 3.7 \times 10^{-7}\,M$. **17.83.** $[H^+] = 0.1\,M$; $[OH^-] = 1 \times 10^{-13}\,M$; $[Cl^-] = 0.05\,M$; $[Cu(CN)_2^-] = 0.05\,M$; $[HCN] = 5 \times 10^{-4}\,M$; $[Cu^+] = 6 \times 10^{-5}\,M$; $[CN^-] = 2 \times 10^{-12}\,M$. **17.84.** 4.5. **17.85.** 1.0×10^{-16}. **17.86.** 3.4×10^{-6}. **17.87.** 7.85 to 12.3.

Chapter 18

18.2. 5.1 kJ. **18.3.** −4250 J. **18.6.** $\Delta E = \Delta H = 0$, $q = 1530$ J, $w = -1530$ J. **18.7.** $\Delta E = \Delta H = 0$, $q = -5230$ J, $w = 5230$ J. **18.8.** $\Delta E = \Delta H = 0$, $q = 3440$ J, $w = -3440$ J. **18.9.** $\Delta E = \Delta H = 0$, $q = -5150$ J, $w = 5150$ J. **18.10.** $\Delta E = \Delta H = 0$, $q = 1730$ J, $w = -1730$ J. **18.13.** $\Delta E = 3120$ J, $\Delta H = 5210$ J, $q = 3120$ J, $w = 0$. **18.14.** $\Delta H = q = 5200$ J, $w = -2080$ J, $\Delta E = 3120$ J. **18.15(a).** −1860 J; **(b).** −2470 J; **(c).** −2260 J; **(d).** −2680 J. **18.16.** $\Delta H = q = -8180$ J, $\Delta E = -4910$ J, $w = 3270$ J. **18.17(a).** $w = -1350$ J, $q = 100$ J, $\Delta E = -1250$ J, $\Delta H = -2080$ J; **(b).** $w = -828$ J, $q = -423$ J, $\Delta E = -1250$ J, $\Delta H = -2080$ J; **(c).** $w = -895$ J, $q = -351$ J, $\Delta E = -1250$ J; $\Delta H = -2080$ J. **18.18.** $\Delta H = q = 2260$ kJ, $\Delta E = 2090$ kJ, $w = -172$ kJ. **18.19.** $\Delta H = q = 42.4$ kJ, $w = -3.18$ kJ, $\Delta E = 39.2$ kJ. **18.20.** $w = 0$, $\Delta H = \Delta E = q = -5610$ J. **18.21.** −57.3 kJ. **18.22(a).** 800 J; **(b).** 800 J; **(c).** 0. **18.23.** 600 J. **18.25.** $\Delta S = 25.5$ J/K, $\Delta G = -13.1$ kJ. **18.26.** $\Delta S = 0.79$ J/K, $\Delta G = 1300$ J. **18.28.** 2503 K. **18.29.** 114 J/K. **18.30.** $\Delta S = 0.8$ J/K, $\Delta G = -210$ J. **18.31.** q: +, +, −, +, +; w: −, −, 0, 0, 0; ΔE: 0, +, −, +, +; ΔH: 0, +, −, +, +; ΔS: +, +, −, +, +.

18.34. −106 kJ, $K_p = 3.9 \times 10^{18}$; −109 kJ, $K_p = 1.3 \times 10^{19}$. **18.35.** −1331 kJ, $K_p = 4.85 \times 10^{233}$. **18.37.** 1.26×10^{33}. **18.38.** 1.49×10^{-17}. **18.39.** −16.6 kJ/mol. **18.40(a).** −92 kJ; **(b).** −4 kJ; **(c).** −242 kJ. **18.41.** −213 kJ. **18.42.** 55.5 kJ.

Chapter 19

19.33. 1.6 g. **19.34.** 96 cm^3. **19.35.** 0.91 g. **19.36(a).** 0.600 *N*; **(b).** 0.0120; **(c).** 0.0120; **(d).** 42.5; **(e).** 0.510 g; **(f).** 50.0%. **19.37.** 0.134 *M*. **19.38.** 88.9%. **19.39.** 0.0202 *M*. **19.40.** 8.46%. **19.44.** 12.2 g. **19.45.** 557,000 C. **19.46.** 73.2 g. **19.47.** 2.4 g. **19.48.** 24 g. **19.49.** 101 g. **19.50.** 1.60×10^{-19} C. **19.51.** 140 cm^3. **19.53.** 55.9 g. **19.54.** 386 C. **19.55(a).** 9650 s; **(b).** 2.6 L; **(c).** 1.2×10^{23}; **(d).** 800 cm^3. **19.56(a).** $Y_2(SO_4)_3$; **(b).** 27 min; **(c).** 189 cm^3. **19.57.** MCl_3. **19.58.** 114. **19.59.** 126.9. **19.60.** 65.4. **19.62.** +4. **19.63.** 435 s. **19.64(a).** 0.513 V; **(b).** 1.997 V; **(c).** 0.517 V; **(d).** 0.739 V. **19.65(a).** 0.96 V; **(b).** 1.71 V; **(c).** 1.12 V. **19.66(a).** yes; **(b).** yes; **(c).** no; **(d).** yes; **(e).** yes. **19.67(a).** no; **(b).** no; **(c).** yes; **(d).** no; **(e).** yes. **19.68.** 2.99 V, Be anode, Au cathode. **19.69.** 2.715 V, Cu cathode, Mg anode. **19.70.** 0.793 V. **19.71.** − 0.806 V. **19.72(a).** 1.09 V; **(b).** 1.00 V; **(c).** 0.885 V; **(d).** 1.315. **19.73(a).** 0.103; **(b).** 0.354. **19.74(a).** 0.0827 V; **(b).** − 0.200 V. **19.75.** $E°_{cell} = 0.24$ V, $E°_{anode} = -0.24$ V. **19.76.** At pH = 2, − 0.12 V; at pH = 4, − 0.24 V; at pH = 7, − 0.41 V. **19.77.** Anode $[Zn^{2+}] = 0.10$ *M*, cathode $[Zn^{2+}] = 1.0$ *M*, 0.030 V. **19.78.** 0.047 V. **19.79.** 0.097 *M*. **19.80.** 7.3 *M*. **19.81(a).** 1×10^{10}; **(b).** 1×10^{8}; **(c).** 1×10^{120}; **(d).** 1×10^{37}. **19.82.** 4.82×10^{-13}. **19.83.** −88.66 kJ. **19.84.** −110 kJ. **19.85(a).** 0.824 V; **(b).** −79.5 kJ, 0.824 V. **19.86(a).** −0.948 V; **(b).** $Cu(s)|Cu^{2+}(1\ M)\|Au^{+}(1\ M), Au^{3+}(1\ M)|Pt$; **(c).** 0.948 V. **19.87.** 0.36 V. **19.88.** 1.47 V. **19.89.** 0.40 V. **19.91.** yes; $3H_2SO_3 \rightleftarrows S + 2SO_4^{2-} + 4H^+ + H_2O$. **19.92(a).** −0.48 V; **(b)** yes; $3ClO^- \rightleftarrows ClO_3^- + 2Cl^-$. **19.93.** yes. **19.94.** −0.140 V. **19.95.** −1. **19.96.** 0.56 V. **19.97(a).** 0.06 V; **(b).** no; **(c).** yes; **(d).** no; **(e1).** + 2; **(e2).** + 4; **(e3).** + 3; **(f).** no. **19.98.** 1.41×10^{9}.

INDEX